BMW 3 and 5 Series Automotive Repair Manual

by Larry Warren and John H Haynes
Member of the Guild of Motoring Writers

Models covered:
3-series
   318i (1984, 1985), 325, 325e, 325es (1984 through 1988), 325i, 325is, 325iC (1987 through 1990)

5-series
   525i (1989, 1990), 528e, (1982 through 1988), 533i (1983, 1984), 535i, 535is (1985 through 1992)

(11W2 - 2020)

ABCDE
FGHIJ
KLMNO
PQ

Haynes Publishing Group
Sparkford Nr Yeovil
Somerset BA22 7JJ England

Haynes North America, Inc
861 Lawrence Drive
Newbury Park
California 91320 USA

Acknowledgements

We are grateful to the Champion Spark Plug Company, who supplied the illustrations of various spark plug conditions. Technical writers who contributed to this project include Robert Maddox, Mark Ryan and Mike Stubblefield.

© Haynes North America, Inc. 1993

With permission from J.H. Haynes & Co. Ltd.

A book in the Haynes Automotive Repair Manual Series

Printed in the U.S.A.

All rights reserved. No part of this book may be reproduced or transmitted in any form or by any means, electronic or mechanical, including photocopying, recording or by any information storage or retrieval system, without permission in writing from the copyright holder.

ISBN 1 56392 020 4

Library of Congress Catalog Card Number 93-78147

While every attempt is made to ensure that the information in this manual is correct, no liability can be accepted by the authors or publishers for loss, damage or injury caused by any errors in, or omissions from, the information given.

Contents

Introductory pages

About this manual	0-5
Introduction to the BMW 3 and 5 Series	0-5
Vehicle identification numbers	0-6
Buying parts	0-6
Anti-theft audio system	0-7
Instrument panel language display	0-7
Maintenance techniques, tools and working facilities	0-7
Jacking and towing	0-14
Booster battery (jump) starting	0-15
Automotive chemicals and lubricants	0-16
Safety first!	0-17
Conversion factors	0-18
Troubleshooting	0-19

Chapter 1
Tune-up and routine maintenance — 1-1

Chapter 2 Part A
Engines — 2A-1

Chapter 2 Part B
General engine overhaul procedures — 2B-1

Chapter 3
Cooling, heating and air conditioning systems — 3-1

Chapter 4
Fuel and exhaust systems — 4-1

Chapter 5
Engine electrical systems — 5-1

Chapter 6
Emissions and engine control systems — 6-1

Chapter 7 Part A
Manual transmission — 7A-1

Chapter 7 Part B
Automatic transmission — 7B-1

Chapter 8
Clutch and driveline — 8-1

Chapter 9
Brakes — 9-1

Chapter 10
Suspension and steering systems — 10-1

Chapter 11
Body — 11-1

Chapter 12
Chassis electrical system — 12-1

Wiring diagrams — 12-11

Index — IND-1

Haynes mechanic, author and photographer with 1989 BMW 525i

About this manual

Its purpose
The purpose of this manual is to help you get the best value from your vehicle. It can do so in several ways. It can help you decide what work must be done, even if you choose to have it done by a dealer service department or a repair shop; it provides information and procedures for routine maintenance and servicing; and it offers diagnostic and repair procedures to follow when trouble occurs.

We hope you use the manual to tackle the work yourself. For many simpler jobs, doing it yourself may be quicker than arranging an appointment to get the vehicle into a shop and making the trips to leave it and pick it up. More importantly, a lot of money can be saved by avoiding the expense the shop must pass on to you to cover its labor and overhead costs. An added benefit is the sense of satisfaction and accomplishment that you feel after doing the job yourself.

Using the manual
The manual is divided into Chapters. Each Chapter is divided into numbered Sections, which are headed in bold type between horizontal lines. Each Section consists of consecutively numbered paragraphs.

At the beginning of each numbered Section you will be referred to any illustrations which apply to the procedures in that Section. The reference numbers used in illustration captions pinpoint the pertinent Section and the Step within that Section. That is, illustration 3.2 means the illustration refers to Section 3 and Step (or paragraph) 2 within that Section.

Procedures, once described in the text, are not normally repeated. When it's necessary to refer to another Chapter, the reference will be given as Chapter and Section number. Cross references given without use of the word "Chapter" apply to Sections and/or paragraphs in the same Chapter. For example, "see Section 8" means in the same Chapter.

References to the left or right side of the vehicle assume you are sitting in the driver's seat, facing forward.

Even though we have prepared this manual with extreme care, neither the publisher nor the author can accept responsibility for any errors in, or omissions from, the information given.

NOTE
A **Note** provides information necessary to properly complete a procedure or information which will make the procedure easier to understand.

CAUTION
A **Caution** provides a special procedure or special steps which must be taken while completing the procedure where the Caution is found. Not heeding a Caution can result in damage to the assembly being worked on.

WARNING
A **Warning** provides a special procedure or special steps which must be taken while completing the procedure where the Warning is found. Not heeding a Warning can result in personal injury.

Introduction to the BMW 3 and 5 Series

This manual covers service and repair operations for the BMW 3 and 5 Series models produced from 1984 through 1991 for 3-Series vehicles and 1982 through 1992 for 5-Series models.

These models are equipped with single overhead cam inline four- and six-cylinder engines. These engines feature a computer controlled ignition system and electronic fuel injection. Transmissions are a five-speed manual and three- or four-speed automatic. The transmission is mounted to the back of the engine, and power is transmitted to the fully-independent rear axle through a two-piece driveshaft. The differential is bolted solidly to a frame crossmember and drives the wheels through driveaxles equipped with inner and outer constant velocity joints.

The front suspension is MacPherson strut, with the coil spring/shock absorber unit making up the upper suspension link. The rear suspension is made up of MacPherson struts or coil springs and conventional design shock absorbers, depending on model.

Power assisted front and rear disc brakes are standard equipment except on 1985 and earlier 318i models which use drum rear brakes. Some later models are equipped with Anti-lock Brake Systems (ABS).

Vehicle identification numbers

Vehicle Identification Number

The Vehicle Identification Number (VIN) is located on the cowl in the engine compartment, on the driver's side door, and on a plate on top of the dash, just inside the windshield (see illustrations). It contains valuable information such as where and when the vehicle was manufactured, the model year and the body style. This number can be used to cross-check the registration and license. On later models the number is etched on the glass and affixed to most body parts in accordance with federal law.

Engine serial number

The engine serial number is located on the lower left side of the block, toward the front.

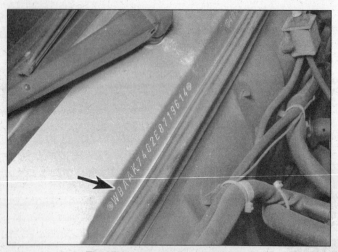

The VIN is stamped on the firewall

The VIN is also present on the end of the driver's door

Buying parts

Replacement parts are available from many sources, which generally fall into one of two categories - authorized dealer parts departments and independent retail auto parts stores. Our advice concerning these parts is as follows:

Retail auto parts stores: Good auto parts stores will stock frequently needed components which wear out relatively fast, such as clutch components, exhaust systems, brake parts, tune-up parts, etc. These stores often supply new or reconditioned parts on an exchange basis, which can save a considerable amount of money. Discount auto parts stores are often very good places to buy materials and parts needed for general vehicle maintenance such as oil, grease, filters, spark plugs, belts, touch-up paint, bulbs, etc. They also usually sell tools and general accessories, have convenient hours, charge lower prices and can often be found not far from home.

Authorized dealer parts department: This is the best source for parts which are unique to the vehicle and not generally available elsewhere (such as major engine parts, transmission parts, trim pieces, etc.).

Warranty information: If the vehicle is still covered under warranty, be sure that any replacement parts purchased - regardless of the source - do not invalidate the warranty!

To be sure of obtaining the correct parts, have engine and chassis numbers available and, if possible, take the old parts along for positive identification.

Anti-theft audio system

General information
1 Some models are equipped with an audio system which includes an anti-theft feature that will render the stereo inoperative if stolen. If the power source to the stereo is cut, the stereo won't work even if the power source is immediately re-connected. If your vehicle is equipped with this anti-theft system, do not disconnect the battery or remove the stereo unless you have the individual code number for the stereo.
2 Refer to your vehicle's owner's manual for more complete information on this audio system and its anti-theft feature.

Unlocking the stereo after a power loss
3 Turn on the radio. The word "CODE" should appear on the display.
4 Using the station preset selector buttons, enter the five-digit code. If you make a mistake when entering the code, continue the five digit sequence anyway. If you hear a "beep," however, stop immediately and start the sequence over again. **Note:** *You have three attempts to enter the correct code. If the correct code isn't entered in three trys, you'll have to wait one hour, with the radio on, before you enter the codes again.*
5 Once the code has been entered correctly, the word "CODE" should disappear from the display and the radio should play (you'll have to tune-in and enter your preset stations, however).
6 If you have lost your code number, contact a BMW dealer service department.

Instrument panel language display

Disconnecting the battery will cause the instrument panel display to default to the German language. It is necessary to reset the correct language after the battery is reconnected. With all the doors shut and the ignition key ON (engine not running), press the Trip Reset button until the panel displays the desired language. There are eight languages available. If you wish to bypass a particular selection, release the Reset button and press again - this will cause the display to advance to the next language. Once the correct language has been selected, continue holding the Reset button until the display reads "I.O. Version 2.0". Continue holding the button until it reads "H.P. Version 3.4", then release the button.

Maintenance techniques, tools and working facilities

Maintenance techniques
There are a number of techniques involved in maintenance and repair that will be referred to throughout this manual. Application of these techniques will enable the home mechanic to be more efficient, better organized and capable of performing the various tasks properly, which will ensure that the repair job is thorough and complete.

Fasteners
Fasteners are nuts, bolts, studs and screws used to hold two or more parts together. There are a few things to keep in mind when working with fasteners. Almost all of them use a locking device of some type, either a lockwasher, locknut, locking tab or thread adhesive. All threaded fasteners should be clean and straight, with undamaged threads and undamaged corners on the hex head where the wrench fits. Develop the habit of replacing all damaged nuts and bolts with new ones. Special locknuts with nylon or fiber inserts can only be used once. If they are removed, they lose their locking ability and must be replaced with new ones.

Rusted nuts and bolts should be treated with a penetrating fluid to ease removal and prevent breakage. Some mechanics use turpentine in a spout-type oil can, which works quite well. After applying the rust penetrant, let it work for a few minutes before trying to loosen the nut or bolt. Badly rusted fasteners may have to be chiseled or sawed off or removed with a special nut breaker, available at tool stores.

If a bolt or stud breaks off in an assembly, it can be drilled and removed with a special tool commonly available for this purpose. Most automotive machine shops can perform this task, as well as other repair procedures, such as the repair of threaded holes that have been stripped out.

Flat washers and lockwashers, when removed from an assembly, should always be replaced exactly as removed. Replace any damaged washers with new ones. Never use a lockwasher on any soft metal surface (such as aluminum), thin sheet metal or plastic.

Maintenance techniques, tools and working facilities

Fastener sizes

For a number of reasons, automobile manufacturers are making wider and wider use of metric fasteners. Therefore, it is important to be able to tell the difference between standard (sometimes called U.S. or SAE) and metric hardware, since they cannot be interchanged.

All bolts, whether standard or metric, are sized according to diameter, thread pitch and length. For example, a standard 1/2 - 13 x 1 bolt is 1/2 inch in diameter, has 13 threads per inch and is 1 inch long. An M12 - 1.75 x 25 metric bolt is 12 mm in diameter, has a thread pitch of 1.75 mm (the distance between threads) and is 25 mm long. The two bolts are nearly identical, and easily confused, but they are not interchangeable.

In addition to the differences in diameter, thread pitch and length, metric and standard bolts can also be distinguished by examining the bolt heads. To begin with, the distance across the flats on a standard bolt head is measured in inches, while the same dimension on a metric bolt is sized in millimeters (the same is true for nuts). As a result, a standard wrench should not be used on a metric bolt and a metric wrench should not be used on a standard bolt. Also, most standard bolts have slashes radiating out from the center of the head to denote the grade or strength of the bolt, which is an indication of the amount of torque that can be applied to it. The greater the number of slashes, the greater the strength of the bolt. Grades 0 through 5 are commonly used on automobiles. Metric bolts have a property class (grade) number, rather than a slash, molded into their heads to indicate bolt strength. In this case, the higher the number, the stronger the bolt. Property class numbers 8.8, 9.8 and 10.9 are commonly used on automobiles.

Strength markings can also be used to distinguish standard hex nuts from metric hex nuts. Many standard nuts have dots stamped into one side, while metric nuts are marked with a number. The greater the number of dots, or the higher the number, the greater the strength of the nut.

Metric studs are also marked on their ends according to property class (grade). Larger studs are numbered (the same as metric bolts), while smaller studs carry a geometric code to denote grade.

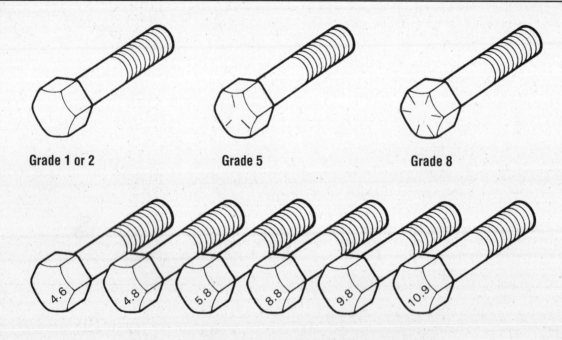

Bolt strength markings (top - standard/SAE/USS; bottom - metric)

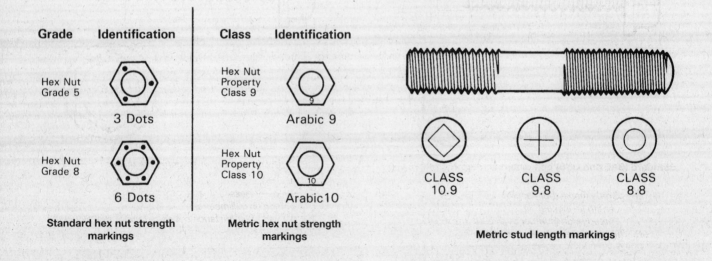

Standard hex nut strength markings

Metric hex nut strength markings

Metric stud length markings

Maintenance techniques, tools and working facilities

It should be noted that many fasteners, especially Grades 0 through 2, have no distinguishing marks on them. When such is the case, the only way to determine whether it is standard or metric is to measure the thread pitch or compare it to a known fastener of the same size.

Standard fasteners are often referred to as SAE, as opposed to metric. However, it should be noted that SAE technically refers to a non-metric fine thread fastener only. Coarse thread non-metric fasteners are referred to as USS sizes.

Since fasteners of the same size (both standard and metric) may have different strength ratings, be sure to reinstall any bolts, studs or nuts removed from your vehicle in their original locations. Also, when replacing a fastener with a new one, make sure that the new one has a strength rating equal to or greater than the original.

Tightening sequences and procedures

Most threaded fasteners should be tightened to a specific torque value (torque is the twisting force applied to a threaded component such as a nut or bolt). Overtightening the fastener can weaken it and cause it to break, while undertightening can cause it to eventually come loose. Bolts, screws and studs, depending on the material they are made of and their thread diameters, have specific torque values, many of which are noted in the Specifications at the beginning of each Chapter. Be sure to follow the torque recommendations closely. For fasteners not assigned a specific torque, a general torque value chart is presented here as a guide. These torque values are for dry (unlubricated) fasteners threaded into steel or cast iron (not aluminum). As was previously mentioned, the size and grade of a fastener determine the amount of torque that can safely be applied to it. The figures listed

	Ft-lbs	Nm
Metric thread sizes		
M-6	6 to 9	9 to 12
M-8	14 to 21	19 to 28
M-10	28 to 40	38 to 54
M-12	50 to 71	68 to 96
M-14	80 to 140	109 to 154
Pipe thread sizes		
1/8	5 to 8	7 to 10
1/4	12 to 18	17 to 24
3/8	22 to 33	30 to 44
1/2	25 to 35	34 to 47
U.S. thread sizes		
1/4 - 20	6 to 9	9 to 12
5/16 - 18	12 to 18	17 to 24
5/16 - 24	14 to 20	19 to 27
3/8 - 16	22 to 32	30 to 43
3/8 - 24	27 to 38	37 to 51
7/16 - 14	40 to 55	55 to 74
7/16 - 20	40 to 60	55 to 81
1/2 - 13	55 to 80	75 to 108

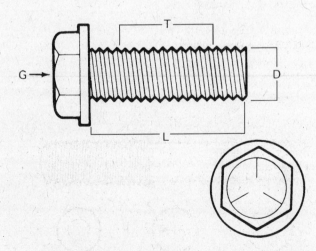

Standard (SAE and USS) bolt dimensions/grade marks

- G Grade marks (bolt length)
- L Length (in inches)
- T Thread pitch (number of threads per inch)
- D Nominal diameter (in inches)

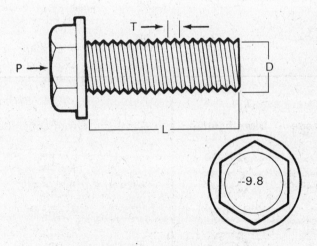

Metric bolt dimensions/grade marks

- P Property class (bolt strength)
- L Length (in millimeters)
- T Thread pitch (distance between threads in millimeters)
- D Diameter

here are approximate for Grade 2 and Grade 3 fasteners. Higher grades can tolerate higher torque values.

Fasteners laid out in a pattern, such as cylinder head bolts, oil pan bolts, differential cover bolts, etc., must be loosened or tightened in sequence to avoid warping the component. This sequence will normally be shown in the appropriate Chapter. If a specific pattern is not given, the following procedures can be used to prevent warping.

Initially, the bolts or nuts should be assembled finger-tight only. Next, they should be tightened one full turn each, in a criss-cross or diagonal pattern. After each one has been tightened one full turn, return to the first one and tighten them all one-half turn, following the same pattern. Finally, tighten each of them one-quarter turn at a time until each fastener has been tightened to the proper torque. To loosen and remove the fasteners, the procedure would be reversed.

Component disassembly

Component disassembly should be done with care and purpose to help ensure that the parts go back together properly. Always keep track of the sequence in which parts are removed. Make note of special characteristics or marks on parts that can be installed more than one way, such as a grooved thrust washer on a shaft. It is a good idea to lay the disassembled parts out on a clean surface in the order that they were removed. It may also be helpful to make sketches or take instant photos of components before removal.

When removing fasteners from a component, keep track of their locations. Sometimes threading a bolt back in a part, or putting the washers and nut back on a stud, can prevent mix-ups later. If nuts and bolts cannot be returned to their original locations, they should be kept in a compartmented box or a series of small boxes. A cupcake or muffin tin is ideal for this purpose, since each cavity can hold the bolts and nuts from a particular area (i.e. oil pan bolts, valve cover bolts, engine mount bolts, etc.). A pan of this type is especially helpful when working on assemblies with very small parts, such as the carburetor, alternator, valve train or interior dash and trim pieces. The cavities can be marked with paint or tape to identify the contents.

Whenever wiring looms, harnesses or connectors are separated, it is a good idea to identify the two halves with numbered pieces of masking tape so they can be easily reconnected.

Gasket sealing surfaces

Throughout any vehicle, gaskets are used to seal the mating surfaces between two parts and keep lubricants, fluids, vacuum or pressure contained in an assembly.

Many times these gaskets are coated with a liquid or paste-type gasket sealing compound before assembly. Age, heat and pressure can sometimes cause the two parts to stick together so tightly that they are very difficult to separate. Often, the assembly can be loosened by striking it with a soft-face hammer near the mating surfaces. A regular hammer can be used if a block of wood is placed between the hammer and the part. Do not hammer on cast parts or parts that could be easily damaged. With any particularly stubborn part, always recheck to make sure that every fastener has been removed.

Avoid using a screwdriver or bar to pry apart an assembly, as they can easily mar the gasket sealing surfaces of the parts, which must remain smooth. If prying is absolutely necessary, use an old broom handle, but keep in mind that extra clean up will be necessary if the wood splinters.

After the parts are separated, the old gasket must be carefully scraped off and the gasket surfaces cleaned. Stubborn gasket material can be soaked with rust penetrant or treated with a special chemical to soften it so it can be easily scraped off. A scraper can be fashioned from a piece of copper tubing by flattening and sharpening one end. Copper is recommended because it is usually softer than the surfaces to be scraped, which reduces the chance of gouging the part. Some gaskets can be removed with a wire brush, but regardless of the method used, the mating surfaces must be left clean and smooth. If for some reason the gasket surface is gouged, then a gasket sealer thick enough to fill scratches will have to be used during reassembly of the components. For most applications, a non-drying (or semi-drying) gasket sealer should be used.

Hose removal tips

Warning: *If the vehicle is equipped with air conditioning, do not disconnect any of the A/C hoses without first having the system depressurized by a dealer service department or a service station.*

Hose removal precautions closely parallel gasket removal precautions. Avoid scratching or gouging the surface that the hose mates against or the connection may leak. This is especially true for radiator hoses. Because of various chemical reactions, the rubber in hoses can bond itself to the metal spigot that the hose fits over. To remove a hose, first loosen the hose clamps that secure it to the spigot. Then, with slip-joint pliers, grab the hose at the clamp and rotate it around the spigot. Work it back and forth until it is completely free, then pull it off. Silicone or other lubricants will ease removal if they can be applied between the hose and the outside of the spigot. Apply the same lubricant to the inside of the hose and the outside of the spigot to simplify installation.

As a last resort (and if the hose is to be replaced with a new one anyway), the rubber can be slit with a knife and the hose peeled from the spigot. If this must be done, be careful that the metal connection is not damaged.

If a hose clamp is broken or damaged, do not reuse it. Wire-type clamps usually weaken with age, so it is a good idea to replace them with screw-type clamps whenever a hose is removed.

Tools

A selection of good tools is a basic requirement for anyone who plans to maintain and repair his or her own vehicle. For the owner who has few tools, the initial investment might seem high, but when compared to the spiraling costs of professional auto maintenance and repair, it is a wise one.

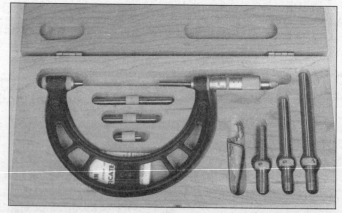

Micrometer set

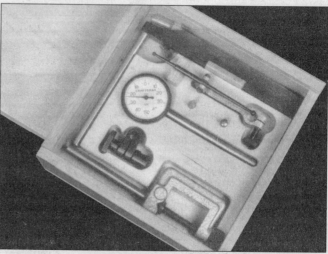

Dial indicator set

Maintenance techniques, tools and working facilities

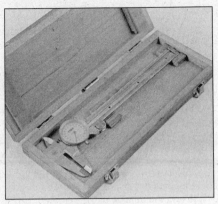
Dial caliper

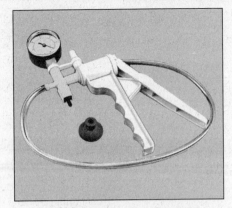
Hand-operated vacuum pump

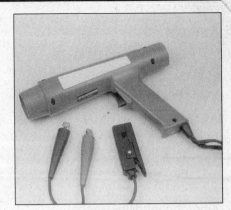
Timing light

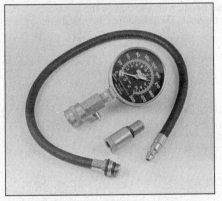
Compression gauge with spark plug hole adapter

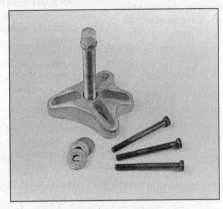
Damper/steering wheel puller

General purpose puller

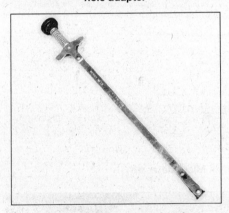
Hydraulic lifter removal tool

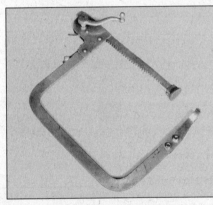
Valve spring compressor

Valve spring compressor

Ridge reamer

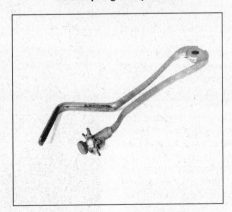
Piston ring groove cleaning tool

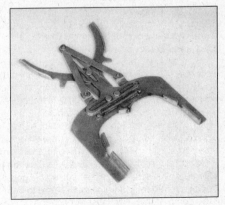
Ring removal/installation tool

Maintenance techniques, tools and working facilities

Ring compressor

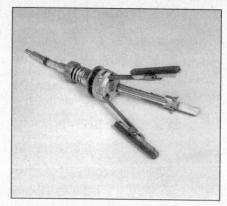
Cylinder hone

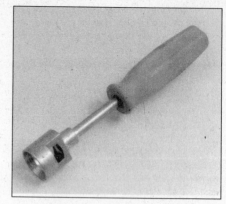
Brake hold-down spring tool

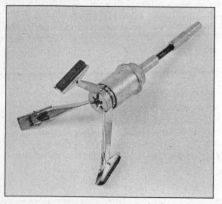
Brake cylinder hone

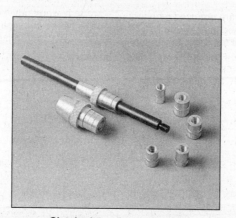
Clutch plate alignment tool

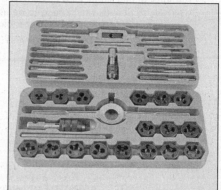
Tap and die set

To help the owner decide which tools are needed to perform the tasks detailed in this manual, the following tool lists are offered: *Maintenance and minor repair*, *Repair/overhaul* and *Special*.

The newcomer to practical mechanics should start off with the *maintenance and minor repair* tool kit, which is adequate for the simpler jobs performed on a vehicle. Then, as confidence and experience grow, the owner can tackle more difficult tasks, buying additional tools as they are needed. Eventually the basic kit will be expanded into the *repair and overhaul* tool set. Over a period of time, the experienced do-it-yourselfer will assemble a tool set complete enough for most repair and overhaul procedures and will add tools from the special category when it is felt that the expense is justified by the frequency of use.

Maintenance and minor repair tool kit

The tools in this list should be considered the minimum required for performance of routine maintenance, servicing and minor repair work. We recommend the purchase of combination wrenches (box-end and open-end combined in one wrench). While more expensive than open end wrenches, they offer the advantages of both types of wrench.

Combination wrench set (1/4-inch to 1 inch or 6 mm to 19 mm)
Adjustable wrench, 8 inch
Spark plug wrench with rubber insert
Spark plug gap adjusting tool
Feeler gauge set
Brake bleeder wrench
Standard screwdriver (5/16-inch x 6 inch)
Phillips screwdriver (No. 2 x 6 inch)
Combination pliers - 6 inch
Hacksaw and assortment of blades
Tire pressure gauge
Grease gun
Oil can
Fine emery cloth
Wire brush

Battery post and cable cleaning tool
Oil filter wrench
Funnel (medium size)
Safety goggles
Jackstands (2)
Drain pan

Note: If basic tune-ups are going to be part of routine maintenance, it will be necessary to purchase a good quality stroboscopic timing light and combination tachometer/dwell meter. Although they are included in the list of special tools, it is mentioned here because they are absolutely necessary for tuning most vehicles properly.

Repair and overhaul tool set

These tools are essential for anyone who plans to perform major repairs and are in addition to those in the maintenance and minor repair tool kit. Included is a comprehensive set of sockets which, though expensive, are invaluable because of their versatility, especially when various extensions and drives are available. We recommend the 1/2-inch drive over the 3/8-inch drive. Although the larger drive is bulky and more expensive, it has the capacity of accepting a very wide range of large sockets. Ideally, however, the mechanic should have a 3/8-inch drive set and a 1/2-inch drive set.

Socket set(s)
Reversible ratchet
Extension - 10 inch
Universal joint
Torque wrench (same size drive as sockets)
Ball peen hammer - 8 ounce
Soft-face hammer (plastic/rubber)
Standard screwdriver (1/4-inch x 6 inch)
Standard screwdriver (stubby - 5/16-inch)
Phillips screwdriver (No. 3 x 8 inch)
Phillips screwdriver (stubby - No. 2)
Pliers - vise grip
Pliers - lineman's

Maintenance techniques, tools and working facilities

Pliers - needle nose
Pliers - snap-ring (internal and external)
Cold chisel - 1/2-inch
Scribe
Scraper (made from flattened copper tubing)
Centerpunch
Pin punches (1/16, 1/8, 3/16-inch)
Steel rule/straightedge - 12 inch
Allen wrench set (1/8 to 3/8-inch or 4 mm to 10 mm)
A selection of files
Wire brush (large)
Jackstands (second set)
Jack (scissor or hydraulic type)

Note: *Another tool which is often useful is an electric drill with a chuck capacity of 3/8-inch and a set of good quality drill bits*

Special tools

The tools in this list include those which are not used regularly, are expensive to buy, or which need to be used in accordance with their manufacturer's instructions. Unless these tools will be used frequently, it is not very economical to purchase many of them. A consideration would be to split the cost and use between yourself and a friend or friends. In addition, most of these tools can be obtained from a tool rental shop on a temporary basis.

This list primarily contains only those tools and instruments widely available to the public, and not those special tools produced by the vehicle manufacturer for distribution to dealer service departments. Occasionally, references to the manufacturer's special tools are included in the text of this manual. Generally, an alternative method of doing the job without the special tool is offered. However, sometimes there is no alternative to their use. Where this is the case, and the tool cannot be purchased or borrowed, the work should be turned over to the dealer service department or an automotive repair shop.

Valve spring compressor
Piston ring groove cleaning tool
Piston ring compressor
Piston ring installation tool
Cylinder compression gauge
Cylinder ridge reamer
Cylinder surfacing hone
Cylinder bore gauge
Micrometers and/or dial calipers
Hydraulic lifter removal tool
Balljoint separator
Universal-type puller
Impact screwdriver
Dial indicator set
Stroboscopic timing light (inductive pick-up)
Hand operated vacuum/pressure pump
Tachometer/dwell meter
Universal electrical multimeter
Cable hoist
Brake spring removal and installation tools
Floor jack

Buying tools

For the do-it-yourselfer who is just starting to get involved in vehicle maintenance and repair, there are a number of options available when purchasing tools. If maintenance and minor repair is the extent of the work to be done, the purchase of individual tools is satisfactory. If, on the other hand, extensive work is planned, it would be a good idea to purchase a modest tool set from one of the large retail chain stores. A set can usually be bought at a substantial savings over the individual tool prices, and they often come with a tool box. As additional tools are needed, add-on sets, individual tools and a larger tool box can be purchased to expand the tool selection. Building a tool set gradually allows the cost of the tools to be spread over a longer period of time and gives the mechanic the freedom to choose only those tools that will actually be used.

Tool stores will often be the only source of some of the special tools that are needed, but regardless of where tools are bought, try to avoid cheap ones, especially when buying screwdrivers and sockets, because they won't last very long. The expense involved in replacing cheap tools will eventually be greater than the initial cost of quality tools.

Care and maintenance of tools

Good tools are expensive, so it makes sense to treat them with respect. Keep them clean and in usable condition and store them properly when not in use. Always wipe off any dirt, grease or metal chips before putting them away. Never leave tools lying around in the work area. Upon completion of a job, always check closely under the hood for tools that may have been left there so they won't get lost during a test drive.

Some tools, such as screwdrivers, pliers, wrenches and sockets, can be hung on a panel mounted on the garage or workshop wall, while others should be kept in a tool box or tray. Measuring instruments, gauges, meters, etc. must be carefully stored where they cannot be damaged by weather or impact from other tools.

When tools are used with care and stored properly, they will last a very long time. Even with the best of care, though, tools will wear out if used frequently. When a tool is damaged or worn out, replace it. Subsequent jobs will be safer and more enjoyable if you do.

Working facilities

Not to be overlooked when discussing tools is the workshop. If anything more than routine maintenance is to be carried out, some sort of suitable work area is essential.

It is understood, and appreciated, that many home mechanics do not have a good workshop or garage available, and end up removing an engine or doing major repairs outside. It is recommended, however, that the overhaul or repair be completed under the cover of a roof.

A clean, flat workbench or table of comfortable working height is an absolute necessity. The workbench should be equipped with a vise that has a jaw opening of at least four inches.

As mentioned previously, some clean, dry storage space is also required for tools, as well as the lubricants, fluids, cleaning solvents, etc. which soon become necessary.

Sometimes waste oil and fluids, drained from the engine or cooling system during normal maintenance or repairs, present a disposal problem. To avoid pouring them on the ground or into a sewage system, pour the used fluids into large containers, seal them with caps and take them to an authorized disposal site or recycling center. Plastic jugs, such as old antifreeze containers, are ideal for this purpose.

Always keep a supply of old newspapers and clean rags available. Old towels are excellent for mopping up spills. Many mechanics use rolls of paper towels for most work because they are readily available and disposable. To help keep the area under the vehicle clean, a large cardboard box can be cut open and flattened to protect the garage or shop floor.

Whenever working over a painted surface, such as when leaning over a fender to service something under the hood, always cover it with an old blanket or bedspread to protect the finish. Vinyl covered pads, made especially for this purpose, are available at auto parts stores.

Jacking and towing

Jacking

The jack supplied with the vehicle should be used only for raising the vehicle when changing a tire or placing jackstands under the frame. **Warning**: *Never crawl under the vehicle or start the engine when this jack is being used as the only means of support.*

The vehicle should be on level ground with the wheels blocked and the transmission in Park (automatic) or Reverse (manual). Pry off the hub cap (if equipped) using the tapered end of the lug wrench. Loosen the lug bolts one-half turn and leave them in place until the wheel is raised off the ground.

Position the head of the jack under the side of the vehicle, making sure it engages with the pocket made for this purpose (just behind the front wheel or forward of the rear wheel). Turn the jack handle clockwise until the wheel is raised off the ground. Remove the lug bolts, pull off the wheel and replace it with the spare.

Install the lug bolts and tighten them until snug. Lower the vehicle by turning the jackscrew counterclockwise. Remove the jack and tighten the bolts in a diagonal pattern to the torque listed in the Chapter 1 Specifications. If a torque wrench is not available, have the torque checked by a service station as soon as possible. Replace the hubcap.

Towing

Manual transmission-equipped vehicles can be towed with all four wheels on the ground. Automatic transmission-equipped vehicles can be towed with all four wheels on the ground if speeds do not exceed 35 mph and the distance is not over 50 miles, otherwise transmission damage can result.

Towing equipment specifically designed for this purpose should be used and should be attached to the main structural members of the vehicle, not the bumper or brackets. Sling-type towing equipment must **not** be used on these vehicles.

Safety is a major consideration while towing and all applicable state and local laws must be obeyed. A safety chain system must be used for all towing.

While towing, the parking brake should be released and the transmission and transfer case should be in Neutral. The steering must be unlocked (ignition switch in the Off position). Remember that power steering and power brakes will not work with the engine off.

Booster battery (jump) starting

Observe these precautions when using a booster battery to start a vehicle:

a) Before connecting the booster battery, make sure the ignition switch is in the Off position.
b) Turn off the lights, heater and other electrical loads.
c) Your eyes should be shielded. Safety goggles are a good idea.
d) Make sure the booster battery is the same voltage as the dead one in the vehicle.
e) The two vehicles MUST NOT TOUCH each other!
f) Make sure the transmission is in Neutral (manual) or Park (automatic).
g) If the booster battery is not a maintenance-free type, remove the vent caps and lay a cloth over the vent holes.

On engine compartment-mounted batteries, connect the red jumper cable to the positive (+) terminals of each battery **(see illustration)**. On vehicles with remotely-mounted batteries (in the trunk or under the rear seat), connect the red jumper cable to the jumper terminal in the engine compartment **(see illustrations)**.

Connect one end of the black jumper cable to the negative (-) terminal of the booster battery. The other end of this cable should be connected to a good ground on the vehicle to be started, such as a bolt or bracket on the body.

Start the engine using the booster battery, then, with the engine running at idle speed, disconnect the jumper cables in the reverse order of connection.

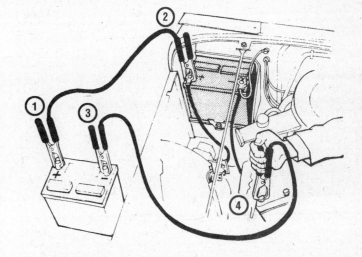

Make the booster battery cable connections in the numerical order shown (note that the negative cable of the booster battery is NOT attached to the negative terminal of the dead battery.

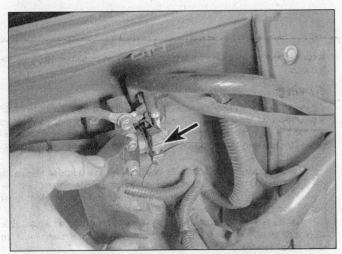

On 3-series models, remove the plastic cover and connect the positive (red) jumper cable to this junction (arrow)

Some 5-series models also have a remotely-mounted battery - connect the positive (red) jumper cable to this terminal in the left rear corner of the engine compartment

Automotive chemicals and lubricants

A number of automotive chemicals and lubricants are available for use during vehicle maintenance and repair. They include a wide variety of products ranging from cleaning solvents and degreasers to lubricants and protective sprays for rubber, plastic and vinyl.

Cleaners

Carburetor cleaner and choke cleaner is a strong solvent for gum, varnish and carbon. Most carburetor cleaners leave a dry-type lubricant film which will not harden or gum up. Because of this film it is not recommended for use on electrical components

Brake system cleaner is used to remove grease and brake fluid from the brake system, where clean surfaces are absolutely necessary. It leaves no residue and often eliminates brake squeal caused by contaminants.

Electrical cleaner removes oxidation, corrosion and carbon deposits from electrical contacts, restoring full current flow. It can also be used to clean spark plugs, carburetor jets, voltage regulators and other parts where an oil-free surface is desired.

Demoisturants remove water and moisture from electrical components such as alternators, voltage regulators, electrical connectors and fuse blocks. They are non-conductive, non-corrosive and non-flammable.

Degreasers are heavy-duty solvents used to remove grease from the outside of the engine and from chassis components. They can be sprayed or brushed on and, depending on the type, are rinsed off either with water or solvent.

Lubricants

Motor oil is the lubricant formulated for use in engines. It normally contains a wide variety of additives to prevent corrosion and reduce foaming and wear. Motor oil comes in various weights (viscosity ratings) from 5 to 80. The recommended weight of the oil depends on the season, temperature and the demands on the engine. Light oil is used in cold climates and under light load conditions. Heavy oil is used in hot climates and where high loads are encountered. Multi-viscosity oils are designed to have characteristics of both light and heavy oils and are available in a number of weights from 5W-20 to 20W-50.

Gear oil is designed to be used in differentials, manual transmissions and other areas where high-temperature lubrication is required.

Chassis and wheel bearing grease is a heavy grease used where increased loads and friction are encountered, such as for wheel bearings, balljoints, tie-rod ends and universal joints.

High-temperature wheel bearing grease is designed to withstand the extreme temperatures encountered by wheel bearings in disc brake equipped vehicles. It usually contains molybdenum disulfide (moly), which is a dry-type lubricant.

White grease is a heavy grease for metal-to-metal applications where water is a problem. White grease stays soft under both low and high temperatures (usually from -100 to +190-degrees F), and will not wash off or dilute in the presence of water.

Assembly lube is a special extreme pressure lubricant, usually containing moly, used to lubricate high-load parts (such as main and rod bearings and cam lobes) for initial start-up of a new engine. The assembly lube lubricates the parts without being squeezed out or washed away until the engine oiling system begins to function.

Silicone lubricants are used to protect rubber, plastic, vinyl and nylon parts.

Graphite lubricants are used where oils cannot be used due to contamination problems, such as in locks. The dry graphite will lubricate metal parts while remaining uncontaminated by dirt, water, oil or acids. It is electrically conductive and will not foul electrical contacts in locks such as the ignition switch.

Moly penetrants loosen and lubricate frozen, rusted and corroded fasteners and prevent future rusting or freezing.

Heat-sink grease is a special electrically non-conductive grease that is used for mounting electronic ignition modules where it is essential that heat is transferred away from the module.

Sealants

RTV sealant is one of the most widely used gasket compounds. Made from silicone, RTV is air curing, it seals, bonds, waterproofs, fills surface irregularities, remains flexible, doesn't shrink, is relatively easy to remove, and is used as a supplementary sealer with almost all low and medium temperature gaskets.

Anaerobic sealant is much like RTV in that it can be used either to seal gaskets or to form gaskets by itself. It remains flexible, is solvent resistant and fills surface imperfections. The difference between an anaerobic sealant and an RTV-type sealant is in the curing. RTV cures when exposed to air, while an anaerobic sealant cures only in the absence of air. This means that an anaerobic sealant cures only after the assembly of parts, sealing them together.

Thread and pipe sealant is used for sealing hydraulic and pneumatic fittings and vacuum lines. It is usually made from a Teflon compound, and comes in a spray, a paint-on liquid and as a wrap-around tape.

Chemicals

Anti-seize compound prevents seizing, galling, cold welding, rust and corrosion in fasteners. High-temperature ant-seize, usually made with copper and graphite lubricants, is used for exhaust system and exhaust manifold bolts.

Anaerobic locking compounds are used to keep fasteners from vibrating or working loose and cure only after installation, in the absence of air. Medium strength locking compound is used for small nuts, bolts and screws that may be removed later. High-strength locking compound is for large nuts, bolts and studs which aren't removed on a regular basis.

Oil additives range from viscosity index improvers to chemical treatments that claim to reduce internal engine friction. It should be noted that most oil manufacturers caution against using additives with their oils.

Gas additives perform several functions, depending on their chemical makeup. They usually contain solvents that help dissolve gum and varnish that build up on carburetor, fuel injection and intake parts. They also serve to break down carbon deposits that form on the inside surfaces of the combustion chambers. Some additives contain upper cylinder lubricants for valves and piston rings, and others contain chemicals to remove condensation from the gas tank.

Miscellaneous

Brake fluid is specially formulated hydraulic fluid that can withstand the heat and pressure encountered in brake systems. Care must be taken so this fluid does not come in contact with painted surfaces or plastics. An opened container should always be resealed to prevent contamination by water or dirt.

Weatherstrip adhesive is used to bond weatherstripping around doors, windows and trunk lids. It is sometimes used to attach trim pieces.

Undercoating is a petroleum-based, tar-like substance that is designed to protect metal surfaces on the underside of the vehicle from corrosion. It also acts as a sound-deadening agent by insulating the bottom of the vehicle.

Waxes and polishes are used to help protect painted and plated surfaces from the weather. Different types of paint may require the use of different types of wax and polish. Some polishes utilize a chemical or abrasive cleaner to help remove the top layer of oxidized (dull) paint on older vehicles. In recent years many non-wax polishes that contain a wide variety of chemicals such as polymers and silicones have been introduced. These non-wax polishes are usually easier to apply and last longer than conventional waxes and polishes.

Safety first

Regardless of how enthusiastic you may be about getting on with the job at hand, take the time to ensure that your safety is not jeopardized. A moment's lack of attention can result in an accident, as can failure to observe certain simple safety precautions. The possibility of an accident will always exist, and the following points should not be considered a comprehensive list of all dangers. Rather, they are intended to make you aware of the risks and to encourage a safety conscious approach to all work you carry out on your vehicle.

Essential DOs and DON'Ts

DON'T rely on a jack when working under the vehicle. Always use approved jackstands to support the weight of the vehicle and place them under the recommended lift or support points.

DON'T attempt to loosen extremely tight fasteners (i.e. wheel lug nuts) while the vehicle is on a jack - it may fall.

DON'T start the engine without first making sure that the transmission is in Neutral (or Park where applicable) and the parking brake is set.

DON'T remove the radiator cap from a hot cooling system - let it cool or cover it with a cloth and release the pressure gradually.

DON'T attempt to drain the engine oil until you are sure it has cooled to the point that it will not burn you.

DON'T touch any part of the engine or exhaust system until it has cooled sufficiently to avoid burns.

DON'T siphon toxic liquids such as gasoline, antifreeze and brake fluid by mouth, or allow them to remain on your skin.

DON'T inhale brake lining dust - it is potentially hazardous (see *Asbestos* below)

DON'T allow spilled oil or grease to remain on the floor - wipe it up before someone slips on it.

DON'T use loose fitting wrenches or other tools which may slip and cause injury.

DON'T push on wrenches when loosening or tightening nuts or bolts. Always try to pull the wrench toward you. If the situation calls for pushing the wrench away, push with an open hand to avoid scraped knuckles if the wrench should slip.

DON'T attempt to lift a heavy component alone - get someone to help you.

DON'T rush or take unsafe shortcuts to finish a job.

DON'T allow children or animals in or around the vehicle while you are working on it.

DO wear eye protection when using power tools such as a drill, sander, bench grinder, etc. and when working under a vehicle.

DO keep loose clothing and long hair well out of the way of moving parts.

DO make sure that any hoist used has a safe working load rating adequate for the job.

DO get someone to check on you periodically when working alone on a vehicle.

DO carry out work in a logical sequence and make sure that everything is correctly assembled and tightened.

DO keep chemicals and fluids tightly capped and out of the reach of children and pets.

DO remember that your vehicle's safety affects that of yourself and others. If in doubt on any point, get professional advice.

Asbestos

Certain friction, insulating, sealing, and other products - such as brake linings, brake bands, clutch linings, torque converters, gaskets, etc. - contain asbestos. Extreme care must be taken to avoid inhalation of dust from such products, since it is hazardous to health. If in doubt, assume that they do contain asbestos.

Fire

Remember at all times that gasoline is highly flammable. Never smoke or have any kind of open flame around when working on a vehicle. But the risk does not end there. A spark caused by an electrical short circuit, by two metal surfaces contacting each other, or even by static electricity built up in your body under certain conditions, can ignite gasoline vapors, which in a confined space are highly explosive. Do not, under any circumstances, use gasoline for cleaning parts. Use an approved safety solvent.

Always disconnect the battery ground (-) cable at the battery before working on any part of the fuel system or electrical system. Never risk spilling fuel on a hot engine or exhaust component. It is strongly recommended that a fire extinguisher suitable for use on fuel and electrical fires be kept handy in the garage or workshop at all times. Never try to extinguish a fuel or electrical fire with water.

Fumes

Certain fumes are highly toxic and can quickly cause unconsciousness and even death if inhaled to any extent. Gasoline vapor falls into this category, as do the vapors from some cleaning solvents. Any draining or pouring of such volatile fluids should be done in a well ventilated area.

When using cleaning fluids and solvents, read the instructions on the container carefully. Never use materials from unmarked containers.

Never run the engine in an enclosed space, such as a garage. Exhaust fumes contain carbon monoxide, which is extremely poisonous. If you need to run the engine, always do so in the open air, or at least have the rear of the vehicle outside the work area.

If you are fortunate enough to have the use of an inspection pit, never drain or pour gasoline and never run the engine while the vehicle is over the pit. The fumes, being heavier than air, will concentrate in the pit with possibly lethal results.

The battery

Never create a spark or allow a bare light bulb near a battery. They normally give off a certain amount of hydrogen gas, which is highly explosive.

Always disconnect the battery ground (-) cable at the battery before working on the fuel or electrical systems.

If possible, loosen the filler caps or cover when charging the battery from an external source (this does not apply to sealed or maintenance-free batteries). Do not charge at an excessive rate or the battery may burst.

Take care when adding water to a non maintenance-free battery and when carrying a battery. The electrolyte, even when diluted, is very corrosive and should not be allowed to contact clothing or skin.

Always wear eye protection when cleaning the battery to prevent the caustic deposits from entering your eyes.

Household current

When using an electric power tool, inspection light, etc., which operates on household current, always make sure that the tool is correctly connected to its plug and that, where necessary, it is properly grounded. Do not use such items in damp conditions and, again, do not create a spark or apply excessive heat in the vicinity of fuel or fuel vapor.

Secondary ignition system voltage

A severe electric shock can result from touching certain parts of the ignition system (such as the spark plug wires) when the engine is running or being cranked, particularly if components are damp or the insulation is defective. In the case of an electronic ignition system, the secondary system voltage is much higher and could prove fatal.

Conversion factors

Length (distance)
Inches (in)	X	25.4	= Millimetres (mm)	X 0.0394	= Inches (in)
Feet (ft)	X	0.305	= Metres (m)	X 3.281	= Feet (ft)
Miles	X	1.609	= Kilometres (km)	X 0.621	= Miles

Volume (capacity)
Cubic inches (cu in; in^3)	X	16.387	= Cubic centimetres (cc; cm^3)	X 0.061	= Cubic inches (cu in; in^3)
Imperial pints (Imp pt)	X	0.568	= Litres (l)	X 1.76	= Imperial pints (Imp pt)
Imperial quarts (Imp qt)	X	1.137	= Litres (l)	X 0.88	= Imperial quarts (Imp qt)
Imperial quarts (Imp qt)	X	1.201	= US quarts (US qt)	X 0.833	= Imperial quarts (Imp qt)
US quarts (US qt)	X	0.946	= Litres (l)	X 1.057	= US quarts (US qt)
Imperial gallons (Imp gal)	X	4.546	= Litres (l)	X 0.22	= Imperial gallons (Imp gal)
Imperial gallons (Imp gal)	X	1.201	= US gallons (US gal)	X 0.833	= Imperial gallons (Imp gal)
US gallons (US gal)	X	3.785	= Litres (l)	X 0.264	= US gallons (US gal)

Mass (weight)
Ounces (oz)	X	28.35	= Grams (g)	X 0.035	Ounces (oz)
Pounds (lb)	X	0.454	= Kilograms (kg)	X 2.205	= Pounds (lb)

Force
Ounces-force (ozf; oz)	X	0.278	= Newtons (N)	X 3.6	= Ounces-force (ozf; oz)
Pounds-force (lbf; lb)	X	4.448	= Newtons (N)	X 0.225	= Pounds-force (lbf; lb)
Newtons (N)	X	0.1	= Kilograms-force (kgf; kg)	X 9.81	= Newtons (N)

Pressure
Pounds-force per square inch (psi; lbf/in^2; lb/in^2)	X	0.070	= Kilograms-force per square centimetre (kgf/cm^2; kg/cm^2)	X 14.223	= Pounds-force per square inch (psi; lbf/in^2; lb/in^2)
Pounds-force per square inch (psi; lbf/in^2; lb/in^2)	X	0.068	= Atmospheres (atm)	X 14.696	= Pounds-force per square inch (psi; lbf/in^2; lb/in^2)
Pounds-force per square inch (psi; lbf/in^2; lb/in^2)	X	0.069	= Bars	X 14.5	= Pounds-force per square inch (psi; lbf/in^2; lb/in^2)
Pounds-force per square inch (psi; lbf/in^2; lb/in^2)	X	6.895	= Kilopascals (kPa)	X 0.145	= Pounds-force per square inch (psi; lbf/in^2; lb/in^2)
Kilopascals (kPa)	X	0.01	= Kilograms-force per square centimetre (kgf/cm^2; kg/cm^2)	X 98.1	= Kilopascals (kPa)

Torque (moment of force)
Pounds-force inches (lbf in; lb in)	X	1.152	= Kilograms-force centimetre (kgf cm; kg cm)	X 0.868	= Pounds-force inches (lbf in; lb in)
Pounds-force inches (lbf in; lb in)	X	0.113	= Newton metres (Nm)	X 8.85	= Pounds-force inches (lbf in; lb in)
Pounds-force inches (lbf in; lb in)	X	0.083	= Pounds-force feet (lbf ft; lb ft)	X 12	= Pounds-force inches (lbf in; lb in)
Pounds-force feet (lbf ft; lb ft)	X	0.138	= Kilograms-force metres (kgf m; kg m)	X 7.233	= Pounds-force feet (lbf ft; lb ft)
Pounds-force feet (lbf ft; lb ft)	X	1.356	= Newton metres (Nm)	X 0.738	= Pounds-force feet (lbf ft; lb ft)
Newton metres (Nm)	X	0.102	= Kilograms-force metres (kgf m; kg m)	X 9.804	= Newton metres (Nm)

Power
Horsepower (hp)	X	745.7	= Watts (W)	X 0.0013	= Horsepower (hp)

Velocity (speed)
Miles per hour (miles/hr; mph)	X	1.609	= Kilometres per hour (km/hr; kph)	X 0.621	= Miles per hour (miles/hr; mph)

Fuel consumption*
Miles per gallon, Imperial (mpg)	X	0.354	= Kilometres per litre (km/l)	X 2.825	= Miles per gallon, Imperial (mpg)
Miles per gallon, US (mpg)	X	0.425	= Kilometres per litre (km/l)	X 2.352	= Miles per gallon, US (mpg)

Temperature

Degrees Fahrenheit = (°C x 1.8) + 32

Degrees Celsius (Degrees Centigrade; °C) = (°F - 32) x 0.56

*It is common practice to convert from miles per gallon (mpg) to litres/100 kilometres (l/100km), where mpg (Imperial) x l/100 km = 282 and mpg (US) x l/100 km = 235

Troubleshooting

Contents

Symptom	Section
Engine	
Engine backfires	15
Engine diesels (continues to run) after switching off	18
Engine hard to start when cold	3
Engine hard to start when hot	4
Engine lacks power	14
Engine lopes while idling or idles erratically	8
Engine misses at idle speed	9
Engine misses throughout driving speed range	10
Engine rotates but will not start	2
Engine runs with oil pressure light on	17
Engine stalls	13
Engine starts but stops immediately	6
Engine stumbles on acceleration	11
Engine surges while holding accelerator steady	12
Engine will not rotate when attempting to start	1
Oil puddle under engine	7
Pinging or knocking engine sounds during acceleration or uphill	16
Starter motor noisy or excessively rough in engagement	5
Engine electrical system	
Discharge warning light fails to go out	20
Battery will not hold a charge	19
Discharge warning light fails to come on when key is turned on	21
Fuel system	
Excessive fuel consumption	22
Fuel leakage and/or fuel odor	23
Cooling system	
Coolant loss	28
External coolant leakage	26
Internal coolant leakage	27
Overcooling	25
Overheating	24
Poor coolant circulation	29
Clutch	
Clutch pedal stays on floor	38
Clutch slips (engine speed increases with no increase in vehicle speed)	35
Fluid in area of master cylinder dust cover and on pedal	31
Fluid on slave cylinder	32
Grabbing (chattering) as clutch is engaged	36
High pedal effort	39
Noise in clutch area	37
Pedal feels "spongy" when depressed	33
Pedal travels to floor - no pressure or very little resistance	30
Unable to select gears	34
Manual transmission	
Leaks lubricant	45
Noisy in all gears	43
Noisy in Neutral with engine running	41
Noisy in one particular gear	42
Slips out of gear	44
Vibration	40
Automatic transmission	
Engine will start in gears other than Park or Neutral	50
Fluid leakage	46
General shift mechanism problems	48
Transmission fluid brown or has a burned smell	47
Transmission slips, shifts roughly, is noisy or has no drive in forward or reverse gears	51
Transmission will not downshift with accelerator pedal pressed to the floor	49
Brakes	
Brake pedal feels spongy when depressed	59
Brake pedal travels to the floor with little resistance	60
Brake roughness or chatter (pedal pulsates)	54
Dragging brakes	57
Excessive brake pedal travel	56
Excessive pedal effort required to stop vehicle	55
Grabbing or uneven braking action	58
Noise (high-pitched squeal when the brakes are applied)	53
Parking brake does not hold	61
Vehicle pulls to one side during braking	52
Suspension and steering systems	
Abnormal or excessive tire wear	63
Abnormal noise at the front end	68
Cupped tires	73
Erratic steering when braking	70
Excessive pitching and/or rolling around corners or during braking	71
Excessive play or looseness in steering system	77
Excessive tire wear on inside edge	75
Excessive tire wear on outside edge	74
Hard steering	66
Poor returnability of steering to center	67
Rattling or clicking noise in rack-and-pinion	78
Shimmy, shake or vibration	65
Suspension bottoms	72
Tire tread worn in one place	76
Vehicle pulls to one side	62
Wander or poor steering stability	69
Wheel makes a "thumping" noise	64

Troubleshooting

This Section provides an easy reference guide to the more common problems which may occur during the operation of your vehicle. These problems and their possible causes are grouped under headings denoting various components or systems, such as Engine, Cooling system, etc. They also refer you to the Chapter and/or Section which deals with the problem.

Remember that successful troubleshooting is not a mysterious black art practiced only by professional mechanics. It is simply the result of the right knowledge combined with an intelligent, systematic approach to the problem. Always work by a process of elimination, starting with the simplest solution and working through to the most complex - and never overlook the obvious. Anyone can run the gas tank dry or leave the lights on overnight, so don't assume that you are exempt from such oversights.

Finally, always establish a clear idea of why a problem has occurred and take steps to ensure that it doesn't happen again. If the electrical system fails because of a poor connection, check all other connections in the system to make sure that they don't fail as well. If a particular fuse continues to blow, find out why - don't just replace one fuse after another. Remember, failure of a small component can often be indicative of potential failure or incorrect functioning of a more important component or system.

Engine

1 Engine will not rotate when attempting to start

1 Battery terminal connections loose or corroded (Chapter 1).
2 Battery discharged or faulty (Chapter 1).
3 Automatic transmission not completely engaged in Park (Chapter 7) or clutch not completely depressed (Chapter 8).
4 Broken, loose or disconnected wiring in the starting circuit (Chapters 5 and 12).
5 Starter motor pinion jammed in flywheel ring gear (Chapter 5).
6 Starter solenoid faulty (Chapter 5).
7 Starter motor faulty (Chapter 5).
8 Ignition switch faulty (Chapter 12).
9 Starter pinion or flywheel teeth worn or broken (Chapter 5).
10 Internal engine problem (Chapter 2B).

2 Engine rotates but will not start

1 Fuel tank empty.
2 Battery discharged (engine rotates slowly) (Chapter 5).
3 Battery terminal connections loose or corroded (Chapter 1).
4 Leaking fuel injector(s), faulty fuel pump, pressure regulator, etc. (Chapter 4).
5 Fuel not reaching fuel injection system (Chapter 4).
6 Ignition components damp or damaged (Chapter 5).
7 Fuel injector stuck open (Chapter 4).
8 Worn, faulty or incorrectly gapped spark plugs (Chapter 1).
9 Broken, loose or disconnected wiring in the starting circuit (Chapter 5).
10 Loose distributor is changing ignition timing (Chapter 1).
11 Broken, loose or disconnected wires at the ignition coil or faulty coil (Chapter 5).

3 Engine hard to start when cold

1 Battery discharged or low (Chapter 1).
2 Fuel system malfunctioning (Chapter 4).
3 Injector(s) leaking (Chapter 4).
4 Distributor rotor carbon tracked (Chapter 5).

4 Engine hard to start when hot

1 Air filter clogged (Chapter 1).
2 Fuel not reaching the fuel injection system (Chapter 4).
3 Corroded battery connections, especially ground (Chapter 1).

5 Starter motor noisy or excessively rough in engagement

1 Pinion or flywheel gear teeth worn or broken (Chapter 5).
2 Starter motor mounting bolts loose or missing (Chapter 5).

6 Engine starts but stops immediately

1 Loose or faulty electrical connections at distributor, coil or alternator (Chapter 5).
2 Insufficient fuel reaching the fuel injector(s) (Chapters 1 and 4).
3 Damaged fuel injection system speed sensors (Chapter 5).
4 Faulty fuel injection relays (Chapter 5).

7 Oil puddle under engine

1 Oil pan gasket and/or oil pan drain bolt seal leaking (Chapter 2).
2 Oil pressure sending unit leaking (Chapter 2).
3 Valve cover gaskets leaking (Chapter 2).
4 Engine oil seals leaking (Chapter 2).

8 Engine lopes while idling or idles erratically

1 Vacuum leakage (Chapter 4).
2 Air filter clogged (Chapter 1).
3 Fuel pump not delivering sufficient fuel to the fuel injection system (Chapter 4).
4 Leaking head gasket (Chapter 2).
5 Timing belt/chain and/or sprockets worn (Chapter 2).
6 Camshaft lobes worn (Chapter 2).
7 Faulty charcoal canister on some 325i models produced in December 1986 (Chapter 6).

9 Engine misses at idle speed

1 Spark plugs worn or not gapped properly (Chapter 1).
2 Faulty spark plug wires (Chapter 1).
3 Vacuum leaks (Chapter 1).
4 Incorrect ignition timing (Chapter 5).
5 Uneven or low compression (Chapter 2).
6 Faulty charcoal canister on some 325i models produced in December 1986 (Chapter 6).

10 Engine misses throughout driving speed range

1 Fuel filter clogged and/or impurities in the fuel system (Chapter 1).
2 Low fuel output at the injectors (Chapter 4).
3 Faulty or incorrectly gapped spark plugs (Chapter 1).
4 Incorrect ignition timing (Chapter 5).
5 Cracked distributor cap, disconnected distributor wires or damaged distributor components (Chapter 1).
6 Leaking spark plug wires (Chapter 1).
7 Faulty emission system components (Chapter 6).
8 Low or uneven cylinder compression pressures (Chapter 2).
9 Weak or faulty ignition system (Chapter 5).
10 Vacuum leak in fuel injection system, intake manifold or vacuum hoses (Chapter 4).

Troubleshooting   0-21

11 Engine stumbles on acceleration

1. Spark plugs fouled (Chapter 1).
2. Fuel injection system malfunctioning (Chapter 4).
3. Fuel filter clogged (Chapters 1 and 4).
4. Incorrect ignition timing (Chapter 5).
5. Intake manifold air leak (Chapter 4).

12 Engine surges while holding accelerator steady

1. Intake air leak (Chapter 4).
2. Fuel pump faulty (Chapter 4).
3. Loose fuel injector harness connections (Chapters 4 and 6).
4. Defective ECU (Chapter 5).

13 Engine stalls

1. Idle speed incorrect (Chapter 1).
2. Fuel filter clogged and/or water and impurities in the fuel system (Chapter 1).
3. Distributor components damp or damaged (Chapter 5).
4. Faulty emissions system components (Chapter 6).
5. Faulty or incorrectly gapped spark plugs (Chapter 1).
6. Faulty spark plug wires (Chapter 1).
7. Vacuum leak in the fuel injection system, intake manifold or vacuum hoses (Chapter 4).

14 Engine lacks power

1. Incorrect ignition timing (Chapter 5).
2. Excessive play in distributor shaft (Chapter 5).
3. Worn rotor, distributor cap or wires (Chapters 1 and 5).
4. Faulty or incorrectly gapped spark plugs (Chapter 1).
5. Fuel injection system malfunctioning (Chapter 4).
6. Faulty coil (Chapter 5).
7. Brakes binding (Chapter 1).
8. Automatic transmission fluid level incorrect (Chapter 1).
9. Clutch slipping (Chapter 8).
10. Fuel filter clogged and/or impurities in the fuel system (Chapter 1).
11. Emission control system not functioning properly (Chapter 6).
12. Low or uneven cylinder compression pressures (Chapter 2).

15 Engine backfires

1. Emissions system not functioning properly (Chapter 6).
2. Ignition timing incorrect (Chapter 1).
3. Faulty secondary ignition system (cracked spark plug insulator, faulty plug wires, distributor cap and/or rotor) (Chapters 1 and 5).
4. Fuel injection system malfunctioning (Chapter 4).
5. Vacuum leak at fuel injector(s), intake manifold or vacuum hoses (Chapter 4).

16 Pinging or knocking engine sounds during acceleration or uphill

1. Incorrect grade of fuel.
2. Ignition timing incorrect (Chapter 5).
3. Fuel injection system in need of adjustment (Chapter 4).
4. Improper or damaged spark plugs or wires (Chapter 1).
5. Worn or damaged distributor components (Chapter 5).
6. Faulty emission system (Chapter 6).
7. Vacuum leak (Chapter 4).

17 Engine runs with oil pressure light on

1. Low oil level (Chapter 1).
2. Idle rpm too low (Chapter 1).
3. Short in wiring circuit (Chapter 12).
4. Faulty oil pressure sending unit (Chapter 2).
5. Worn engine bearings and/or oil pump (Chapter 2).

18 Engine diesels (continues to run) after switching off

1. Idle speed too high (Chapter 1).
2. Excessive engine operating temperature (Chapter 3).
3. Incorrect fuel octane grade.

Engine electrical system

19 Battery will not hold a charge

1. Alternator drivebelt defective or not adjusted properly (Chapter 1).
2. Electrolyte level low (Chapter 1).
3. Battery terminals loose or corroded (Chapter 1).
4. Alternator not charging properly (Chapter 5).
5. Loose, broken or faulty wiring in the charging circuit (Chapter 5).
6. Short in vehicle wiring (Chapters 5 and 12).
7. Internally defective battery (Chapters 1 and 5).
8. Burned-out dashboard warning light bulb on some 1982 through 1986 models (Chapter 5).

20 Discharge warning light fails to go out

1. Faulty alternator or charging circuit (Chapter 5).
2. Alternator drivebelt defective or out of adjustment (Chapter 1).
3. Alternator voltage regulator inoperative (Chapter 5).

21 Discharge warning light fails to come on when key is turned on

1. Warning light bulb defective (Chapter 12).
2. Fault in the printed circuit, dash wiring or bulb holder (Chapter 12).

Fuel system

22 Excessive fuel consumption

1. Dirty or clogged air filter element (Chapter 1).
2. Incorrectly set ignition timing (Chapter 5).
3. Emissions system not functioning properly (Chapter 6).
4. Fuel injection internal parts excessively worn or damaged (Chapter 4).
5. Low tire pressure or incorrect tire size (Chapter 1).

23 Fuel leakage and/or fuel odor

1. Leak in a fuel feed or vent line (Chapter 4).
2. Tank overfilled.
3. Fuel injector internal parts excessively worn (Chapter 4).

Troubleshooting

Cooling system

24 Overheating

1 Insufficient coolant in system (Chapter 1).
2 Water pump drivebelt defective or out of adjustment (Chapter 1).
3 Radiator core blocked or grille restricted (Chapter 3).
4 Thermostat faulty (Chapter 3).
5 Radiator cap not maintaining proper pressure (Chapter 3).
6 Ignition timing incorrect (Chapter 5).

25 Overcooling

1 Faulty thermostat (Chapter 3).

26 External coolant leakage

1 Deteriorated/damaged hoses; loose clamps (Chapters 1 and 3).
2 Water pump seal defective (Chapters 1 and 3).
3 Leakage from radiator core or header tank (Chapter 3).
4 Engine drain or water jacket core plugs leaking (Chapter 2).

27 Internal coolant leakage

1 Leaking cylinder head gasket (Chapter 2).
2 Cracked cylinder bore or cylinder head (Chapter 2).

28 Coolant loss

1 Too much coolant in system (Chapter 1).
2 Coolant boiling away because of overheating (Chapter 3).
3 Internal or external leakage (Chapter 3).
4 Faulty radiator cap (Chapter 3).

29 Poor coolant circulation

1 Inoperative water pump (Chapter 3).
2 Restriction in cooling system (Chapters 1 and 3).
3 Water pump drivebelt defective/out of adjustment (Chapter 1).
4 Thermostat sticking (Chapter 3).

Clutch

30 Pedal travels to floor - no pressure or very little resistance

1 Master or slave cylinder faulty (Chapter 8).
2 Hose/pipe burst or leaking (Chapter 8).
3 Connections leaking (Chapter 8).
4 No fluid in reservoir (Chapter 1).
5 If fluid is present in master cylinder dust cover, rear master cylinder seal has failed (Chapter 8).
6 Broken release bearing or fork (Chapter 8).

31 Fluid in area of master cylinder dust cover and on pedal

Rear seal failure in master cylinder (Chapter 8).

32 Fluid on slave cylinder

Slave cylinder plunger seal faulty (Chapter 8).

33 Pedal feels "spongy" when depressed

Air in system (Chapter 8).

34 Unable to select gears

1 Faulty transmission (Chapter 7).
2 Faulty clutch plate (Chapter 8).
3 Fork and bearing not assembled properly (Chapter 8).
4 Faulty pressure plate (Chapter 8).
5 Pressure plate-to-flywheel bolts loose (Chapter 8).

35 Clutch slips (engine speed increases with no increase in vehicle speed)

1 Clutch plate worn (Chapter 8).
2 Clutch plate is oil soaked by leaking rear main seal (Chapter 8).
3 Clutch plate not seated. It may take 30 or 40 normal starts for a new one to seat.
4 Warped pressure plate or flywheel (Chapter 8).
5 Weak diaphragm spring (Chapter 8).
6 Clutch plate overheated. Allow to cool.

36 Grabbing (chattering) as clutch is engaged

1 Oil on clutch plate lining, burned or glazed facings (Chapter 8).
2 Worn or loose engine or transmission mounts (Chapters 2 and 7).
3 Worn splines on clutch plate hub (Chapter 8).
4 Warped pressure plate or flywheel (Chapter 8).

37 Noise in clutch area

1 Fork improperly installed (Chapter 8).
2 Faulty release bearing (Chapter 8).

38 Clutch pedal stays on floor

1 Fork binding in housing (Chapter 8).
2 Broken release bearing or fork (Chapter 8).

39 High pedal effort

1 Fork binding in housing (Chapter 8).
2 Pressure plate faulty (Chapter 8).
3 Incorrect size master or slave cylinder installed (Chapter 8).

Manual transmission

40 Vibration

1 Rough wheel bearing (Chapters 1 and 10).
2 Damaged axleshaft (Chapter 8).
3 Out-of-round tires (Chapter 1).
4 Tire out-of-balance (Chapters 1 and 10).
5 Worn U-joint (Chapter 8).

41 Noisy in Neutral with engine running

Damaged clutch release bearing (Chapter 8).
Damaged transmission input shaft bearing (Chapter 7A).

Troubleshooting 0-23

42 Noisy in one particular gear

1. Damaged or worn constant mesh gears.
2. Damaged or worn synchronizers.

43 Noisy in all gears

1. Insufficient lubricant (Chapter 1).
2. Damaged or worn bearings.
3. Worn or damaged input gear shaft and/or output gear shaft.

44 Slips out of gear

1. Worn or improperly adjusted linkage (Chapter 7).
2. Transmission loose on engine (Chapter 7).
3. Shift linkage does not work freely, binds (Chapter 7).
4. Input shaft bearing retainer broken or loose (Chapter 7).
5. Dirt between bellhousing and engine block (Chapter 7).
6. Worn shift fork (Chapter 7).

45 Leaks lubricant

1. Excessive amount of lubricant in transmission (Chapters 1 and 7).
2. Loose or broken input shaft bearing retainer (Chapter 7).
3. Input shaft bearing retainer O-ring and/or lip seal damaged (Chapter 7).

Automatic transmission

Note: *Due to the complexity of the automatic transmission, it is difficult for the home mechanic to properly diagnose and service this component. For problems other than the following, the vehicle should be taken to a dealer or transmission shop.*

46 Fluid leakage

1. Automatic transmission fluid is a deep red color. Fluid leaks should not be confused with engine oil, which can easily be blown by air flow to the transmission.
2. To pinpoint a leak, first remove all built-up dirt and grime from the transmission housing with degreasing agents and/or steam cleaning. Then drive the vehicle at low speeds so air flow will not blow the leak far from its source. Raise the vehicle and determine where the leak is coming from. Common areas of leakage are:
   a) Pan (Chapters 1 and 7)
   b) Filler pipe (Chapter 7)
   c) Transmission fluid cooler lines (Chapter 7)
   d) Speedometer sensor (Chapter 7)

47 Transmission fluid brown or has a burned smell

Transmission fluid burned (Chapter 1).

48 General shift mechanism problems

1. Chapter 7 Part B deals with checking and adjusting the shift linkage on automatic transmissions. Common problems which may be attributed to poorly adjusted linkage are:
   a) Engine starting in gears other than Park or Neutral.
   b) Indicator on shifter pointing to a gear other than the one actually being used.
   c) Vehicle moves when in Park.
2. Refer to Chapter 7 Part B for the shift linkage adjustment procedure.

49 Transmission will not downshift with accelerator pedal pressed to the floor

Throttle valve cable out of adjustment (Chapter 7).

50 Engine will start in gears other than Park or Neutral

Neutral start switch malfunctioning (Chapter 7).

51 Transmission slips, shifts roughly, is noisy or has no drive in forward or reverse gears

There are many probable causes for the above problems, but the home mechanic should be concerned with only one possibility - fluid level. Before taking the vehicle to a repair shop, check the level and condition of the fluid as described in Chapter 1. Correct the fluid level as necessary or change the fluid if needed. If the problem persists, have a professional diagnose the probable cause.

Brakes

Note: *Before assuming that a brake problem exists, make sure that:*
   a) The tires are in good condition and properly inflated (Chapter 1).
   b) The front end alignment is correct (Chapter 10).
   c) The vehicle is not loaded with weight in an unequal manner.

52 Vehicle pulls to one side during braking

1. Incorrect tire pressures (Chapter 1).
2. Front end out of line (have the front end aligned).
3. Unmatched tires on same axle.
4. Restricted brake lines or hoses (Chapter 9).
5. Malfunctioning caliper assembly (Chapter 9).
6. Loose suspension parts (Chapter 10).
7. Loose calipers (Chapter 9).

53 Noise (high-pitched squeal when the brakes are applied)

Front and/or rear disc brake pads worn out. The noise comes from the wear sensor rubbing against the disc. Replace pads with new ones immediately (Chapter 9).

54 Brake roughness or chatter (pedal pulsates)

1. Excessive lateral disc runout (Chapter 9).
2. Parallelism not within specifications (Chapter 9).
3. Uneven pad wear caused by caliper not sliding due to improper clearance or dirt (Chapter 9).
4. Defective disc (Chapter 9).

55 Excessive pedal effort required to stop vehicle

1. Malfunctioning power brake booster (Chapter 9).
2. Partial system failure (Chapter 9).
3. Excessively worn pads (Chapter 9).
4. Piston in caliper stuck or sluggish (Chapter 9).
5. Brake pads contaminated with oil or grease (Chapter 9).
6. New pads installed and not yet seated. It will take a while for the new material to seat against the rotor.

Troubleshooting

56 Excessive brake pedal travel

1 Partial brake system failure (Chapter 9).
2 Insufficient fluid in master cylinder (Chapters 1 and 9).
3 Air trapped in system (Chapters 1 and 9).

57 Dragging brakes

1 Master cylinder pistons not returning correctly (Chapter 9).
2 Restricted brakes lines or hoses (Chapters 1 and 9).
3 Incorrect parking brake adjustment (Chapter 9).

58 Grabbing or uneven braking action

1 Malfunction of power brake booster unit (Chapter 9).
2 Binding brake pedal mechanism (Chapter 9).

59 Brake pedal feels spongy when depressed

1 Air in hydraulic lines (Chapter 9).
2 Master cylinder mounting bolts loose (Chapter 9).
3 Master cylinder defective (Chapter 9).

60 Brake pedal travels to the floor with little resistance

Little or no fluid in the master cylinder reservoir caused by leaking caliper piston(s), loose, damaged or disconnected brake lines (Chapter 9).

61 Parking brake does not hold

Parking brake linkage improperly adjusted (Chapter 9).

Suspension and steering systems

Note: *Before attempting to diagnose the suspension and steering systems, perform the following preliminary checks:*
a) Tires for wrong pressure and uneven wear.
b) Steering universal joints from the column to the steering gear for loose connector0s or wear.
c) Front and rear suspension and the rack and pinion assembly for loose or damaged parts.
d) Out-of-round or out-of-balance tires, bent rims and loose and/or rough wheel bearings.

62 Vehicle pulls to one side

1 Mismatched or uneven tires (Chapter 10).
2 Broken or sagging springs (Chapter 10).
3 Front wheel or rear wheel alignment (Chapter 10).
4 Front brakes dragging (Chapter 9).

63 Abnormal or excessive tire wear

1 Front wheel or rear wheel alignment (Chapter 10).
2 Sagging or broken springs (Chapter 10).
3 Tire out-of-balance (Chapter 10).
4 Worn shock absorber (Chapter 10).
5 Overloaded vehicle.
6 Tires not rotated regularly.

64 Wheel makes a "thumping" noise

1 Blister or bump on tire (Chapter 10).
2 Improper shock absorber action (Chapter 10).

65 Shimmy, shake or vibration

1 Tire or wheel out-of-balance or out-of-round (Chapter 10).
2 Loose, worn or out-of-adjustment wheel bearings (Chapter 1).
3 Worn tie-rod ends (Chapter 10).
4 Worn balljoints (Chapter 10).
5 Excessive wheel runout (Chapter 10).
6 Blister or bump on tire (Chapter 10).

66 Hard steering

1 Lack of lubrication at balljoints, tie-rod ends and rack-and-pinion assembly (Chapter 1).
2 Front wheel alignment (Chapter 10).
3 Low tire pressure(s) (Chapter 1).

67 Poor returnability of steering to center

1 Lack of lubrication at balljoints and tie-rod ends (Chapter 1).
2 Binding in balljoints (Chapter 10).
3 Binding in steering column (Chapter 10).
4 Lack of lubricant in rack-and-pinion assembly (Chapter 10).
5 Front wheel alignment (Chapter 10).

68 Abnormal noise at the front end

1 Lack of lubrication at balljoints and tie-rod ends (Chapter 1).
2 Damaged shock absorber mounting (Chapter 10).
3 Worn control arm bushings or tie-rod ends (Chapter 10).
4 Loose stabilizer bar (Chapter 10).
5 Loose wheel nuts (Chapter).
6 Loose suspension bolts (Chapter 10).

69 Wander or poor steering stability

1 Mismatched or uneven tires (Chapter 10).
2 Lack of lubrication at balljoints and tie-rod ends (Chapter 1).
3 Worn shock absorbers (Chapter 10).
4 Loose stabilizer bar (Chapter 10).
5 Broken or sagging springs (Chapter 10).
6 Front or rear wheel alignment (Chapter 10).

70 Erratic steering when braking

1 Wheel bearings worn (Chapter 1).
2 Broken or sagging springs (Chapter 10).
3 Leaking wheel cylinder or caliper (Chapter 9).
4 Warped discs (Chapter 9).

Troubleshooting

71 Excessive pitching and/or rolling around corners or during braking

1. Loose stabilizer bar (Chapter 10).
2. Worn shock absorbers or mounts (Chapter 10).
3. Broken or sakgging springs (Chapter 10).
4. Overloaded vehicle.

72 Suspension bottoms

1. Overloaded vehicle.
2. Worn shock absorbers (Chapter 10).
3. Incorrect, broken or sagging springs (Chapter 10).

73 Cupped tires

1. Front wheel or rear wheel alignment (Chapter 10).
2. Worn shock absorbers (Chapter 10).
3. Wheel bearings worn (Chapter 10).
4. Excessive tire or wheel runout (Chapter 10).
5. Worn balljoints (Chapter 10).

74 Excessive tire wear on outside edge

1. Inflation pressures incorrect (Chapter 1).
2. Excessive speed in turns.
3. Front end alignment incorrect (excessive toe-in). Have professionally aligned.
4. Suspension arm bent or twisted (Chapter 10).

75 Excessive tire wear on inside edge

1. Inflation pressures incorrect (Chapter 1).
2. Front end alignment incorrect (toe-out). Have professionally aligned.
3. Loose or damaged steering components (Chapter 10).

76 Tire tread worn in one place

1. Tires out-of-balance.
2. Damaged or buckled wheel. Inspect and replace if necessary.
3. Defective tire (Chapter 1).

77 Excessive play or looseness in steering system

1. Wheel bearing(s) worn (Chapter 10.
2. Tie-rod end loose or worn (Chapter 10).
3. Rack and pinion loose (Chapter 10).

78 Rattling or clicking noise in rack-and-pinion

1. Insufficient or improper lubricant in rack-and-pinion assembly (Chapter 10).
2. Steering gear or rack-and-pinion mount loose (Chapter 10).

Troubleshooting

Notes

Chapter 1 Tune-up and routine maintenance

Contents

Air filter replacement	19
Automatic transmission fluid and filter change	27
Automatic transmission fluid level check	8
Battery check, maintenance and charging	12
Brake system check	25
Cooling system check	21
Cooling system servicing (draining, flushing and refilling)	28
Differential lubricant change	31
Differential lubricant level check	16
Driveaxle boot check	24
Drivebelt check, adjustment and replacement	11
Engine oil and filter change	6
Engine timing belt replacement	See Chapter 2
Exhaust system check	22
Evaporative Emissions Control (EVAP) system check	32
Fluid level checks	4
Fuel filter replacement	29
Fuel system check	20
Introduction	1
Maintenance schedule	2
Manual transmission lubricant change	30
Manual transmission lubricant level check	15
Oxygen sensor replacement	33
Power steering fluid level check	7
Service light resetting	34
Spark plug check and replacement	13
Spark plug wire, distributor cap and rotor check and replacement	14
Steering and suspension check	23
Throttle linkage - check and lubrication	18
Tire and tire pressure checks	5
Tire rotation	9
Tune-up general information	3
Underhood hose check and replacement	10
Valve clearance check and adjustment	17
Wiper blade check and replacement	26

Specifications

Recommended lubricants and fluids

Engine oil	
Type	API grade SG
Viscosity	See chart on next page
Power steering fluid type	DEXRON II automatic transmission fluid
Brake fluid type	DOT 3 heavy duty brake fluid
Automatic transmission fluid type	DEXRON II automatic transmission fluid
Manual transmission lubricant type	
Red label	DEXRON II automatic transmission fluid
Green label	Synthetic gear oil
No label	API-GL-4 SAE 80-non-hypoid gear lubricant or single-grade API SE or SF HD engine oil
Differential lubricant	API GL-5 SAE 90-hypoid gear lubricant
Coolant type	50/50 mixture of ethylene glycol-based antifreeze and water

Four-cylinder engines

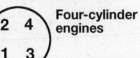

Six-cylinder engines

Front

Cylinder location and distributor rotation

Capacities *

Engine oil with filter change	5 qts
Manual transmission	1.7 qts
Automatic transmission (drain and refill)	
1987 and earlier	2.1 qts
1988 and later	3.2 qts
Cooling system	
Four-cylinder engines	9 qts
Six-cylinder engines	11 qts

All capacities approximate. Add as necessary to bring to appropriate level.

Chapter 1 Tune-up and routine maintenance

Brakes
Disc brake pad thickness (minimum) .... 1/8 inch
Drum brake shoe lining thickness (minimum) .... 1/16 inch

Spark plug type and gap
318i
   Type .... Bosch WR9DP
   Gap .... 0.036 inch
325e and 325es (1988 and earlier)
   Type .... Bosch WR9LP
   Gap .... 0.028 inch
325i, 325is and 325iC (1989 and later)
   Type .... Bosch WR8LPR
   Gap .... 0.028 inch
528e
   1987 and earlier
      Type .... Bosch WR9LP
      Gap .... 0.028 inch
   1988
      Type .... Bosch WR8LPR
      Gap .... 0.028 inch
525i
   1989 and later
      Type .... Bosch WR8LPR
      Gap .... 0.028 inch
   1991
      Type .... Bosch F03DAR
      Gap .... 0.036 inch
533i, 535i and 535is
   1988 and earlier
      Type .... Bosch WR8LP
      Gap .... 0.028 inch
   1989 and later
      Type .... Bosch W8LPR
      Gap .... 0.028 inch

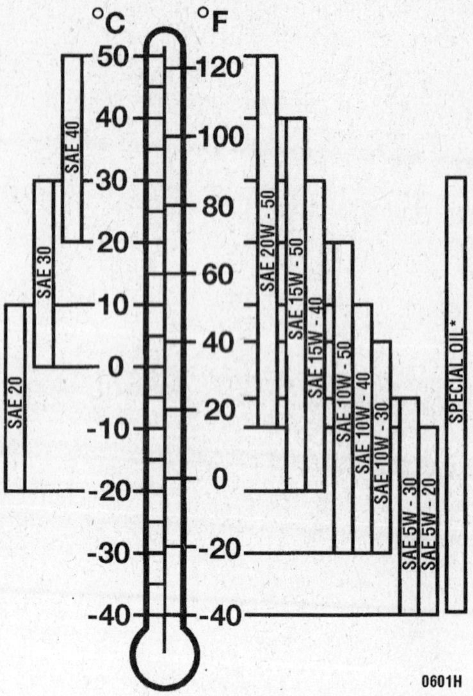

Valve clearances (intake and exhaust)
Four-cylinder engines
   Cold .... 0.008 inch
   Warm .... 0.010 inch
Six-cylinder engines
   Cold .... 0.010 inch
   Warm .... 0.012 inch

Torque specifications
Ft-lbs (unless otherwise indicated)
Automatic transmission pan bolts
   Three-speed .... 72 to 84 in-lbs
   Four-speed .... 48 to 60 in-lbs
Spark plugs .... 15 to 22
Oxygen sensor .... 41 to 44
Wheel lug bolts 81

Oil viscosity chart

1 Introduction

This Chapter is designed to help the home mechanic maintain his or her vehicle with the goals of maximum performance, economy, safety and reliability in mind. Included is a master maintenance schedule (page 1-6), followed by procedures dealing specifically with each item on the schedule. Visual checks, adjustments, component replacement and other helpful items are included. Refer to the accompanying illustrations of the engine compartment and the underside of the vehicle for the locations of various components. Servicing the vehicle, in accordance with the mileage/time maintenance schedule and the step-by-step procedures will result in a planned maintenance program that should produce a long and reliable service life. Keep in mind that it is a comprehensive plan, so maintaining some items but not others at specified intervals will not produce the same results.

As you service the vehicle, you will discover that many of the procedures can - and should - be grouped together because of the nature of the particular procedure you're performing or because of the close proximity of two otherwise unrelated components to one another. For example, if the vehicle is raised for chassis lubrication, you should inspect the exhaust, suspension, steering and fuel systems while you're under the vehicle. When you're rotating the tires, it makes good sense to check the brakes since the wheels are already removed. Finally, let's suppose you have to borrow or rent a torque wrench. Even if you only need it to tighten the spark plugs, you might as well check the torque of as many critical fasteners as time allows.

The first step in this maintenance program is to prepare yourself before the actual work begins. Read through all the procedures you're planning to do, then gather up all the parts and tools needed. If it looks like you might run into problems during a particular job, seek advice from a mechanic or an experienced do-it-yourselfer.

Chapter 1 Tune-up and routine maintenance

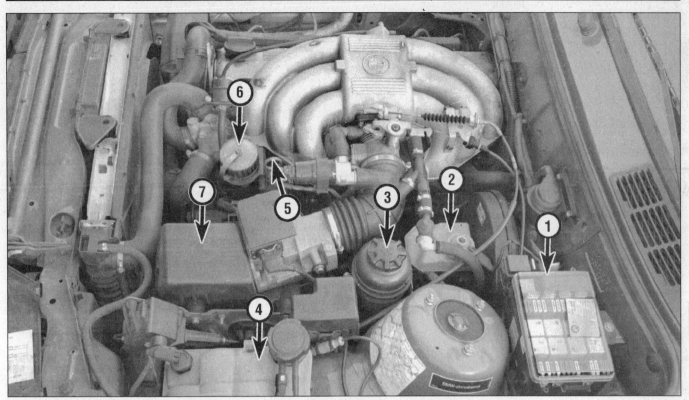

Typical engine compartment components - left side (1988 325is)

1. Main fuse block
2. Brake fluid reservoir
3. Power steering fluid reservoir
4. Coolant reservoir (expansion tank)
5. Engine oil dipstick
6. Check service connector
7. Air cleaner housing

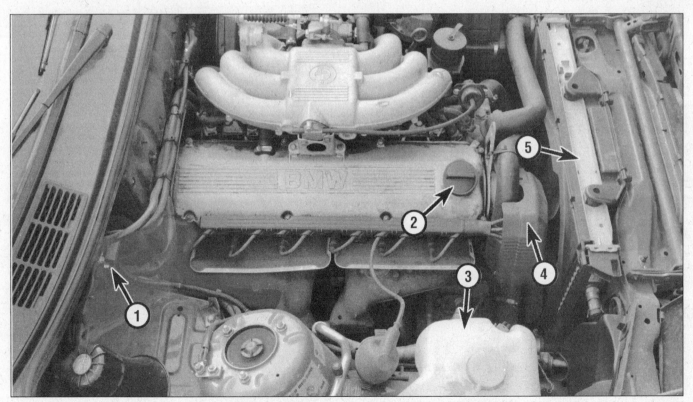

Typical engine compartment components - right side (1988 325is)

1. Remote battery positive terminal for jump starting
2. Engine oil filler cap
3. Windshield washer fluid reservoir
4. Distributor
5. Radiator

Chapter 1 Tune-up and routine maintenance

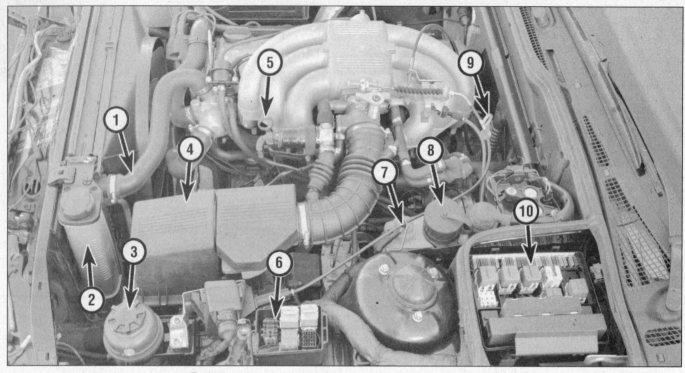

Typical engine compartment components - left side (1989 525i)

1. Radiator hose
2. Radiator coolant reservoir (expansion tank)
3. Power steering fluid reservoir
4. Air cleaner housing
5. Engine oil dipstick
6. Relay box
7. Brake fluid reservoir
8. Check service connector
9. Automatic transmission fluid dipstick
10. Main fuse block

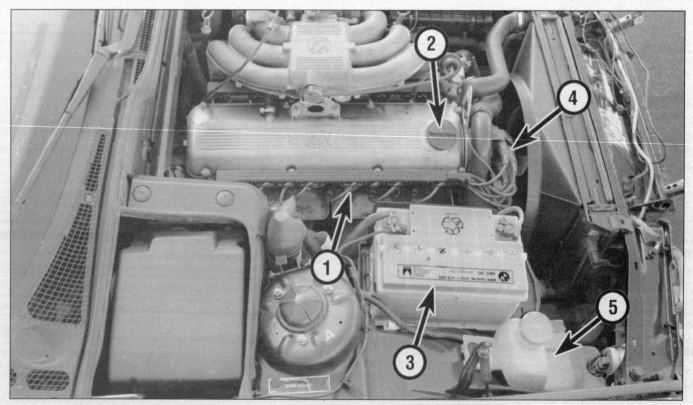

Typical engine compartment components - right side (1989 525i)

1. Spark plug and wire
2. Oil filler cap
3. Battery
4. Distributor
5. Windshield washer fluid reservoir

Chapter 1 Tune-up and routine maintenance

1-5

Typical engine compartment underside components (3-Series model)

1. Engine oil drain plug
2. Steering gear boot
3. Strut assembly
4. Front disc brake
5. Front suspension balljoint

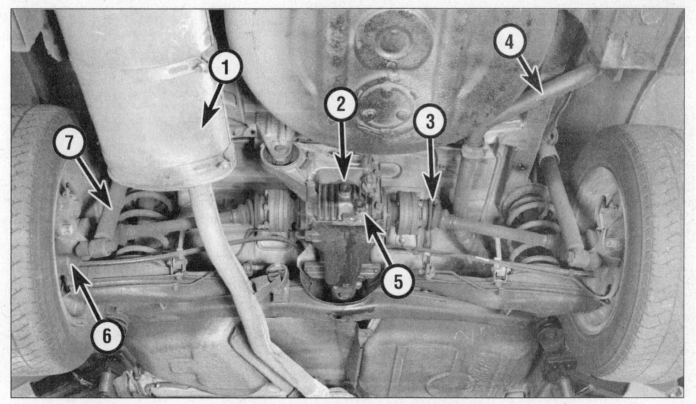

Typical rear underside components (3-Series model)

1. Exhaust system
2. Differential fill/check plug
3. Driveaxle boot
4. Fuel tank filler tube
5. Differential drain plug
6. Rear brake
7. Rear shock absorber

2 BMW 3 and 5-Series maintenance schedule

The following maintenance intervals are based on the assumption that the vehicle owner will be doing the maintenance or service work, as opposed to having a dealer service department do the work. Although the time/mileage intervals are loosely based on factory recommendations, most have been shortened to ensure, for example, that such items as lubricants and fluids are checked/changed at intervals that promote maximum engine/driveline service life. Also, subject to the preference of the individual owner interested in keeping his or her vehicle in peak condition at all times, and with the vehicle's ultimate resale in mind, many of the maintenance procedures may be performed more often than recommended in the following schedule. We encourage such owner initiative.

When the vehicle is new it should be serviced initially by a factory authorized dealer service department to protect the factory warranty. In many cases the initial maintenance check is done at no cost to the owner (check with your dealer service department for more information).

Every 250 miles or weekly, whichever comes first

Check the engine oil level (Section 4)
Check the engine coolant level (Section 4)
Check the brake fluid level (Section 4)
Check the clutch fluid level (Section 4)
Check the washer fluid level (Section 4)
Check the tires and tire pressures (Section 5)

Every 3,000 miles or 3 months, whichever comes first

All items listed above, plus:
Change the engine oil and oil filter (Section 6)
Check the power steering fluid level (Section 7)
Check the automatic transmission fluid level (Section 8)

Every 6,000 miles or 6 months, whichever comes first

All items listed above, plus:
Rotate the tires (Section 9)
Inspect/replace the underhood hoses (Section 10)
Check/adjust the drivebelts (Section 11)

Every 15,000 miles or 12 months, whichever comes first

All items listed above, plus:
Check/service the battery (Section 12)
Check/regap the spark plugs (Section 13)
Check/replace the spark plug wires, distributor cap and rotor (Section 14)
Check/replenish the manual transmission lubricant (Section 15)
Check the differential lubricant level (Section 16)
Check and adjust if necessary, the valve clearances (Section 17)

Check and lubricate the throttle linkage (Section 18)
Replace the air filter (Section 19)
Check the fuel system (Section 20)
Inspect the cooling system (Section 21)
Inspect the exhaust system (Section 22)
Inspect the steering and suspension components (Section 23)
Check the driveaxle boot(s) (Section 24)
Inspect the brakes (Section 25)
Inspect/replace the windshield wiper blades (Section 26)

Every 30,000 miles or 24 months, whichever comes first

All items listed above plus:
Change the automatic transmission fluid and filter (Section 27)
Service the cooling system (drain, flush and refill) (Section 28)
Replace the spark plugs (Section 13)
Check/replace the spark plug wires (Section 14)
Replace the fuel filter (Section 29)
Change the manual transmission lubricant (Section 30)
Change the differential lubricant (Section 31)
Check the evaporative emissions system (Section 32)
Reset the service lights (Section 34)

Oxygen sensor replacement intervals (Section 33)

Every 30,000 miles:

1984 3-series models with four-cylinder engines
5-series models with unheated (single-wire) oxygen sensors

Every 50,000 miles:

1985 3-series with four-cylinder engine
All six-cylinder 3-series models
5-series models with heated (multi-wire) oxygen sensors

Chapter 1 Tune-up and routine maintenance

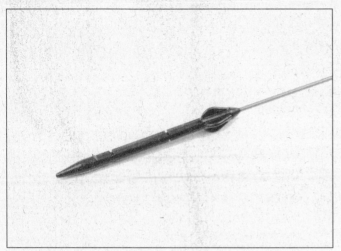

4.4 The oil level should be kept between the two notches, at or near the upper one - if it isn't, add enough oil to bring the level to the upper notch (it takes one full quart to raise the level from the lower to the upper mark)

4.6 The threaded oil filler cap is located in the valve cover - always make sure the area around the opening is clean before unscrewing the cap

3 Tune-up general information

The term tune-up is used in this manual to represent a combination of individual operations rather than one specific procedure.

If, from the time the vehicle is new, the routine maintenance schedule is followed closely and frequent checks are made of fluid levels and high wear items, as suggested throughout this manual, the engine will be kept in relatively good running condition and the need for additional work will be minimized.

More likely than not, however, there will be times when the engine is running poorly due to a lack of regular maintenance. This is even more likely if a used vehicle, which has not received regular and frequent maintenance checks, is purchased. In such cases, an engine tune-up will be needed outside of the regular maintenance intervals.

The first step in any tune-up or diagnostic procedure to help correct a poor running engine is a cylinder compression check. A compression check (see Chapter 2, Part B) will help determine the condition of internal engine components and should be used as a guide for tune-up and repair procedures. If, for instance, a compression check indicates serious internal engine wear, a conventional tune-up will not improve the performance of the engine and would be a waste of time and money. Because of its importance, the compression check should be done by someone with the right equipment and the knowledge to use it properly.

The following procedures are those most often needed to bring a generally poor running engine back into a proper state of tune.

Minor tune-up
Check all engine related fluids (Section 4)
Check all underhood hoses (Section 10)
Check and adjust the drivebelts (Section 11)
Clean, inspect and test the battery (Section 12)
Replace the spark plugs (Section 13)
Inspect the spark plug wires, distributor cap and rotor (Section 14)
Check the air filter (Section 19)
Check the cooling system (Section 21)

Major tune-up
All items listed under minor tune-up, plus . . .
Check the ignition system (see Chapter 5)
Check the charging system (see Chapter 5)
Check the fuel system (see Chapter 4)
Replace the spark plug wires, distributor cap and rotor (Section 14)

4 Fluid level checks (every 250 miles or weekly)

Refer to illustrations 4.4, 4.6, 4.9a, 4.9b, 4.9c, 4.16 and 4.22
Note: *The following are fluid level checks to be done on a 250 mile or weekly basis. Additional fluid level checks can be found in specific maintenance procedures which follow. Regardless of intervals, be alert to fluid leaks under the vehicle which would indicate a fault to be corrected immediately.*

1 Fluids are an essential part of the lubrication, cooling, brake and windshield washer systems. Because the fluids gradually become depleted and/or contaminated during normal operation of the vehicle, they must be periodically replenished. See *Recommended lubricants and fluids* at the beginning of this Chapter before adding fluid to any of the following components. **Note:** *The vehicle must be on level ground when fluid levels are checked.*

Engine oil

2 Engine oil is checked with a dipstick, which is located on the side of the engine (refer to the underhood illustration at the front of this Chapter for dipstick location). The dipstick extends through a metal tube down into the oil pan.

3 The engine oil should be checked before the vehicle has been driven, or about 15 minutes after the engine has been shut off. If the oil is checked immediately after driving the vehicle, some of the oil will remain in the upper part of the engine, resulting in an inaccurate reading on the dipstick.

4 Pull the dipstick out of the tube and wipe all of the oil away from the end with a clean rag or paper towel. Insert the clean dipstick all the way back into the tube and pull it out again. Note the oil at the end of the dipstick. At its highest point, the oil should be between the two **(see illustration)**.

5 It takes one quart of oil to raise the level from the lower mark to the upper mark on the dipstick. Do not allow the level to drop below the lower mark or oil starvation may cause engine damage. Conversely, overfilling the engine (adding oil above the upper mark) may cause oil fouled spark plugs, oil leaks or oil seal failures.

6 To add oil, remove the filler cap located on the valve cover **(see illustration)**. After adding oil, wait a few minutes to allow the level to stabilize, then pull the dipstick out and check the level again. Add more oil if required. Install the filler cap and tighten it by hand only.

7 Checking the oil level is an important preventive maintenance step. A consistently low oil level indicates oil leakage through damaged seals, defective gaskets or past worn rings or valve guides. The condition of the oil should also be noted. If the oil looks milky in color or has water droplets in it, the cylinder head gasket may be blown or the head or block may be cracked. The engine should be repaired immediately.

1-8 Chapter 1 Tune-up and routine maintenance

4.9a On some models the expansion tank (coolant reservoir) is mounted on the radiator - make sure the level is kept at or near the Max mark (arrow)

4.9b On some models the expansion tank (coolant reservoir) is located on the side of the engine compartment - remove the cap to add coolant

4.9c On some 5-Series models the expansion tank (coolant reservoir) is located on the firewall

Whenever you check the oil level, slide your thumb and index finger up the dipstick before wiping off the oil. If you see small dirt or metal particles clinging to the dipstick, the oil should be changed (see Section 6).

Engine coolant

Warning: *Do not allow antifreeze to come in contact with your skin or painted surfaces of the vehicle. Rinse off spills immediately with plenty of water. Antifreeze is highly toxic if ingested. Never leave antifreeze lying around in an open container or in puddles on the floor; children and pets are attracted by it's sweet smell and may drink it. Check with local authorities about disposing of used antifreeze. Many communities have collection centers which will see that antifreeze is disposed of safely.*

8   All vehicles covered by this manual are equipped with a pressurized coolant recovery system. On most models, a white plastic expansion tank (or coolant reservoir) located in the engine compartment is connected by a hose to the radiator. As the engine heats up during operation, the expanding coolant fills the tank. As the engine cools, the coolant is automatically drawn back into the cooling system to maintain the correct level.

9   The coolant level in the reservoir **(see illustrations)** should be checked regularly. **Warning:** *Do not remove the expansion tank cap or radiator cap to check the coolant level when the engine is warm! The level in the reservoir varies with the temperature of the engine. When* the engine is cold, the coolant level should be above the LOW mark on the reservoir. Once the engine has warmed up, the level should be at or near the FULL mark. If it isn't, allow the engine to cool, then remove the cap from the reservoir and add a 50/50 mixture of ethylene glycol based antifreeze and water. Don't use rust inhibitors or additives.

10 Drive the vehicle and recheck the coolant level. If only a small amount of coolant is required to bring the system up to the proper level, water can be used. However, repeated additions of water will dilute the antifreeze and water solution. In order to maintain the proper ratio of antifreeze and water, always top up the coolant level with the correct mixture. An empty plastic milk jug or bleach bottle makes an excellent container for mixing coolant.

11 If the coolant level drops consistently, there may be a leak in the system. Inspect the radiator, hoses, filler cap, drain plugs and water pump (see Section 28). If no leaks are noted, have the expansion tank cap or radiator cap pressure tested by a service station.

12 If you have to remove the cap, wait until the engine has cooled completely, then wrap a thick cloth around the cap and turn it to the first stop. If coolant or steam escapes, let the engine cool down longer, then remove the cap.

13 Check the condition of the coolant as well. It should be relatively clear. If it's brown or rust colored, the system should be drained, flushed and refilled. Even if the coolant appears to be normal, the corrosion inhibitors wear out, so it must be replaced at the specified intervals.

Brake and clutch fluid

Warning: *Brake fluid can harm your eyes and damage painted surfaces, so use extreme caution when handling or pouring it. Do not use brake fluid that has been standing open or is more than one year old. Brake fluid absorbs moisture from the air, which can cause a dangerous loss of brake effectiveness. Use only the specified type of brake fluid. Mixing different types (such as DOT 3 or 4 and DOT 5) can cause brake failure.*

14 The brake master cylinder is mounted at the left (driver's side) rear corner of the engine compartment. The clutch fluid reservoir (used on models with manual transmissions) is mounted adjacent to it.

15 To check the clutch fluid level, observe the level through the translucent reservoir. The level should be at or near the step molded into the reservoir. If the level is low, remove the reservoir cap to add the specified fluid.

16 The brake fluid level is checked by looking through the plastic reservoir mounted on the master cylinder **(see illustration)**. The fluid level should be between the MAX and MIN lines on the reservoir. If the fluid level is low, wipe the top of the reservoir and the cap with a clean rag to prevent contamination of the system as the cap is unscrewed. Top up with the recommended brake fluid, but do not overfill.

Chapter 1 Tune-up and routine maintenance

4.16 The brake fluid level should be kept above the MIN mark on the translucent reservoir - unscrew the cap to add fluid

4.22 The windshield washer fluid reservoir is located in the right front corner of the engine compartment on most models

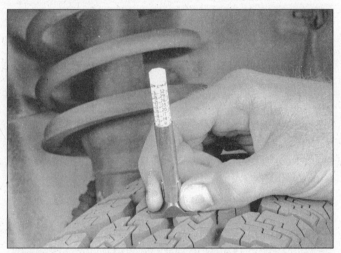

5.2 Use a tire tread depth indicator to monitor tire wear - they are available at auto parts stores and service stations and cost very little

antifreeze, available at any auto parts store, to lower the freezing point of the fluid. This comes in concentrated or pre-mixed form. If you purchase concentrated antifreeze, mix the antifreeze with water in accordance with the manufacturer's directions on the container. **Caution:** *Do not use cooling system antifreeze - it will damage the vehicle's paint.*

5 Tire and tire pressure checks (every 250 miles or weekly)

Refer to illustrations 5.2, 5.3, 5.4a, 5.4b and 5.8

1 Periodic inspection of the tires may save you the inconvenience of being stranded with a flat tire. It can also provide you with vital information regarding possible problems in the steering and suspension systems before major damage occurs.

2 Tires are equipped with 1/2-inch wide bands that will appear when tread depth reaches 1/16-inch, at which time the tires can be considered worn out. Tread wear can be monitored with a simple, inexpensive device known as a tread depth indicator **(see illustration)**.

3 Note any abnormal tire wear **(see illustration)**. Tread pattern irregularities such as cupping, flat spots and more wear on one side that the other are indications of front end alignment and/or balance problems. If any of these conditions are noted, take the vehicle to a tire shop or service station to correct the problem.

4 Look closely for cuts, punctures and embedded nails or tacks. Sometimes a tire will hold air pressure for a short time or leak down very slowly after a nail has embedded itself in the tread. If a slow leak persists, check the valve stem core to make sure it is tight **(see illustration)**. Examine the tread for an object that may have embedded itself in the tire or for a "plug" that may have begun to leak (radial tire punctures are repaired with a plug that is installed in the puncture). If a puncture is suspected, it can be easily verified by spraying a solution of soapy water onto the puncture **(see illustration)**. The soapy solution will bubble if there is a leak. Unless the puncture is unusually large, a tire shop or service station can usually repair the tire.

5 Carefully inspect the inner sidewall of each tire for evidence of brake fluid leakage. If you see any, inspect the brakes immediately.

6 Correct air pressure adds miles to the lifespan of the tires, improves mileage and enhances overall ride quality. Tire pressure cannot be accurately estimated by looking at a tire, especially if it's a radial. A tire pressure gauge is essential. Keep an accurate gauge in the glove compartment. The pressure gauges attached to the nozzles of air hoses at gas stations are often inaccurate.

7 Always check tire pressure when the tires are cold. Cold, in this case, means the vehicle has not been driven over a mile in the three hours preceding a tire pressure check. A pressure rise of four to eight pounds is not uncommon once the tires are warm.

17 While the reservoir cap is off, check the master cylinder reservoir for contamination. If rust deposits, dirt particles or water droplets are present, the system should be drained and refilled by a dealer service department or repair shop.

18 After filling the reservoir to the proper level, make sure the cap is seated to prevent fluid leakage and/or contamination.

19 The fluid level in the master cylinder will drop slightly as the disc brake pads wear. A very low level may indicate worn brake pads. Check for wear (see Section 25).

20 If the brake fluid level drops consistently, check the entire system for leaks immediately. Examine all brake lines, hoses and connections, along with the calipers, wheel cylinders and master cylinder (see Section 25).

21 When checking the fluid level, if you discover one or both reservoirs empty or nearly empty, the brake or clutch hydraulic system should be checked for leaks and bled (see Chapters 8 and 9).

Windshield washer fluid

22 Fluid for the windshield washer system is stored in a plastic reservoir in the engine compartment **(see illustration)**.

23 In milder climates, plain water can be used in the reservoir, but it should be kept no more than two-thirds full to allow for expansion if the water freezes. In colder climates, use windshield washer system

Condition	Probable cause	Corrective action	Condition	Probable cause	Corrective action
Shoulder wear	• Underinflation (both sides wear) • Incorrect wheel camber (one side wear) • Hard cornering • Lack of rotation	• Measure and adjust pressure. • Repair or replace axle and suspension parts. • Reduce speed. • Rotate tires.	Toe wear (Feathered edge)	• Incorrect toe	• Adjust toe-in.
Center wear	• Overinflation • Lack of rotation	• Measure and adjust pressure. • Rotate tires.	Uneven wear	• Incorrect camber or caster • Malfunctioning suspension • Unbalanced wheel • Out-of-round brake drum • Lack of rotation	• Repair or replace axle and suspension parts. • Repair or replace suspension parts. • Balance or replace. • Turn or replace. • Rotate tires.

5.3 This chart will help you determine the condition of the tires, the probable cause(s) of abnormal wear and the corrective action necessary

5.4a If a tire loses air on a steady basis, check the valve core first to make sure it's snug (special inexpensive wrenches are commonly available at auto parts stores)

5.4b If the valve core is tight, raise the corner of the vehicle with the low tire and spray a soapy water solution onto the tread as the tire is turned slowly - leaks will cause small bubbles to appear

8 Unscrew the valve cap protruding from the wheel or hubcap and push the gauge firmly onto the valve stem **(see illustration)**. Note the reading on the gauge and compare the figure to the recommended tire pressure shown in your owner's manual or on the tire placard on the passenger side door or door pillar. Be sure to reinstall the valve cap to keep dirt and moisture out of the valve stem mechanism. Check all four tires and, if necessary, add enough air to bring them to the recommended pressure.

9 Don't forget to keep the spare tire inflated to the specified pressure (refer to your owner's manual or the placard attached to the door pillar).

5.8 To extend the life of the tires, check the air pressure at least once a week with an accurate gauge (don't forget the spare!)

6 **Engine oil and filter change (every 3000 miles or 3 months)**

Refer to illustrations 6.2, 6.7, 6.16 and 6.17

1 Frequent oil changes are the most important preventive maintenance procedures that can be done by the home mechanic. As engine

Chapter 1 Tune-up and routine maintenance

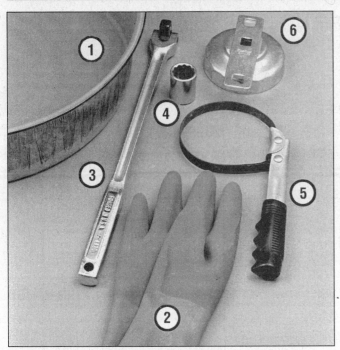

6.2 These tools are required when changing the engine oil and filter

1. **Drain pan** - It should be fairly shallow in depth, but wide to prevent spills
2. **Rubber gloves** - When removing the drain plug and filter, you will get oil on your hands (the gloves will prevent burns)
3. **Breaker bar** - Sometimes the oil drain plug is tight, and a long breaker bar is needed to loosen it
4. **Socket** - To be used with the breaker bar or a ratchet (must be the correct size to fit the drain plug - six-point preferred)
5. **Filter wrench** - This is a metal band-type wrench, which requires clearance around the filter to be effective
6. **Filter wrench** - This type fits on the bottom of the filter and can be turned with a ratchet or breaker bar (different-size wrenches are available for different types of filters)

6.7 Use a proper size box-end wrench or socket to remove the oil drain plug to avoid rounding it off

6.16 Lubricate the oil filter gasket with clean engine oil before installing the filter on the engine

oil ages, it becomes diluted and contaminated, which leads to premature engine wear.

2 Make sure that you have all the necessary tools before you begin this procedure **(see illustration)**. You should also have plenty of rags or newspapers handy for mopping up oil spills

3 Start the engine and allow it to reach normal operating temperature - oil and sludge will flow more easily when warm. If new oil, a filter or tools are needed, use the vehicle to go get them and warm up the engine oil at the same time. Park on a level surface and shut off the engine when it's warmed up. Remove the oil filler cap from the valve cover.

4 Access to the oil drain plug and filter will be improved if the vehicle can be lifted on a hoist, driven onto ramps or supported by jackstands. **Warning:** *DO NOT work under a vehicle supported only by a bumper, hydraulic or scissors-type jack - always use jackstands!*

5 Raise the vehicle and support it on jackstands. Make sure it is safely supported!

6 If you haven't changed the oil on this vehicle before, get under it and locate the drain plug and the oil filter. The exhaust components will be hot as you work, so note how they are routed to avoid touching them.

7 Being careful not to touch the hot exhaust components, position a drain pan under the plug in the bottom of the engine. Clean the area around the plug, then remove the plug **(see illustration)**. It's a good idea to wear a rubber glove while unscrewing the plug the final few turns to avoid being scalded by hot oil. It will also help to hold the drain plug against the threads as you unscrew it, then pull it away from the drain hole suddenly. This will place your arm out of the way of the hot oil, as well as reducing the chances of dropping the drain plug into the drain pan.

8 It may be necessary to move the drain pan slightly as oil flow slows to a trickle. Inspect the old oil for the presence of metal particles.

9 After all the oil has drained, wipe off the drain plug with a clean rag. Any small metal particles clinging to the plug would immediately contaminate the new oil.

10 Reinstall the plug and tighten it securely, but don't strip the threads.

11 Move the drain pan into position under the oil filter.

Spin-off oil filter

12 Loosen the spin-off type oil filter by turning it counterclockwise with a filter wrench. Any standard filter wrench will work

13 Sometimes the spin-off type oil filter is screwed on so tightly that it can't be loosened. If it is, punch a metal bar or long screwdriver directly through it and use it as a T-bar to turn the filter. Be prepared for oil to spurt out of the canister as it's punctured.

14 Once the filter is loose, use your hands to unscrew it from the block. Just as the filter is detached from the block, immediately tilt the open end up to prevent oil inside the filter from spilling out.

15 Using a clean rag, wipe off the mounting surface on the block. Also, make sure that none of the old gasket remains stuck to the mounting surface. It can be removed with a scraper if necessary.

16 Compare the old filter with the new one to make sure they are the same type. Smear some engine oil on the rubber gasket of the new filter and screw it into place **(see illustration)**. Overtightening the filter will damage the gasket, so don't use a filter wrench. Most filter manufacturers recommend tightening the filter by hand only. Normally, they should be tightened 3/4-turn after the gasket contacts the block, but be sure to follow the directions on the filter or container.

6.17 Remove the bolt and lift off the cover for access to the the filter cartridge

7.2 The power steering fluid reservoir (arrow) is located on the left side of the engine compartment

Cartridge-type oil filter

17 Some models are equipped with a cartridge-type oil filter. Remove the bolt and lift the filter out **(see illustration)**.
18 Compare the new cartridge with the old one to make sure they are the same type, then lower it into the housing.
19 Using a clean rag, wipe off the mounting surface of the housing and cover. Smear some clean oil on the gasket and install the cover and bolt. Tighten the bolt securely.

All models

20 Remove all tools and materials from under the vehicle, being careful not to spill the oil in the drain pan, then lower the vehicle.
21 Add new oil to the engine through the oil filler cap in the valve over. Use a funnel to prevent oil from spilling onto the top of the engine. Pour four quarts of fresh oil into the engine. Wait a few minutes to allow the oil to drain into the pan, then check the level on the dipstick (see Section 4 if necessary). If the oil level is in the SAFE range, install the filler cap.
22 Start the engine and run it for about a minute. While the engine is running, look under the vehicle and check for leaks at the oil pan drain plug and around the oil filter. If either one is leaking, stop the engine and tighten the plug or filter slightly.
23 Wait a few minutes, then recheck the level on the dipstick. Add oil as necessary to bring the level into the SAFE range.
24 During the first few trips after an oil change, make it a point to check frequently for leaks and proper oil level.
25 The old oil drained from the engine cannot be reused in its present state and should be discarded. Oil reclamation centers and some auto repair shops and gas stations will accept the oil, which can be recycled. After the oil has cooled, it can be drained into a container (plastic jugs, bottles, milk cartons, etc.) for transport to a disposal site.

7 Power steering fluid level check (every 3000 miles or 3 months)

Refer to illustrations 7.2 and 7.5

1 Check the power steering fluid level periodically to avoid steering system problems, such as damage to the pump. **Caution:** *DO NOT hold the steering wheel against either stop (extreme left or right turn) for more than five seconds. If you do, the power steering pump could be damaged.*
2 The power steering fluid reservoir is located on the left side of the engine compartment, and is equipped with a twist-off cap with an integral fluid level dipstick **(see illustration)**. Some 533 and 535 models use a hydraulic power steering and brake booster system which combines the fluid in one reservoir located at the right rear corner of the engine compartment.

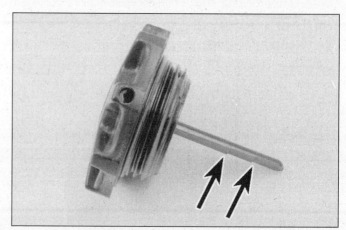

7.5 The power steering fluid level should be kept between the two arrows near the upper step on the dipstick

3 Park the vehicle on level ground and apply the parking brake.
4 On dipstick-equipped models, run the engine until it has reached normal operating temperature. With the engine at idle, turn the steering wheel back-and-forth several times to get any air out of the steering system. Shut the engine off, remove the cap by turning it counterclockwise, wipe the dipstick clean and reinstall the cap. On hydraulic booster models, pump the brake pedal about ten times or until the pedal is firm.
5 On dipstick equipped models, remove the cap again and note the fluid level. It must be between the two lines **(see illustration)**. On hydraulic booster models, remove the nut, lift the cap off and make sure the fluid is within 1/4-inch of the top of the reservoir.
6 Add small amounts of fluid until the level is correct. **Caution:** *Do not overfill the reservoir. If too much fluid is added, remove the excess with a clean syringe or suction pump. Insert the dipstick or install the cap.*
7 Check the power steering hoses and connections for leaks and wear (see Section 10).
8 Check the condition and tension of the drivebelt (see Section 11).

8 Automatic transmission fluid level check (every 3000 miles or 3 months)

Refer to illustrations 8.5 and 8.6
Caution: *The use of transmission fluid other than the type listed in this Chapter's Specifications could result in transmission malfunctions or failure.*

Chapter 1 Tune-up and routine maintenance

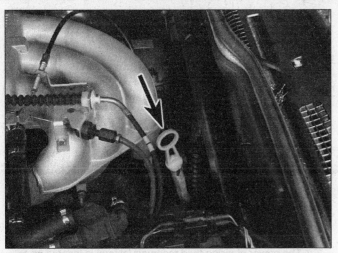

8.5 The automatic transmission fluid dipstick (arrow) is located near the firewall on the left side of the engine compartment

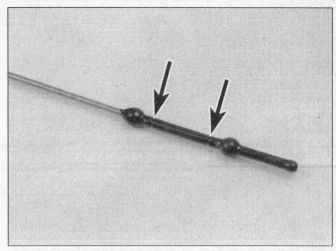

8.6 With the fluid hot, the level should be kept between the two dipstick notches, near the upper one

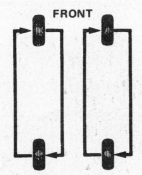

9.2 The tire rotation pattern for these models

1 The automatic transmission fluid should be carefully maintained. Low fluid level can lead to slipping or loss of drive, while overfilling can cause foaming and loss of fluid. Either condition can cause transmission damage.
2 Since transmission fluid expands as it heats up, the fluid level should only be checked when the transmission is warm (at normal operating temperature). If the vehicle has just been driven over 20 miles (32 km), the transmission can be considered warm. **Caution:** *If the vehicle has just been driven for a long time at high speed or in city traffic, in hot weather, or if it has been pulling a trailer, an accurate fluid level reading cannot be obtained. Allow the transmission to cool down for about 30 minutes. You can also check the transmission fluid level when the transmission is cold. If the vehicle has not been driven for over five hours and the fluid is about room temperature (70 to 95-degrees F), the transmission is cold. However, the fluid level is normally checked with the transmission warm to ensure accurate results.*
3 Immediately after driving the vehicle, park it on a level surface, set the parking brake and start the engine. While the engine is idling, depress the brake pedal and move the selector lever through all the gear ranges, beginning and ending in Park.
4 Locate the automatic transmission dipstick tube in the left rear corner of the engine compartment.
5 With the engine still idling, pull the dipstick out of the tube **(see illustration)**, wipe it off with a clean rag, push it all the way back into the tube and withdraw it again, then note the fluid level.
6 The level should be between the two marks **(see illustration)**. If the level is low, add the specified automatic transmission fluid through the dipstick tube - use a clean funnel to prevent spills.
7 Add just enough of the recommended fluid to fill the transmission to the proper level. It takes about one pint to raise the level from the low mark to the high mark when the fluid is hot, so add the fluid a little

at a time and keep checking the level until it's correct.
8 The condition of the fluid should also be checked along with the level. If the fluid is black or a dark reddish-brown color, or if it smells burned, it should be changed (see Section 27). If you are in doubt about its condition, purchase some new fluid and compare the two for color and smell.

9 Tire rotation (every 6000 miles or 6 months)

Refer to illustration 9.2
1 The tires should be rotated at the specified intervals and whenever uneven wear is noticed. Since the vehicle will be raised and the tires checked anyway, check the brakes also (see Section 25). **Note:** *Even if you don't rotate the tires, at least check the lug bolt tightness.*
2 It is recommended that the tires be rotated in a specific pattern **(see illustration)**.
3 Refer to the information in *Jacking and towing* at the front of this manual for the proper procedure to follow when raising the vehicle and changing a tire. If the brakes must be checked, don't apply the parking brake as stated.
4 The vehicle must be raised on a hoist or supported on jackstands to get all four tires off the ground. Make sure the vehicle is safely supported!
5 After the rotation procedure is finished, check and adjust the tire pressures as necessary and be sure to check the lug bolt tightness.

10 Underhood hose check and replacement (every 6,000 miles or 6 months)

Warning: *Replacement of air conditioning hoses must be left to a dealer service department or air conditioning shop that has the equipment to depressurize the system safely. Never disconnect air conditioning hoses or components until the system has been depressurized.*

General
1 High temperatures under the hood can cause deterioration of the rubber and plastic hoses used for engine, accessory and emission systems operation. Periodic inspection should be made for cracks, loose clamps, material hardening and leaks.
2 Information specific to the cooling system can be found in Section 21.
3 Most (but not all) hoses are secured to the fitting with clamps. Where clamps are used, check to be sure they haven't lost their tension, allowing the hose to leak. If clamps aren't used, make sure the hose has not expanded and/or hardened where it slips over the fitting, allowing it to leak.

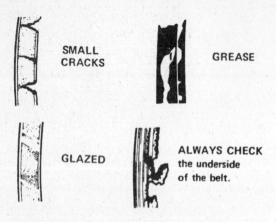

11.3 Here are some of the more common problems associated with drivebelts (check the belts very carefully to prevent an untimely breakdown)

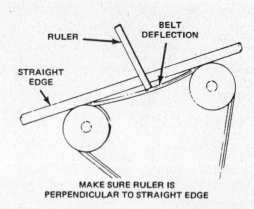

11.4 Measuring drivebelt deflection with a straightedge and ruler

Vacuum hoses

4   It's quite common for vacuum hoses, especially those in the emissions system, to be color coded or identified by colored stripes molded into them. Various systems require hoses with different wall thicknesses, collapse resistance and temperature resistance. When replacing hoses, be sure the new ones are made of the same material.

5   Often the only effective way to check a hose is to remove it completely from the vehicle. If more than one hose is removed, be sure to label the hoses and fittings to ensure correct installation.

6   When checking vacuum hoses, be sure to include any plastic T-fittings in the check. Inspect the fittings for cracks and the hose where it fits over each fitting for distortion, which could cause leakage.

7   A small piece of vacuum hose can be used as a stethoscope to detect vacuum leaks. Hold one end of the hose to your ear and probe around vacuum hoses and fittings, listening for the "hissing" sound characteristic of a vacuum leak. **Warning:** *When probing with the vacuum hose stethoscope, be careful not to come into contact with moving engine components such as the drivebelt, cooling fan, etc.*

Fuel hoses

Warning: *There are certain precautions which must be taken when servicing or inspecting fuel system components. Work in a well ventilated area and do not allow open flames (cigarettes, appliance pilot lights, etc.) or bare light bulbs near the work area. Mop up any spills immediately and do not store fuel-soaked rags where they could ignite.*

8   The fuel lines are usually under pressure, so if any fuel lines are to be disconnected be prepared to catch spilled fuel. **Warning:** *Your vehicle is equipped with fuel injection and you must relieve the fuel system pressure before servicing the fuel lines. Refer to Chapter 4 for the fuel system pressure relief procedure.*

9   Check all rubber fuel lines for deterioration and chafing. Check especially for cracks in areas where the hose bends and just before fittings, such as where a hose attaches to the fuel pump, fuel filter and fuel injection system.

10   Only high quality fuel line, made specifically for use with high-pressure fuel injection systems, should be used for fuel line replacement. Never, under any circumstances, use unreinforced vacuum line, clear plastic tubing or water hose for fuel lines.

11   Band-type clamps are commonly used on fuel lines. These clamps often lose their tension over a period of time, and can be "sprung" during removal. Replace all band-type clamps with screw clamps whenever a hose is replaced

Metal lines

12   Sections of metal line are often used for fuel line between the fuel pump and fuel injection system. Check carefully to make sure the line isn't bent, crimped or cracked.

13   If a section of metal fuel line must be replaced, use seamless steel tubing only, since copper and aluminum tubing do not have the strength necessary to withstand the vibration caused by the engine.

14   Check the metal brake lines where they enter the master cylinder and brake proportioning unit (if used) for cracks in the lines and loose fittings. Any sign of brake fluid leakage calls for an immediate thorough inspection of the brake system.

Power steering hoses

15   Check the power steering hoses for leaks, loose connections and worn clamps. Tighten loose connections. Worn clamps or leaky hoses should be replaced.

11   Drivebelt check, adjustment and replacement (every 6000 miles or 6 months)

Refer to illustrations 11.3, 11.4 and 11.6

Check

1   The drivebelts, sometimes called V-belts or simply "fan" belts, are located at the front of the engine and play an important role in the overall operation of the vehicle and its components. Due to their function and material make up, the belts are prone to failure after a period of time and should be inspected and adjusted periodically to prevent major engine damage.

2   The number of belts used on a particular vehicle depends on the accessories installed. Drivebelts are used to turn the alternator, power steering pump, water pump and air conditioning compressor. Depending on the pulley arrangement, a single belt may be used to drive more than one of these components.

3   With the engine off, open the hood and locate the various belts at the front of the engine. Using your fingers (and a flashlight, if necessary), move along the belts checking for cracks and separation of the belt plies. Also check for fraying and glazing, which gives the belt a shiny appearance **(see illustration)**. Both sides of the belts should be inspected, which means you will have to twist each belt to check the underside.

4   The tension of each belt is checked by pushing firmly with your thumb and see how much the belt moves (deflects). Measure the deflection with a ruler **(see illustration)**. A good rule of thumb is that the belt should deflect 1/4-inch if the distance from pulley center-to-pulley center is between 7 and 11 inches. The belt should deflect 1/2-inch if the distance from pulley center-to-pulley center is between 12 and 16 inches.

Adjustment

5   If it is necessary to adjust the belt tension, either to make the belt tighter or looser, it is done by moving the belt driven accessory on the bracket.

Chapter 1 Tune-up and routine maintenance

11.6 Loosen the nut on the other end of the adjuster bolt (arrow) and turn the bolt to increase or decrease tension on the drivebelt

6 For each component there will be an adjusting bolt and a pivot bolt. Both bolts must be loosened slightly to enable you to move the component. On some components the drivebelt tension can be adjusted by turning an adjusting bolt after loosening the lockbolt **(see illustration)**.

7 After the two bolts have been loosened, move the component away from the engine to tighten the belt or toward the engine to loosen the belt. Hold the accessory in position and check the belt tension. If it is correct, tighten the two bolts until just snug, then recheck the tension. If the tension is correct, tighten the bolts.

8 It will often be necessary to use some sort of prybar to move the accessory while the belt is adjusted. If this must be done to gain the proper leverage, be very careful not to damage the component being moved or the part being pried against.

Replacement

9 To replace a belt, follow the instructions above for adjustment, however completely remove the belt from the pulleys.

10 In some cases you will have to remove more then one belt because of their arrangement on the front of the engine. Due to this and the fact that belts will tend to fail at the same time, it is wise to replace all belts together. Mark each belt and its appropriate pulley groove so all replacement belts can be installed in their proper positions.

11 It is a good idea to take the old belts with you when buying new ones in order to make a direct comparison for length, width and design.

12 Battery check, maintenance and charging (every 15,000 miles or 12 months)

Check and maintenance

Refer to illustrations 12.1, 12.8a, 12.8b, 12.8c and 12.8d
Warning: *Certain precautions must be followed when checking and servicing the battery. Hydrogen gas, which is highly flammable, is always present in the battery cells, so keep lighted tobacco and all other flames and sparks away from it. The electrolyte inside the battery is actually dilute sulfuric acid, which will cause injury if splashed on your skin or in your eyes. It will also ruin clothes and painted surfaces. When removing the battery cables, always detach the negative cable first and hook it up last!*

1 Battery maintenance is an important procedure which will help ensure that you are not stranded because of a dead battery. Several tools are required for this procedure **(see illustration)**.

2 Before servicing the battery, always turn the engine and all accessories off and disconnect the cable from the negative terminal of the battery. **Caution:** *If the radio in your vehicle is equipped with an anti-theft system, make sure you have the correct activation code before disconnecting the battery.* **Note:** *If, after connecting the battery, the*

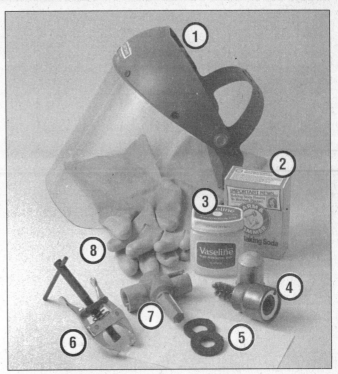

12.1 Tools and materials required for battery maintenance

1 **Face shield/safety goggles** - *When removing corrosion with a brush, the acidic particles can easily fly up into your eyes*
2 **Baking soda** - *A solution of baking soda and water can be used to neutralize corrosion*
3 **Petroleum jelly** - *A layer of this on the battery posts will help prevent corrosion*
4 **Battery post/cable cleaner** - *This wire brush cleaning tool will remove all traces of corrosion from the battery posts and cable clamps*
5 **Treated felt washers** - *Placing one of these on each post, directly under the cable clamps, will help prevent corrosion*
6 **Puller** - *Sometimes the cable clamps are very difficult to pull off the posts, even after the nut/bolt has been completely loosened. This tool pulls the clamp straight up and off the post without damage*
7 **Battery post/cable cleaner** - *Here is another cleaning tool which is a slightly different version of Number 4 above, but it does the same thing*
8 **Rubber gloves** - *Another safety item to consider when servicing the battery; remember that's acid inside the battery!*

wrong language appears on the instrument panel display, refer to page 0-7 for the language resetting procedure.

3 A low maintenance battery is standard equipment. The cell caps can be removed and distilled water can be added, if necessary.

4 Remove the caps and check the electrolyte level in each of the battery cells. It must be above the plates. There's usually a split-ring indicator in each cell to indicate the correct level. If the level is low, add distilled water only, then install the cell caps. **Caution:** *Overfilling the cells may cause electrolyte to spill over during periods of heavy charging, causing corrosion and damage to nearby components.*

5 If the positive terminal and cable clamp on your vehicle's battery is equipped with a rubber protector, make sure that it's not torn or damaged. It should completely cover the terminal.

6 The external condition of the battery should be checked periodically. Look for damage such as a cracked case.

7 Check the tightness of the battery cable clamps to ensure good electrical connections and inspect the entire length of each cable, looking for cracked or abraded insulation and frayed conductors.

8 If corrosion (visible as white, fluffy deposits) is evident, remove the cables from the terminals, clean them with a battery brush and re-

1-16 Chapter 1 Tune-up and routine maintenance

12.8a Battery terminal corrosion usually appears as light, fluffy powder

12.8b Removing a cable from the battery post with a wrench - sometimes special battery pliers are required for this procedure if corrosion has caused deterioration of the hex nut (always remove the ground cable first and hook it up last!)

12.8c Regardless of the type of tool used on the battery posts, a clean, shiny surface should be the result

12.8d When cleaning the cable clamps, all corrosion must be removed (the inside of the clamp is tapered to match the taper on the post, so don't remove too much material)

install them **(see illustrations)**. Corrosion can be kept to a minimum by installing specially treated washers available at auto parts stores or by applying a layer of petroleum jelly or grease to the terminals and cable clamps after they are assembled.

9 Make sure that the battery carrier is in good condition and that the hold-down clamp bolt is tight. If the battery is removed (see Chapter 5 for the removal and installation procedure), make sure that no parts remain in the bottom of the carrier when it's reinstalled. When reinstalling the hold-down clamp, don't overtighten the bolt.

10 Corrosion on the carrier, battery case and surrounding areas can be removed with a solution of water and baking soda. Apply the mixture with a small brush, let it work, then rinse it off with plenty of clean water.

11 Any metal parts of the vehicle damaged by corrosion should be coated with a zinc-based primer, then painted.

12 Additional information on the battery and jump starting can be found in Chapter 5 and the front of this manual.

Charging

Note: *The manufacturer recommends the battery be removed from the vehicle for charging because the gas which escapes during this procedure can damage the paint or interior, depending on the model. Fast charging with the battery cables connected can result in damage to the electrical system.*

13 Remove all of the cell caps (if equipped) and cover the holes with a clean cloth to prevent spattering electrolyte. Disconnect the negative battery cable and hook the battery charger leads to the battery posts (positive to positive, negative to negative), then plug in the charger. Make sure it is set at 12-volts if it has a selector switch. **Caution:** *If the radio in your vehicle is equipped with an anti-theft system, make sure you have the correct activation code before disconnecting the battery.* **Note**: *If, after connecting the battery, the wrong language appears on the instrument panel display, refer to page 0-7 for the language resetting procedure.*

14 If you're using a charger with a rate higher than two amps, check the battery regularly during charging to make sure it doesn't overheat. If you're using a trickle charger, you can safely let the battery charge overnight after you've checked it regularly for the first couple of hours.

15 If the battery has removable cell caps, measure the specific gravity with a hydrometer every hour during the last few hours of the charging cycle. Hydrometers are available inexpensively from auto parts stores - follow the instructions that come with the hydrometer. Consider the battery charged when there's no change in the specific gravity reading for two hours and the electrolyte in the cells is gassing (bubbling) freely. The specific gravity reading from each cell should be very close to the others. If not, the battery probably has a bad cell(s).

16 Some batteries with sealed tops have built-in hydrometers on the top that indicate the state of charge by the color displayed in the the hydrometer window. Normally, a bright-colored hydrometer indicates a full charge and a dark hydrometer indicates the battery still needs charging. Check the battery manufacturer's instructions to be sure you know what the colors mean.

17 If the battery has a sealed top and no built-in hydrometer, you can hook up a voltmeter across the battery terminals to check the charge. A fully charged battery should read 12.6-volts or higher.

18 Further information on the battery and jump starting can be found in Chapter 5 and at the front of this manual.

13 Spark plug check and replacement (every 15,000 miles or 12 months)

Refer to illustrations 13.1, 13.4a, 13.4b, 13.5, 13.7 and 13.9

1 Before beginning, obtain the necessary tools, which will include a spark plug socket and a gap gauge **(see illustration)**.

2 The best procedure to follow when replacing the spark plugs is to purchase the new spark plugs beforehand, adjust them to the proper gap, and then replace each plug one at a time. When buying the new spark plugs it is important to obtain the correct plugs for your specific engine. This information can be found on the Vehicle Emissions Control Information label located under the hood, in the Specifications section in the front of this Chapter or in the owner's manual. If differences

Chapter 1 Tune-up and routine maintenance

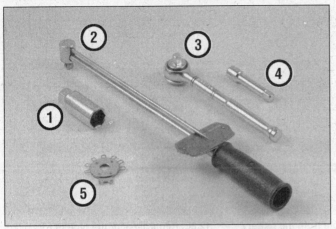

13.1 Tools required for changing spark plugs

1. **Spark plug socket** - This will have special padding inside to protect the spark plug's porcelain insulator
2. **Torque wrench** - Although not mandatory, using this tool is the best way to ensure the plugs are tightened properly
3. **Ratchet** - Standard hand tool to fit the spark plug socket
4. **Extension** - Depending on model and accessories, you may need special extensions and universal joints to reach one or more of the plugs
5. **Spark plug gap gauge** - This gauge for checking the gap comes in a variety of styles. Make sure the gap for your engine is included

13.4a Spark plug manufacturers recommend using a wire-type gauge when checking the gap - if the wire does not slide between the electrodes with a slight drag, adjustment is required

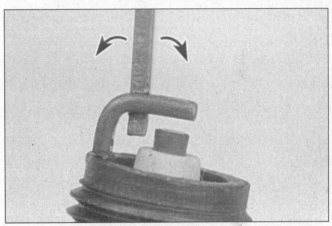

13.4b To change the gap, bend the side electrode only, as indicated by the arrows, and be very careful not to crack or chip the porcelain insulator surrounding the center electrode

13.5 When removing the spark plug wires, pull only on the boot and twist it back-and-forth

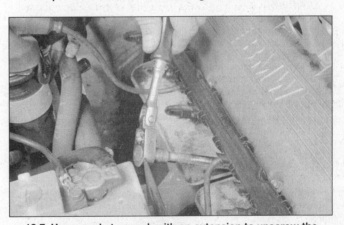

13.7 Use a socket wrench with an extension to unscrew the spark plugs

exist between these sources, purchase the spark plug type specified on the Emissions Control label, because the information was printed for your specific engine.

3 With the new spark plugs at hand, allow the engine to cool completely before attempting plug removal. During this time, each of the new spark plugs can be inspected for defects and the gaps can be checked.

4 The gap is checked by inserting the proper thickness gauge between the electrodes at the tip of the plug **(see illustration)**. The gap between the electrodes should be the same as that given in the Specifications or on the Emissions Control label. The wire should just touch each of the electrodes. If the gap is incorrect, use the notched adjuster to bend the curved side of the electrode slightly until the proper gap is achieved **(see illustration)**. **Note:** *When adjusting the gap of a new plug, bend only the base of the ground electrode, do not touch the tip. If the side electrode is not exactly over the center electrode, use the notched adjuster to align the two. Check for cracks in the porcelain insulator, indicating the spark plug should not be used.*

5 With the engine cool, remove the spark plug wire from one spark plug. Do this by grabbing the boot at the end of the wire, not the wire itself **(see illustration)**. Sometimes it is necessary to use a twisting motion while the boot and plug wire are pulled free.

6 If compressed air is available, use it to blow any dirt or foreign material away from the spark plug area. A common bicycle pump will also work. The idea here is to eliminate the possibility of debris falling into the cylinder as the spark plug is removed.

7 Place the spark plug socket over the plug and remove it from the engine by turning in a counterclockwise direction **(see illustration)**.

For a COLOR version of this spark plug diagnosis page, please see the inside rear cover of this manual

CARBON DEPOSITS
Symptoms: Dry sooty deposits indicate a rich mixture or weak ignition. Causes misfiring, hard starting and hesitation.

Recommendation: Check for a clogged air cleaner, high float level, sticky choke and worn ignition points. Use a spark plug with a longer core nose for greater anti-fouling protection.

NORMAL
Symptoms: Brown to grayish-tan color and slight electrode wear. Correct heat range for engine and operating conditions.

Recommendation: When new spark plugs are installed, replace with plugs of the same heat range.

OIL DEPOSITS
Symptoms: Oily coating caused by poor oil control. Oil is leaking past worn valve guides or piston rings into the combustion chamber. Causes hard starting, misfiring and hesition.

Recommendation: Correct the mechanical condition with necessary repairs and install new plugs.

ASH DEPOSITS
Symptoms: Light brown deposits encrusted on the side or center electrodes or both. Derived from oil and/or fuel additives. Excessive amounts may mask the spark, causing misfiring and hesitation during acceleration.

Recommendation: If excessive deposits accumulate over a short time or low mileage, install new valve guide seals to prevent seepage of oil into the combustion chambers. Also try changing gasoline brands.

TOO HOT
Symptoms: Blistered, white insulator, eroded electrode and absence of deposits. Results in shortened plug life.

Recommendation: Check for the correct plug heat range, over-advanced ignition timing, lean fuel mixture, intake manifold vacuum leaks and sticking valves. Check the coolant level and make sure the radiator is not clogged.

WORN
Symptoms: Rounded electrodes with a small amount of deposits on the firing end. Normal color. Causes hard starting in damp or cold weather and poor fuel economy.

Recommendation: Replace with new plugs of the same heat range.

PREIGNITION
Symptoms: Melted electrodes. Insulators are white, but may be dirty due to misfiring or flying debris in the combustion chamber. Can lead to engine damage.

Recommendation: Check for the correct plug heat range, over-advanced ignition timing, lean fuel mixture, clogged cooling system and lack of lubrication.

DETONATION
Symptoms: Insulators may be cracked or chipped. Improper gap setting techniques can also result in a fractured insulator tip. Can lead to piston damage.

Recommendation: Make sure the fuel anti-knock values meet engine requirements. Use care when setting the gaps on new plugs. Avoid lugging the engine.

HIGH SPEED GLAZING
Symptoms: Insulator has yellowish, glazed appearance. Indicates that combustion chamber temperatures have risen suddenly during hard acceleration. Normal deposits melt to form a conductive coating. Causes misfiring at high speeds.

Recommendation: Install new plugs. Consider using a colder plug if driving habits warrant.

SPLASHED DEPOSITS
Symptoms: After long periods of misfiring, deposits can loosen when normal combustion temperature is restored by an overdue tune-up. At high speeds, deposits flake off the piston and are thrown against the hot insulator, causing misfiring.

Recommendation: Replace the plugs with new ones or clean and reinstall the originals.

GAP BRIDGING
Symptoms: Combustion deposits lodge between the electrodes. Heavy deposits accumulate and bridge the electrode gap. The plug ceases to fire, resulting in a dead cylinder.

Recommendation: Locate the faulty plug and remove the deposits from between the electrodes.

MECHANICAL DAMAGE
Symptoms: May be caused by a foreign object in the combustion chamber or the piston striking an incorrect reach (too long) plug. Causes a dead cylinder and could result in piston damage.

Recommendation: Remove the foreign object from the engine and/or install the correct reach plug.

Chapter 1 Tune-up and routine maintenance   1-19

13.9 Use a length of 3/8-inch ID rubber to hose to start the spark plug into the plug hole

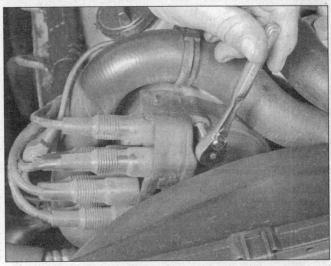

14.11a On later models, loosen the screws and detach the distributor cap up so you can inspect the inside

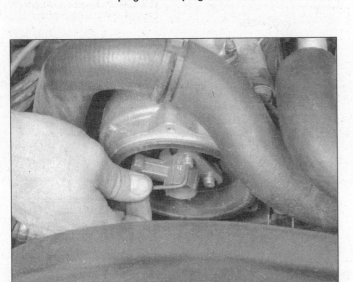

14.11b Use an Allen-head wrench to remove the screws, then lift the rotor off the shaft (later models)

14 Spark plug wire, distributor cap and rotor check and replacement (every 15,000 miles or 12 months)

Refer to illustrations 14.11a, 14.11b, 14.11c and 14.11d

1   The spark plug wires should be checked at the recommended intervals and whenever new spark plugs are installed in the engine.
2   Begin this procedure by making a visual check of the spark plug wires while the engine is running. In a darkened garage (make sure there is ventilation) start the engine and observe each plug wire. Be careful not to come into contact with any moving engine parts. If there is a break in the wire, you will see arcing or a small spark at the damaged area. If arcing is noticed, make a note to obtain new wires, then allow the engine to cool.
3   Disconnect the negative cable from the battery. **Caution:** *If the radio in your vehicle is equipped with an anti-theft system, make sure you have the correct activation code before disconnecting the battery.* **Note**: *If, after connecting the battery, the wrong language appears on the instrument panel display, refer to page 0-7 for the language resetting procedure.*
4   The wires should be inspected one at a time to prevent mixing up the order, which is essential for proper engine operation.
5   Disconnect the plug wire from the spark plug. A removal tool can be used for this purpose or you can grab the plastic boot, twist slightly and pull the wire free. Do not pull on the wire itself, only on the boot.
6   Inspect inside the boot for corrosion, which will look like a white crusty powder. Push the wire and boot back onto the end of the spark plug. It should be a tight fit on the plug end. If it is not, remove the wire and use pliers to carefully crimp the metal connector inside the boot until it fits securely on the end of the spark plug.
7   Using a clean rag, wipe the entire length of the wire to remove any built-up dirt and grease. Once the wire is clean, check for burns, cracks and other damage. Do not bend the wire excessively, since the conductor might break.
8   Disconnect the wire from the distributor. Again, pull only on the boot. Check for corrosion and a tight fit in the same manner as the spark plug end. Replace the wire in the distributor.
9   Check the remaining spark plug wires, making sure they are securely fastened at the distributor and spark plug when the check is complete.
10 If new spark plug wires are required, purchase a set for your specific engine model. Wire sets are available pre-cut, with the boots already installed. Remove and replace the wires one at a time to avoid mix-ups in the firing order.
11 Check the distributor cap and rotor for wear. Loosen the screws or detach the clips and remove the distributor cap **(see illustration)**.

8   Compare the spark plug with those shown in the accompanying photos to get an indication of the overall running condition of the engine.
9   Apply a thin coat of anti-seize compound to the threads of the spark plugs. Install the plug into the head, turning it with your fingers until it no longer turns, then tighten it with the socket. Where there might be difficulty in inserting the spark plugs into the spark plug holes, or the possibility of cross threading them into the head, a short piece of 3/8-inch rubber tubing can be fitted over the end of the spark plug **(see illustration)**. The flexible tubing will act as a universal joint to help align the plug with the plug hole, and should the plug begin to cross thread, the hose will slip on the spark plug, preventing thread damage. If one is available, use a torque wrench to tighten the plug to ensure that it is seated correctly. The correct torque figure is included in this Chapter's Specifications.
10 Before pushing the spark plug wire onto the end of the plug, inspect it following the procedures outlined in Section 14.
11 Attach the plug wire to the new spark plug, again using a twisting motion on the boot until it is firmly seated on the spark plug.
12 Follow the above procedure for the remaining spark plugs, replacing them one at a time to prevent mixing up the spark plug wires.

Chapter 1 Tune-up and routine maintenance

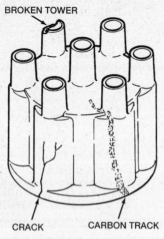

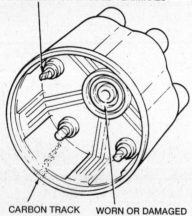

14.11c Shown here are some of the common defects to look for when inspecting the distributor cap (if in doubt about its condition, install a new one)

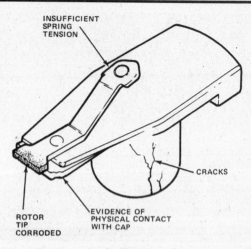

14.11d The ignition rotor should be checked for wear and corrosion as indicated here (if in doubt about its condition, buy a new one)

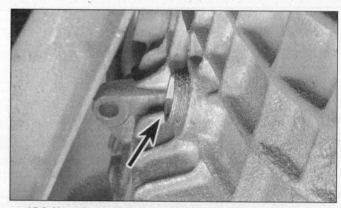

15.2 Use a large Allen wrench to remove the check/fill plug (arrow) and check the lubricant level with your little finger (it should be even with the bottom of the hole) - if it's low, add lubricant

Remove the screw (if equipped) and pull the rotor off the shaft **(see illustration)**. Look for cracks, carbon tracks and worn, burned or loose contacts **(see illustrations)**. Replace the cap and rotor with new parts if defects are found. It is common practice to install a new cap and rotor whenever new spark plug wires are installed. When installing a new cap, remove the wires from the old cap one at a time and attach them to the new cap in the exact same location - do not simultaneously remove all the wires or firing order mix-ups may occur.

15 Manual transmission lubricant level check (every 15,000 miles or 12 months)

Refer to illustration 15.2

1   The transmission has a check/fill plug which must be removed to check the lubricant level. If the vehicle is raised to gain access to the plug, be sure to support it safely on jackstands - DO NOT crawl under a vehicle which is supported only by a jack!
2   Remove the plug from the side of the transmission **(see illustration)** and use your little finger to reach inside plug from the housing and feel the lubricant level. It should be at or very near the bottom of the plug hole.
3   If it isn't, add the recommended lubricant through the plug hole with a syringe or squeeze bottle. Some models have different color stickers on the bellhousing, used to specify the lubricant type. Refer to the Specifications Section at the beginning of this Chapter for the correct lubricant type.

4   Install the plug securely and check for leaks after the first few miles of driving.

16 Differential lubricant level check (every 15,000 miles or 12 months)

Refer to illustration 16.2

1   The differential has a check/fill plug which must be removed to check the lubricant level. If the vehicle is raised to gain access to the plug, be sure to support it safely on jackstands - DO NOT crawl under the vehicle when it's supported only by the jack!
2   Remove the lubricant check/fill plug from the differential **(see illustration)**. Use an Allen wrench to unscrew the plug.
3   Use your little finger as a dipstick to make sure the lubricant level is even with the bottom of the plug hole. If not, use a syringe or squeeze bottle to add the recommended lubricant until it just starts to run out of the opening.
4   Install the plug and tighten it securely.

17 Valve clearance check and adjustment (every 15,000 miles or 12 months)

Refer to illustration 17.6

1   The valve clearances can be checked with the engine hot or cold. If checking and adjusting the valve clearances with the engine hot, start and run the engine until it reaches normal operating temperature,

Chapter 1 Tune-up and routine maintenance

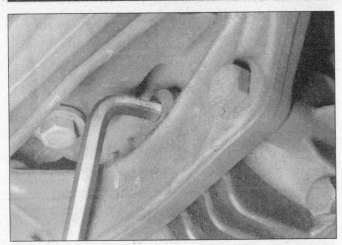

16.2 Remove the differential fill plug (arrow) with an Allen wrench and make sure the lubricant level is even with the bottom of the hole

17.6 The valve clearance is changed by turning the adjusting eccentric with a hooked wire - once the specified clearance is obtained, tighten the locknut with a wrench, then remove the feeler gauge

18.3 Check and lubricate the throttle linkage at the points shown (arrows)

19.1 Detach the duct from the air cleaner housing

then shut it off. **Caution:** *Since the engine will be hot as this procedure is done, extra care must be taken to avoid burns.*
2   Remove the valve cover from the engine (see Chapter 2A).
3   Position the number one (front) piston at Top Dead Center (TDC) on the compression stroke (see Chapter 2A).
4   Check the valve clearances for the number one cylinder. The valve clearances can be found in the Specifications Section at the beginning of this Chapter.
5   The clearance is measured by inserting the specified size feeler gauge between the end of the valve stem and the rocker arm adjusting eccentric. You should feel a slight amount of drag when the feeler gauge is moved back-and-forth.
6   If the gap is too large or too small, loosen the locknut, insert a hook made from large diameter wire and rotate the eccentric to obtain the correct gap **(see illustration)**.
7   Once the gap has been set, hold the eccentric in position with the hook and retighten the locknut securely. Recheck the clearance - sometimes it'll change slightly when the locknut is tightened. If so, readjust until it's correct.
8   On four-cylinder engines the valves are adjusted in the firing order, which is 1-3-4-2. After adjusting the number one cylinder valves rotate the crankshaft 1/2-turn (180-degrees), then check and adjust the valves on the number three cylinder. Repeat the procedure on the remaining cylinders.
9   On six-cylinder engines the valves are adjusted following the firing order, which is 1-5-3-6-2-4. After adjusting the number one cylinder valves, rotate the crankshaft 1/3-turn (120-degrees), then check and

adjust the valves on the number five cylinder. Repeat the procedure for the remaining cylinders.
10   Reinstall the valve cover (use a new gasket) and tighten the mounting nuts evenly and securely.
11   Start the engine and check for oil leakage between the valve cover and the cylinder head.

18 Throttle linkage - check and lubrication (every 15,000 miles or 12 months)

Refer to illustration 18.3
1   The throttle linkage should be checked and lubricated periodically to ensure its proper operation.
2   Check the linkage to make sure it isn't binding.
3   Inspect the linkage joints for looseness and the connections for corrosion and damage, replacing parts as necessary **(see illustration)**.
4   Lubricate the connections with spray lubricant or white lithium grease.

19 Air filter replacement (every 15,000 miles or 12 months)

Refer to illustrations 19.1, 19.2, 19.3a and 19.3b
1   Loosen the clamp on the air intake duct and detach the duct **(see illustration)**.

19.2 Use a screwdriver to detach the air cleaner cover clips

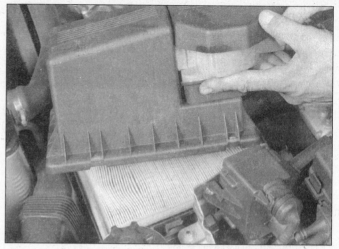

19.3a Rotate the cover up . . .

19.3b . . . and lift the air filter element out

2   Release the retaining clips **(see illustration)**.
3   Rotate the cover up, lift it off and lift the element out, noting the direction it faces **(see illustrations)**.
4   Wipe the inside of the air cleaner housing with a clean cloth. If the element is marked TOP, be sure the marked side faces up.
5   Reinstall the cover and secure the clips.
6   Connect the air hose and tighten the clamp screw.

20 Fuel system check (every 15,000 miles or 12 months)

Warning: *Certain precautions should be observed when inspecting or servicing the fuel system components. Work in a well ventilated area and don't allow open flames (cigarettes, appliance pilot lights, etc.) near the work area. Mop up spills immediately. Do not store fuel soaked rags where they could ignite. It is a good idea to keep a dry chemical (Class B) fire extinguisher near the work area any time the fuel system is being serviced.*

1   If you smell gasoline while driving or after the vehicle has been sitting in the sun, inspect the fuel system immediately.
2   Remove the fuel filler cap and inspect it for damage and corrosion. The gasket should have an unbroken sealing imprint. If the gasket is damaged or corroded, install a new cap.
3   Inspect the fuel feed and return lines for cracks. Make sure that the connections between the fuel lines and the fuel injection system and between the fuel lines and the in-line fuel filter are tight. **Warning:**

The fuel system pressure must be relieved before servicing fuel system components. The fuel system pressure relief procedure is outlined in Chapter 4.
4   Since some components of the fuel system - the fuel tank and some of the fuel feed and return lines, for example - are underneath the vehicle, they can be inspected more easily with the vehicle raised on a hoist. If that's not possible, raise the vehicle and support it on jackstands.
5   With the vehicle raised and safely supported, inspect the gas tank and filler neck for punctures, cracks or other damage. The connection between the filler neck and the tank is particularly critical. Sometimes a rubber filler neck will leak because of loose clamps or deteriorated rubber. Inspect all fuel tank mounting brackets and straps to be sure the tank is securely attached to the vehicle. **Warning:** *Do not, under any circumstances, try to repair a fuel tank (except rubber components). A welding torch or any open flame can easily cause fuel vapors inside the tank to explode.*
6   Carefully check all flexible hoses and metal lines leading away from the fuel tank. Check for loose connections, deteriorated hoses, crimped lines and other damage. Repair or replace damaged sections as necessary (see Chapter 4).

21 Cooling system check (every 15,000 miles or 12 months)

Refer to illustration 21.4

1   Many major engine failures can be attributed to a faulty cooling system. If the vehicle is equipped with an automatic transmission, the cooling system also plays an important role in prolonging transmission life because it cools the fluid.
2   The engine should be cold for the cooling system check, so perform the following procedure before the vehicle is driven for the day or after it has been shut off for at least three hours.
3   Remove the radiator cap and clean it thoroughly, inside and out, with clean water. Also clean the filler neck on the radiator. The presence of rust or corrosion in the filler neck means the coolant should be changed (see Section 28). The coolant inside the radiator should be relatively clean and transparent. If it's rust colored, drain the system and refill with new coolant.
4   Carefully check the radiator hoses and smaller diameter heater hoses. Inspect each coolant hose along its entire length, replacing any hose which is cracked, swollen or deteriorated **(see illustration)**. Cracks will show up better if the hose is squeezed. Pay close attention to hose clamps that secure the hoses to cooling system components. Hose clamps can pinch and puncture hoses, resulting in coolant leaks.
5   Make sure all hose connections are tight. A leak in the cooling system will usually show up as white or rust colored deposits on the area adjoining the leak. If wire-type clamps are used on the hoses, it may be a good idea to replace them with screw-type clamps.

Chapter 1 Tune-up and routine maintenance

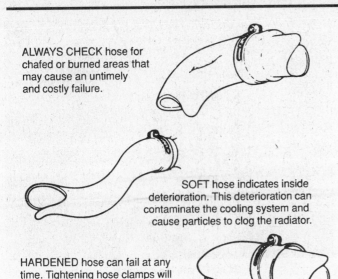

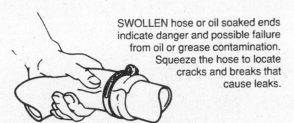

21.4 Hoses, like drivebelts, have a habit of failing at the worst possible time - to prevent the inconvenience of a blown radiator or heater hose, inspect them carefully as shown here

6 Clean the front of the radiator and air conditioning condenser with compressed air, if available, or a soft brush. Remove all bugs, leaves, etc. embedded in the radiator fins. Be extremely careful not to damage the cooling fins or cut your fingers on them.
7 If the coolant level has been dropping consistently and no leaks are detectable, have the radiator cap and cooling system pressure checked at a service station.

22 Exhaust system check (every 15,000 miles or 12 months)

Refer to illustration 22.2
1 With the engine cold (at least three hours after the vehicle has been driven), check the complete exhaust system from the engine to end of the tailpipe. Ideally, the inspection should be done with the vehicle on a hoist to permit unrestricted access. If a hoist isn't available, raise the vehicle and support it securely on jackstands.
2 Check the exhaust pipes and connections for evidence of leaks, severe corrosion and damage. Make sure that all brackets and hangers are in good condition and are tight **(see illustration)**.
3 At the same time, inspect the underside of the body for holes, corrosion, open seams, etc. which may allow exhaust gases to enter the passenger compartment. Seal all body openings with silicone or body putty.
4 Rattles and other noises can often be traced to the exhaust system, especially the mounts, hangers and heat shields. Try to move the pipes, muffler and catalytic converter. If the components can come in contact with the body or suspension parts, secure the exhaust system with new mounts.
5 Check the running condition of the engine by inspecting inside the end of the tailpipe. The exhaust deposits here are an indication of

22.2 Check the exhaust system rubber hangers for cracks

engine state-of-tune. If the pipe is black and sooty or coated with white deposits, the engine may need a tune-up, including a thorough fuel system inspection.

23 Steering and suspension check (every 15,000 miles or 12 months)

Refer to illustration 23.10
Note: *The steering linkage and suspension components should be checked periodically. Worn or damaged suspension and steering linkage components can result in excessive and abnormal tire wear, poor ride quality and vehicle handling and reduced fuel economy. For detailed illustrations of the steering and suspension components, refer to Chapter 10.*

Strut/shock absorber check
1 Park the vehicle on level ground, turn the engine off and set the parking brake. Check the tire pressures.
2 Push down at one corner of the vehicle, then release it while noting the movement of the body. It should stop moving and come to rest in a level position with one or two bounces.
3 If the vehicle continues to move up-and-down or if it fails to return to its original position, a worn or weak strut or shock absorber is probably the reason.
4 Repeat the above check at each of the three remaining corners of the vehicle.
5 Raise the vehicle and support it on jackstands.
6 Check the struts/shock absorbers for evidence of fluid leakage. A light film of fluid is no cause for concern. Make sure that any fluid noted is from the struts/shocks and not from any other source. If leakage is noted, replace the struts or shocks as a set.
7 Check the struts/shock absorbers to be sure that they are securely mounted and undamaged. Check the upper mounts for damage and wear. If damage or wear is noted, replace the struts or shock absorbers as a set.
8 If the struts or shock absorbers must be replaced, refer to Chapter 10 for the procedure.

Steering and suspension check
9 Visually inspect the steering system components for damage and distortion. Look for leaks and damaged seals, boots and fittings.
10 Clean the lower end of the steering knuckle. Have an assistant grasp the lower edge of the tire and move the wheel in-and-out while you look for movement at the steering knuckle-to-axle arm balljoints. Inspect the balljoint boots for tears **(see illustration)**. If there is any movement, or the boots are torn or leaking, the balljoint(s) must be replaced.

Chapter 1 Tune-up and routine maintenance

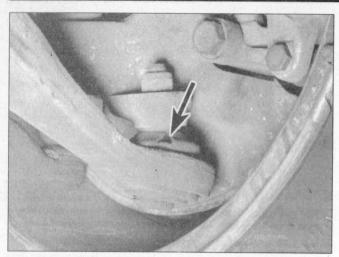

23.10 Inspect the balljoint boots for tears (arrow)

24.2 Push on the driveaxle boots to check for cracks

11 Grasp each front tire at the front and rear edges, push in at the front, pull out at the rear and feel for play in the steering linkage. If any freeplay is noted, check the steering gear mounts and the tie-rod balljoints for looseness. If the steering gear mounts are loose, tighten them. If the tie-rods are loose, the balljoints may be worn (check to make sure the nuts are tight). Additional steering and suspension system information can be found in Chapter 10.

24 Driveaxle boot check (every 15,000 miles or 12 months)

Refer to illustration 24.2

1 The rear driveaxle boots are very important because they prevent dirt, water and foreign material from entering and damaging the constant velocity (CV) joints. Oil and grease can cause the boot material to deteriorate prematurely, so it's a good idea to wash the the boots with soap and water.
2 Inspect the boots for tears and cracks as well as loose clamps (see illustration). If there is any evidence of cracks or leaking lubricant, the boot must be replaced with a new one.

25 Brake system check (every 15,000 miles or 12 months)

Refer to illustrations 25.11 and 25.15
Warning: *Dust produced by lining wear and deposited on brake components may contain asbestos, which is hazardous to your health. DO NOT blow it out with compressed air and DO NOT inhale it! DO NOT use gasoline or solvents to remove the dust. Brake system cleaner should be used to flush the dust into a drain pan. After the brake components are wiped with a damp rag, dispose of the contaminated rag(s) and brake cleaner in a covered and labeled container. Try to use non-asbestos replacement parts whenever possible.*
Note: *In addition to the specified intervals, the brake system should be inspected each time the wheels are removed or a malfunction is indicated. Because of the obvious safety considerations, the following brake system checks are some of the most important maintenance procedures you can perform on your vehicle.*

Symptoms of brake system problems

1 The disc brakes have built-in electrical wear indicators which cause a red warning light on the dash to come on when they're worn to the replacement point. When the light comes on, replace the pads immediately or expensive damage to the brake discs could result.
2 Any of the following symptoms could indicate a potential brake system defect. The vehicle pulls to one side when the brake pedal is depressed, the brakes make squealing or dragging noises when applied, brake travel is excessive, the pedal pulsates and brake fluid leaks are noted (usually on the inner side of the tire or wheel). If any of these conditions are noted, inspect the brake system immediately.

Brake lines and hoses

Note: *Steel tubing is used throughout the brake system, with the exception of flexible, reinforced hoses at the front wheels and as connectors at the rear axle. Periodic inspection of these lines is very important.*
3 Park the vehicle on level ground and turn the engine off.
4 Remove the wheel covers. Loosen, but do not remove, the lug bolts on all four wheels.
5 Raise the vehicle and support it securely on jackstands.
6 Remove the wheels (see *Jacking and towing* at the front of this book, or refer to your owner's manual, if necessary).
7 Check all brake lines and hoses for cracks, chafing of the outer cover, leaks, blisters and distortion. Check the brake hoses at front and rear of the vehicle for softening, cracks, bulging, or wear from rubbing on other components. Check all threaded fittings for leaks and make sure the brake hose mounting bolts and clips are secure.
8 If leaks or damage are discovered, they must be fixed immediately. Refer to Chapter 9 for detailed brake system repair procedures.

Disc brakes

9 If it hasn't already been done, raise the vehicle and support it securely on jackstands. Remove the front wheels.
10 The disc brake calipers, which contain the pads, are now visible. Each caliper has an outer and an inner pad - all pads should be checked.
11 Note the pad thickness by looking through the inspection hole in the caliper (see illustration). If the lining material is 1/8-inch thick or less, or if it is tapered from end-to-end, the pads should be replaced (see Chapter 9). Keep in mind that the lining material is riveted or bonded to a metal plate or shoe - the metal portion is not included in this measurement.
12 Check the condition of the brake disc. Look for score marks, deep scratches and overheated areas (they will appear blue or discolored). If damage or wear is noted, the disc can be removed and resurfaced by an automotive machine shop or replaced with a new one. Refer to Chapter 9 for more detailed inspection and repair procedures.
13 Remove the calipers without disconnecting the brake hoses (see Chapter 9).

Drum brakes

14 Refer to Chapter 9 and remove the rear brake drums.
15 Note the thickness of the lining material on the rear brake shoes and look for signs of contamination by brake fluid or grease (see illustration). If the material is within 1/16-inch of the recessed rivets or metal shoes, replace the brake shoes with new ones. The shoes should also be replaced if they are cracked, glazed (shiny lining surfaces), or contaminated with brake fluid or grease. See Chapter 9 for

Chapter 1 Tune-up and routine maintenance 1-25

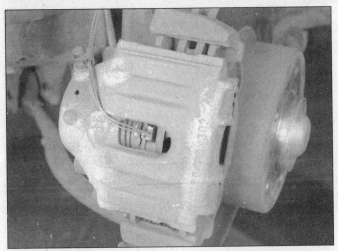

25.11 Look through the caliper inspection window to inspect the brake pads - the pad lining which rubs against the disc can also be inspected by looking at each end of the caliper

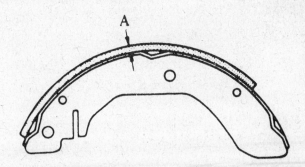

25.15 If the lining is bonded to the brake shoe, measure the lining thickness from the outer surface to the metal shoe, as shown here; if the lining is riveted to the shoe, measure from the lining outer surface to the rivet head

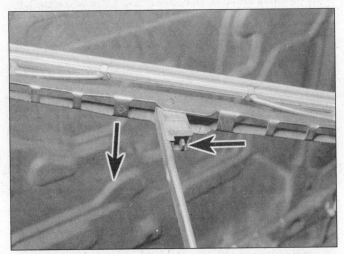

26.5 Press the retaining tab in, then slide the wiper blade assembly down and out of the hook in the end of the wiper arm

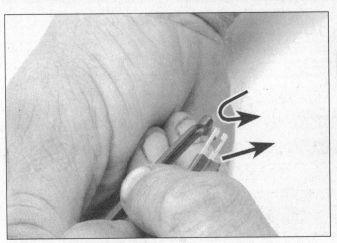

26.6 Detach the end of the wiper element from the end of the frame, then slide the element out

the replacement procedure.

16 Check the shoe return and hold-down springs and the adjusting mechanism to make sure they are installed correctly and in good condition. Deteriorated or distorted springs, if not replaced, could allow the linings to drag and wear prematurely.

17 Check the wheel cylinders for leakage by carefully peeling back the rubber boots. Slight moisture behind the boots is acceptable. If brake fluid is noted behind the boots or if it runs out of the wheel cylinder, the wheel cylinders must be overhauled or replaced (see Chapter 9).

18 Check the drums for cracks, score marks, deep scratches and hard spots, which will appear as small discolored areas. If imperfections cannot be removed with emery cloth, the drums must be resurfaced by an automotive machine shop (see Chapter 9 for more detailed information).

19 Refer to Chapter 9 and install the brake drums.

20 Install the wheels, but don't lower the vehicle yet.

Parking brake

21 The easiest, and perhaps most obvious, method of checking the parking brake is to park the vehicle on a steep hill with the parking brake set and the transmission in Neutral (stay in the vehicle while performing this check). If the parking brake doesn't prevent the vehicle from rolling, refer to Chapter 9 and adjust it.

26 Wiper blade check and replacement (every 15,000 miles or 12 months)

1 Road film can build up on the wiper blades and affect their efficiency, so they should be washed regularly with a mild detergent solution.

Check

2 The wiper and blade assembly should be inspected periodically. Even if you don't use your wipers, the sun and elements will dry out the rubber portions, causing them to crack and break apart. If inspection reveals hardened or cracked rubber, replace the wiper blades. If inspection reveals nothing unusual, wet the windshield, turn the wipers on, allow them to cycle several times, then shut them off. An uneven wiper pattern across the glass or streaks over clean glass indicate that the blades should be replaced.

3 The operation of the wiper mechanism can loosen the fasteners, so they should be checked and tightened, as necessary, at the same time the wiper blades are checked (see Chapter 12 for further information regarding the wiper mechanism).

Wiper blade replacement

Refer to illustrations 26.5 and 26.6

4 Pull the wiper/blade assembly away from the glass.

5 Press the retaining lever and slide the blade assembly down the wiper arm **(see illustration)**.

6 Detach the end of the element from the wiper frame, then slide the element out of the frame **(see illustration)**.

27.5a Unscrew the dipstick tube collar

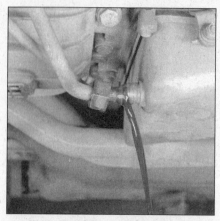
27.5b Detach the tube and let the fluid drain

27.6 Use a socket and extension to remove the bolts and brackets

27.7 Lower the pan from the transmission

27.10a Use a Torx-head driver to remove the filter bolts . . .

27.10b . . . then remove the fluid filter from the transmission

7 Compare the new element with the old for length, design, etc.
8 Slide the new element into place and insert the end in the wiper frame to lock it in place.
9 Reinstall the blade assembly on the arm, wet the glass and check for proper operation.

27 Automatic transmission fluid and filter change (every 30,000 miles or 24 months)

Refer to illustrations 27.5a, 27.5b, 27.6, 27.7, 27.10a, 27.10b and 27.10c

1 At the specified time intervals, the transmission fluid should be drained and replaced. Since the fluid will remain hot long after driving, perform this procedure only after the engine has cooled down completely.
2 Before beginning work, purchase the specified transmission fluid (see *Recommended lubricants and fluids* at the beginning of this Chapter) and a new filter.
3 Other tools necessary for this job include jackstands to support the vehicle in a raised position, a drain pan capable of holding at least eight pints, newspapers and clean rags.
4 Raise the vehicle and support it securely on jackstands.
5 Loosen the dipstick tube collar and let the fluid drain (see illustrations).
6 Remove the pan mounting bolts and brackets (see illustration).
7 Detach the pan from the transmission and lower it, being careful not to spill the remaining fluid (see illustration).
8 Carefully clean the contact surface of the transmission.
9 Drain the fluid from the transmission pan, clean it with solvent and dry it with compressed air. Be sure to clean the metal filings from the

27.10c Remove the O-ring from the transmission, clean it and transfer it to the new fluid filter

magnet, if equipped.
10 Remove the filter from the mount inside the transmission (see illustrations).
11 Install the O-ring and a new filter, being sure to tighten the bolts securely.
12 Make sure the gasket surface on the transmission pan is clean, then install the gasket. Put the pan in place against the transmission and install the brackets and bolts. working around the pan, tighten each bolt a little at a time until the torque listed in this Chapter's Specifications is reached. Don't overtighten the bolts! Connect the dipstick tube and tighten the collar securely.
13 Lower the vehicle and add the specified amount of fluid through the filler tube (see Section 8).

Chapter 1 Tune-up and routine maintenance

1-27

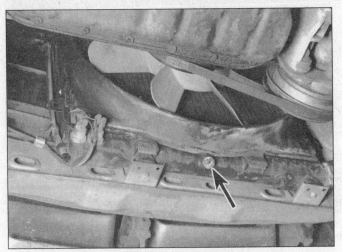

28.4 Radiator drain location (arrow)

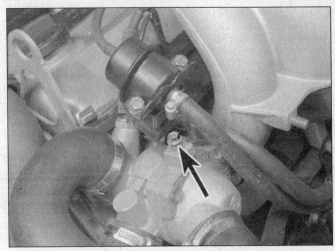

28.16 The bleed screw (arrow) is located on the thermostat housing (six-cylinder models)

14 With the transmission in Park and the parking brake set, run the engine at fast idle, but don't race it.
15 Move the gear selector through each range and back to Park. Check the fluid level.
16 Check under the vehicle for leaks after the first few trips.

28 Cooling system servicing (draining, flushing and refilling) (every 30,000 miles or 24 months)

Refer to illustrations 28.4 and 28.16
Warning: *Do not allow antifreeze to come in contact with your skin or painted surfaces of the vehicle. Rinse off spills immediately with plenty of water. Antifreeze is highly toxic if ingested. Never leave antifreeze lying around in an open container or in puddles on the floor; children and pets are attracted by it's sweet smell and may drink it. Check with local authorities about disposing of used antifreeze. Many communities have collection centers which will see that antifreeze is disposed of safely.*
1 Periodically, the cooling system should be drained, flushed and refilled to replenish the antifreeze mixture and prevent formation of rust and corrosion, which can impair the performance of the cooling system and cause engine damage. When the cooling system is serviced, all hoses and the radiator cap should be checked and replaced if necessary.

Draining

2 Apply the parking brake and block the wheels. If the vehicle has just been driven, wait several hours to allow the engine to cool down before beginning this procedure.
3 Once the engine is completely cool, remove the expansion tank cap or radiator cap.
4 Move a large container under the radiator drain to catch the coolant, then open the drain fitting (a pair of pliers or screwdriver may be required to turn it, depending on the model) **(see illustration)**.
5 While the coolant is draining, check the condition of the radiator hoses, heater hoses and clamps (see Section 21 if necessary).
6 Replace any damaged clamps or hoses (see Chapter 3 for detailed replacement procedures).

Flushing

7 Once the system is completely drained, flush the radiator with fresh water from a garden hose until the water runs clear at the drain. The flushing action of the water will remove sediments from the radiator but will not remove rust and scale from the engine and cooling tube surfaces.
8 These deposits can be removed by a chemical cleaner. Follow the procedure outlined in the manufacturer's instructions. If the radiator is severely corroded, damaged or leaking, it should be removed (see

Chapter 3) and taken to a radiator repair shop.
9 On models so equipped, remove the overflow hose from the coolant recovery reservoir. Drain the reservoir and flush it with clean water, then reconnect the hose.

Refilling

10 Close and tighten the radiator drain. Install and tighten the block drain plug(s).
Place the heater temperature control in the maximum heat position.

Four-cylinder engine

11 Slowly add new coolant (a 50/50 mixture of water and antifreeze) to the radiator until it is full. Add coolant to the reservoir up to the lower mark.
12 Leave the radiator cap off and run the engine in a well-ventilated area until the thermostat opens (coolant will begin flowing through the radiator and the upper radiator hose will become hot).
13 Turn the engine off and let it cool. Add more coolant mixture to bring the coolant level back up to the lip on the radiator filler neck.
14 Squeeze the upper radiator hose to expel air, then add more coolant mixture if necessary. Replace the radiator cap.
15 Start the engine, allow it to reach normal operating temperature and check for leaks.

Six-cylinder engines

16 Loosen the bleed screw in the thermostat housing **(see illustration)**.
17 Fill the radiator with a 50/50 solution of water and antifreeze until it comes out of the bleed screw opening. Tighten the bleed screw.
18 Install the radiator cap and run the engine until the thermostat opens (the upper radiator hose will become hot). Slowly loosen the bleed screw until no bubbles issue from the coolant, then tighten the plug.
19 Repeat the procedure until the air is bled from the system.

29 Fuel filter replacement (every 30,000 miles or 24 months)

Refer to illustration 29.5
Warning: *Gasoline is extremely flammable, so take extra precautions when you work on any part of the fuel system. Don't smoke or allow open flames or bare light bulbs near the work area, and don't work in a garage where a natural gas-type appliance (such as a water heater or clothes dryer) with a pilot light is present. If you spill any fuel on your skin, rinse it off immediately with soap and water. When you perform any kind of work on the fuel system, wear safety glasses and have a Class B type fire extinguisher on hand.*
1 Depressurize the fuel system (see Chapter 4).

Chapter 1 Tune-up and routine maintenance

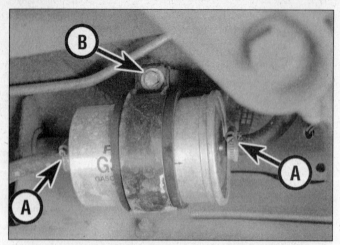

29.5 To replace the fuel filter, disconnect the hoses, then unscrew the nut and detach the filter from the bracket

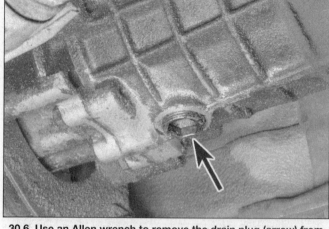

30.6 Use an Allen wrench to remove the drain plug (arrow) from the bottom of the transmission

2   The fuel filter is located in the engine compartment on the firewall (some four-cylinder models) or under the vehicle adjacent to the fuel tank.
3   Because on some four-cylinder models the filter is located adjacent to the starter motor, fuel could leak onto the electrical connections. To avoid this possibility, disconnect the battery negative cable before beginning work. **Caution:** *If the radio in your vehicle is equipped with an anti-theft system, make sure you have the correct activation code before disconnecting the battery.* **Note**: *If, after connecting the battery, the wrong language appears on the instrument panel display, refer to page 0-7 for the language resetting procedure.*
4   Place a pan or rags under the fuel filter to catch any spilled gasoline.
5   Detach the hoses and remove the bracket screws/nuts, then remove the filter and bracket assembly **(see illustration)**.
6   Detach the filter from the bracket.
7   Installation is the reverse of removal. Be sure the arrow on the filter canister points in the direction of fuel flow.

30 Manual transmission lubricant change (every 30,000 miles or 24 months)

Refer to illustration 30.6
1   At the specified time intervals the transmission lubricant should be changed to ensure trouble free operation. Before proceeding, purchase the specified type of lubricant.
2   Tools necessary for this job include jackstands to support the vehicle in a raised position, a wrench to remove the drain plugs, a drain pan, newspapers and clean rags.
3   The lubricant should be drained immediately after the vehicle has been driven. This will remove any contaminants better than if the lubricant were cold. Because of this, it would be wise to wear rubber gloves while removing the drain plug.
4   After the vehicle has been driven to warm up the oil, raise it and place it on jackstands. Make sure it is safely supported and as level as possible.
5   Move the necessary equipment under the vehicle, being careful not to touch any of the hot exhaust components.
6   Place the drain pan under the transmission and remove the check/fill plug from the side of the transmission. Loosen the drain plug **(see illustration)**.
7   Carefully unscrew the plug. Be careful not to burn yourself on the lubricant.
8   Allow the lubricant to drain completely. Clean the drain plug then reinstall and tighten it securely.
9   Refer to Section 15 and fill the transmission with new lubricant, then install the check/fill plug, tightening it securely.

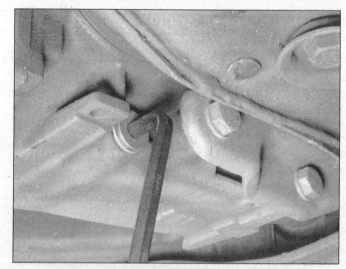

31.4 Remove the differential drain plug (arrow) with an Allen wrench

31 Differential lubricant change (every 30,000 miles or 24 months)

Refer to illustration 31.4
1   Drive the vehicle for several miles to warm up the differential lubricant, then raise the vehicle and support it securely on jackstands.
2   Move a drain pan, rags, newspapers and an Allen wrench under the vehicle.
3   Remove the check/fill plug from the differential. It's the upper of the two plugs.
4   With the drain pan under the differential, use the tool to loosen the drain plug. It's the lower of the two plugs **(see illustration)**.
5   Once loosened, carefully unscrew it with your fingers until you can remove it from the case. Since the lubricant will be hot, wear a rubber glove to prevent burns.
6   Allow all of the oil to drain into the pan, then replace the drain plug and tighten it securely.
7   Refer to Section 16 and fill the differential with lubricant.
8   Reinstall the fill plug and tighten it securely.
9   Lower the vehicle. Check for leaks at the drain plug after the first few miles of driving.

Chapter 1 Tune-up and routine maintenance

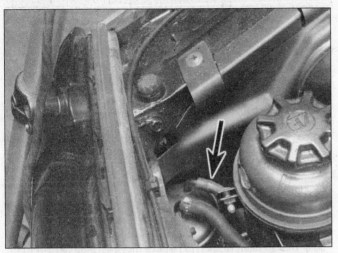

32.2 Inspect the hoses (arrows) that connect to the top of the evaporative emissions charcoal canister for damage

33.2 On most models, the oxygen sensor screws into the exhaust pipe - this one has a heat shield that can be unclipped

34.2a The 15-pin check connector (arrow) is mounted near the front of the engine on earlier models while the 20-pin connector used on later models is located in the left rear corner of the engine compartment

34.2b An inexpensive aftermarket service light tool such as this one can be plugged into the service connector and used to reset the service lights

32 Evaporative Emissions Control (EVAP) system check (every 30,000 miles or 24 months)

Refer to illustration 32.2

1 The function of the Evaporative Emissions Control system is to draw fuel vapors from the tank and fuel system, store them in a charcoal canister and then burn them during normal engine operation.

2 The most common symptom of a fault in the evaporative emissions system is a strong fuel odor in the engine compartment. If a fuel odor is detected, inspect the charcoal canister and system hoses for cracks. The canister is located in the front corner of the engine compartment on most models **(see illustration)**.

3 Refer to Chapter 6 for more information on the evaporative emissions system.

33 Oxygen sensor replacement (every 30,000 or 50,000 miles, depending on model)

Refer to illustration 33.2

1 The oxygen sensor is located in the exhaust pipe. Starting at the oxygen sensor, follow the wire back to the connector. Unplug the oxygen sensor wire at the connector.

2 Use a wrench to unscrew the oxygen sensor **(see illustration)**.

3 Install the new oxygen sensor and tighten it securely.

4 Plug in the electrical connector.

34 Service light resetting

Service lights

Refer to illustrations 34.2a and 34.2b

1 These models are equipped with various service indicator lights on the dash which automatically go on when the mileage interval is reached. These lights can only be turned off by using a special tool which plugs into the service connector located in the engine compartment.

2 Although the service light reset tool can be obtained from a dealer, very reasonably priced tools are also available from aftermarket sources. When obtaining a tool it is important to know the vehicle year and model and whether the service connector has 15 or 20 pins **(see illustrations)**. Once the proper tool is obtained, it is a simple matter to plug it into the service connector and, following the tool manufacturer's instructions, reset the service lights. **Note:** *The brake light will*

not automatically reset if the sensor on the brake pad or its wiring is damaged because it is worn through: it must be repaired first.

3   The service lights are controlled by the Service Indicator (SI) board in the instrument cluster which is powered by rechargeable batteries. Should these batteries fail, problems will develop in the SI board. Symptoms of failed batteries include the inability to reset the service lights and malfunctions affecting the tachometer, temperature gauge and radio operation. Refer to Chapter 12 for more information on the SI board.

Oxygen sensor replacement light

4   Some 1984 four-cylinder models have an oxygen sensor replacement light on the dash. This light comes on only once and cannot be turned off. It can, however be removed.

5   Remove the retaining screw and detach the panel from the dash. Locate the oxygen sensor bulb on the back of the panel and wiggle it back-and-forth until the connector breaks off.

Chapter 2 Part A Engines

Contents

Camshaft - removal, inspection and installation	See Chapter 2B
Compression check	See Chapter 2B
Cylinder head - removal and installation	12
Cylinder head - disassembly and inspection	See Chapter 2B
Drivebelt check, adjustment and replacement	See Chapter 1
Engine mounts - check and replacement	17
Engine oil and filter change	See Chapter 1
Engine overhaul - general information	See Chapter 2B
Engine - removal and installation	See Chapter 2B
Exhaust manifold - removal and installation	6
Flywheel/driveplate - removal and installation	15
Front oil seals - replacement	11
General information	1
Intake manifold - removal and installation	5
Oil pan - removal and installation	13
Oil pump and driveshaft - removal, inspection and installation	14
Rear main oil seal - replacement	16
Repair operations possible with the engine in the vehicle	2
Rocker arm and shaft assembly - disassembly, inspection and assembly	See Chapter 2B
Spark plug replacement	See Chapter 1
Timing belt covers - removal and installation	9
Timing belt and sprockets - removal, inspection and installation	10
Timing chain covers - removal and installation	7
Timing chain and sprockets - removal, inspection and installation	8
Top Dead Center (TDC) for number one piston - locating	3
Valve cover - removal and installation	4
Valves - servicing	See Chapter 2B

Specifications

General

Displacement
- 3-series body style
  - 318i (1984 and 1985) .... 108 cu inches (1.8 Liters)
  - 325i and 325is (1987 through 1990) .... 152 cu inches (2.5 Liters)
  - 325, 325e and 325es (1984 through 1988) .... 164 cu inches (2.7 Liters)
- 5-series body style
  - 525i (1989 and 1990) .... 152 cu inches (2.5 Liters)
  - 528e (1982 through 1988) .... 164 cu inches (2.7 Liters)
  - 533i (1983 and 1984) .... 196 cu inches (3.2 Liters)
  - 535i (1985 through 1990) .... 209 cu inches (3.4 Liters)

Firing order
- Four cylinder engine 1-3-4-2
- Six cylinder engine 1-5-3-6-2-4

Oil pressure at idle (all models) .... 7 to 29 psi

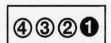

Four-cylinder engines

Six-cylinder engines

Cylinder location and distributor rotation

Front →

Torque specifications

Ft-lbs (unless otherwise indicated)

Timing chain tensioner plug	26
Timing belt tensioner bolts	16
Camshaft sprocket-to-camshaft bolt	
3-series body style	
Four-cylinder engine	5
Six-cylinder engine	48 to 52
5-series body style	
528e	48 to 52
533i and 535i	80 in-lbs
Timing chain or belt covers-to-engine	
Small bolts	8
Large bolts	16
Crankshaft pulley bolts	16
Crankshaft hub bolt or nut	
Four-cylinder engines	190
All six-cylinder engines except those used in 533i and 535i models	302
533i and 535i models	325
Cylinder head bolts	
3-series body style	
Four-cylinder engine	
Step 1	44
Step 2 (wait 15 minutes)	turn an additional 33-degrees
Step 3 (engine at normal operating temperature)	turn an additional 25-degrees
Six-cylinder engine with Torx-head bolts	
Step 1	22
Step 2	turn an additional 90-degrees
Step 3	turn an additional 90-degrees
Six-cylinder engine with hex-head bolts	
Step 1	30
Step 2 (wait 15 minutes)	44
Step 3 (engine at normal operating temperature)	turn an additional 25-degrees
5-series body style	
528e	
Hex-head bolts	
Step 1	30
Step 2 (wait 15 minutes)	44
Step 3 (engine at normal operating temperature)	turn an additional 25-degrees
Torx-head bolts	
Step 1	22
Step 2	turn an additional 90-degrees
Step 3	turn an additional 90-degrees
533i and 535i	
Step 1	44
Step 2 (wait 15 minutes)	59
Step 3 (engine at operating temperature)	turn an additional 35-degrees
Intake manifold-to-cylinder head bolts	22 to 24
Exhaust manifold-to-cylinder head nuts	16 to 18
Flywheel/driveplate bolts	77
Front cover-to-engine bolts (timing belt engines)	
Small bolts	7
Large bolts	16
Intermediate shaft sprocket bolt (timing belt engines)	44
Oil pan-to-block bolts	80 to 97 in-lbs
Oil pump bolts	16
Oil pump sprocket bolts (timing chain engines)	18 to 22
Rear crankshaft oil seal housing-to-block bolts	
Small bolts	80 in-lbs
Large bolts	16

1 General information

This part of Chapter 2 is devoted to in-vehicle engine repair procedures. All information concerning engine removal and installation and engine block and cylinder head overhaul can be found in Chapter 2B.

The following repair procedures are based on the assumption that the engine is installed in the vehicle. If the engine has been removed from the vehicle and mounted on a stand, many of the steps outlined in this part of Chapter 2 will not apply.

The Specifications included in this Part of Chapter 2 apply only to the procedures contained in this Part. Chapter 2B contains the Specifications necessary for cylinder head and engine block rebuilding.

3.8 Align the notch in the pulley with the notch on the timing plate, then check to see if the distributor rotor is pointing to the number 1 cylinder (if not, the camshaft is 180-degrees out of time - the crankshaft will have to be rotated 360-degrees)

The single overhead camshaft four- and six-cylinder engines covered in this manual are very similar in design. Where there are differences, they will be pointed out.

Some engines are equipped with belts to drive the overhead camshaft, while others are equipped with chains. 3-series body types with four-cylinder engines use a timing chain, while six-cylinder 3-series models are equipped with timing belts. 528e models are equipped with a timing belt, but 533i and 535i models are equipped with a timing chain.

2 Repair operations possible with the engine in the vehicle

Many major repair operations can be accomplished without removing the engine from the vehicle.

Clean the engine compartment and the exterior of the engine with some type of degreaser before any work is done. It will make the job easier and help keep dirt out of the internal areas of the engine.

Depending on the components involved, it may be helpful to remove the hood to improve access to the engine as repairs are performed (see Chapter 11 if necessary). Cover the fenders to prevent damage to the paint. Special pads are available, but an old bedspread or blanket will also work.

If vacuum, exhaust, oil or coolant leaks develop, indicating a need for gasket or seal replacement, the repairs can generally be made with the engine in the vehicle. The intake and exhaust manifold gaskets, oil pan gasket, crankshaft oil seals and cylinder head gasket are all accessible with the engine in place.

Exterior engine components, such as the intake and exhaust manifolds, the oil pan, the oil pump, the water pump, the starter motor, the alternator, the distributor and the fuel system components can be removed for repair with the engine in place.

The cylinder head can be removed without pulling the engine, so this procedure covered in this Part of Chapter 2. Camshaft, rocker arm and valve component servicing is most easily accomplished with the cylinder head removed; these procedures are covered in Part B of this Chapter.

In extreme cases caused by a lack of necessary equipment, repair or replacement of piston rings, pistons, connecting rods and rod bearings is possible with the engine in the vehicle. However, this practice is not recommended because of the cleaning and preparation work that must be done to the components involved.

3 Top Dead Center (TDC) for number one piston - locating

Refer to illustration 3.8

Note 1: *The following procedure is based on the assumption that the distributor (if equipped) is correctly installed. If you are trying to locate TDC to install the distributor correctly, piston position must be deter-* mined by feeling for compression at the number one spark plug hole, then aligning the ignition timing marks as described in step 8.

Note 2: *The number one cylinder is the one closest to the radiator.*

1 Top Dead Center (TDC) is the highest point in the cylinder that each piston reaches as it travels up-and-down when the crankshaft turns. Each piston reaches TDC on the compression stroke and again on the exhaust stroke, but TDC generally refers to piston position on the compression stroke.

2 Positioning the piston(s) at TDC is an essential part of many procedures such as timing belt or chain removal and distributor removal.

3 Before beginning this procedure, be sure to place the transmission in Neutral and apply the parking brake or block the rear wheels. Also, disable the ignition system by detaching the coil wire from the center terminal of the distributor cap and grounding it on the block with a jumper wire. Remove the spark plugs (see Chapter 1).

4 In order to bring any piston to TDC, the crankshaft must be turned using one of the methods outlined below. When looking at the front of the engine, normal crankshaft rotation is clockwise.
  a) The preferred method is to turn the crankshaft with a socket and ratchet attached to the bolt threaded into the front of the crankshaft.
  b) A remote starter switch, which may save some time, can also be used. Follow the instructions included with the switch. Once the piston is close to TDC, use a socket and ratchet as described in the previous paragraph.
  c) If an assistant is available to turn the ignition switch to the Start position in short bursts, you can get the piston close to TDC without a remote starter switch. Make sure your assistant is out of the vehicle, away from the ignition switch, then use a socket and ratchet as described in Paragraph a) to complete the procedure.

5 Note the position of the terminal for the number one spark plug wire on the distributor cap. If the terminal isn't marked, follow the plug wire from the number one cylinder spark plug to the cap.

6 Use a felt-tip pen or chalk to make a mark directly below the number one terminal on the distributor body or timing cover.

7 Detach the distributor cap and set it aside (see Chapter 1 if necessary).

8 Turn the crankshaft (see Step 4 above) until the timing marks (located at the front of the engine) are aligned **(see illustration)**.

9 Look at the distributor rotor - it should be pointing directly at the mark you made on the distributor body or timing cover.

10 If the rotor is 180 degrees off, the number one piston is at TDC on the exhaust stroke.

11 To get the piston to TDC on the compression stroke, turn the crankshaft one complete turn (360-degrees) clockwise. The rotor should now be pointing at the mark on the distributor or timing cover. When the rotor is pointing at the number one spark plug wire terminal in the distributor cap and the ignition timing marks are aligned, the number one piston is at TDC on the compression stroke. **Note:** *If it's impossible to align the ignition timing marks when the rotor is pointing at the mark, the timing belt or chain may have jumped the teeth on the sprockets or may have been installed incorrectly.*

12 After the number one piston has been positioned at TDC on the compression stroke, TDC for any of the remaining pistons can be located by turning the crankshaft and following the firing order. Mark the remaining spark plug wire terminal locations just like you did for the number one terminal, then number the marks to correspond with the cylinder numbers. As you turn the crankshaft, the rotor will also turn. When it's pointing directly at one of the marks on the distributor, the piston for that particular cylinder is at TDC on the compression stroke.

4 Valve cover - removal and installation

Removal

Refer to illustrations 4.6a and 4.6b

1 Disconnect the cable from the negative battery terminal. **Caution:** *If the radio in your vehicle is equipped with an anti-theft system, make sure you have the correct activation code before disconnecting the battery.* **Note:** *If, after connecting the battery, the wrong language ap-*

4.6a Valve cover bolt locations on four-cylinder engines

4.6b Valve cover bolt locations on six-cylinder engines

pears on the instrument panel display, refer to Page 0-7 for the language resetting procedure.

2 Detach the breather hose from the valve cover.

3 On 528e models and all 3-series models with six-cylinder engines, unbolt and remove the intake manifold support bracket and, if equipped, the bracket for the engine sensors or idle air stabilizer (it will probably be necessary to disconnect the electrical connectors from the sensors and stabilizer).

4 On 533i and 535i models, disconnect the electrical connector for the air flow sensor and unclip the electrical harness, moving it out of the way.

5 On 533i and 535i models, remove the hoses and fittings from the intake air boot, then loosen the clamp and separate the boot from the throttle body. Unscrew the mounting nuts for the air cleaner housing and remove the housing together with the air boot and air flow sensor.

6 Remove the valve cover retaining nuts and washers **(see illustrations)**. Disconnect the spark plug wire clip or cover from the stud(s) and set it aside. It will usually not be necessary to disconnect the wires from the spark plugs.

7 Remove the valve cover and gasket. Discard the old gasket. If equipped, remove the semi-circular rubber seal from the cutout at the front of the cylinder head.

Installation

8 Using a scraper, remove all traces of old gasket material from the sealing surfaces of the valve cover and cylinder head. **Caution:** *Be very careful not to scratch or gouge the delicate aluminum surfaces.* Gasket removal solvents are available at auto parts stores and may prove helpful. After all gasket material has been removed, the gasket surfaces can be degreased by wiping them with a rag dampened with lacquer thinner or acetone.

9 If equipped, place a new semi-circular rubber seal in the cutout at the front of the cylinder head, then apply RTV-type gasket sealant to the joints between the seal and the mating surface for the valve cover gasket. **Note:** *After the sealant is applied, you should install the valve cover and tighten the nuts within ten minutes.*

10 Install the valve cover and a new gasket. Install the washers and nuts and tighten the nuts evenly and securely. Don't over tighten these nuts - they should be tight enough to prevent oil from leaking past the gasket, but not so tight that they warp the valve cover.

11 The remainder of installation is the reverse of removal.

5 Intake manifold - removal and installation

Removal

Refer to illustration 5.11

1 Allow the engine to cool completely, then relieve the fuel pressure (see Chapter 4).

2 Disconnect the cable from the negative battery terminal. **Caution:** *If the radio in your vehicle is equipped with an anti-theft system, make sure you have the correct activation code before disconnecting the battery.* **Note:***If, after connecting the battery, the wrong language appears on the instrument panel display, refer to Page 0-7 for the language resetting procedure.*

3 Drain the engine coolant (see Chapter 1) below the level of the intake manifold. If the coolant is in good condition, it can be saved and reused.

4 Loosen the hose clamp and disconnect the large air inlet hose from the throttle body. It may also be necessary to remove the entire air cleaner/inlet hose assembly to provide enough working room (see Chapter 4).

5 Disconnect the coolant hoses from the throttle body/intake manifold assembly.

6 Disconnect the throttle cable and, if equipped, cruise control cable from the throttle body (see Chapter 4).

7 Remove the EGR valve and line (see Chapter 6).

8 Disconnect the vacuum hose from the fuel pressure regulator and disconnect the electrical connectors from the fuel injectors (see Chapter 4).

9 Disconnect the fuel lines from the fuel rail (see Chapter 4).

10 Disconnect all remaining hoses and wires attached between the intake manifold/throttle body assembly and the engine or chassis.

11 Remove the bolts and nuts that attach the manifold to the cylinder head **(see illustration)**. Start at the ends and work toward the middle, loosening each one a little at a time until they can be removed by hand. Support the manifold while removing the fasteners so it doesn't fall. **Note:***You can remove the manifold without removing the throttle body, injectors, vacuum/thermo valves and fuel pressure regulator. If you're replacing the manifold, transfer the components (see Chapter 4) and lines to the new manifold before it is installed on the cylinder head.*

12 Move the manifold up and down to break the gasket seal, then lift it away from the head.

Installation

13 Remove the old gasket, then carefully scrape all traces of sealant off of the head and the manifold mating surfaces. Be very careful not to nick or scratch the delicate aluminum mating surfaces. Gasket removal solvents are available at auto parts stores and may prove helpful. Make sure the surfaces are perfectly clean and free of dirt and oil.

14 Check the manifold for corrosion (at the coolant passages), cracks, warping and other damage. Cracks and warping normally show up near the gasket surface, around the stud holes. If defects are found, have the manifold repaired or replaced as necessary.

15 When installing the manifold, use a new gasket and apply a thin, uniform layer of RTV-type gasket sealant to both sides of the gasket.

16 Install the nuts and bolts and tighten them gradually, working

Chapter 2 Part A Engines

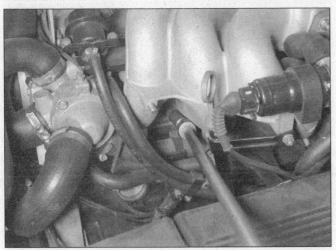

5.11 Remove the intake manifold nuts with a socket, ratchet and long extension

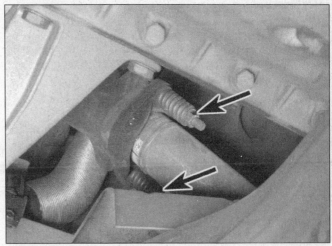

6.5 Remove the exhaust manifold nuts from the exhaust pipe - be sure to soak the nuts with penetrating oil before attempting to remove them

7.5 Place a socket and ratchet on the center bolt to keep the pulley stationary and use another socket and ratchet to remove the smaller bolts attaching the pulley to the vibration damper

from the center out to the ends, to the torque listed in this Chapter's Specifications.
17 The remainder of the installation is the reverse of removal.

6 Exhaust manifold - removal and installation

Refer to illustration 6.5
Warning: *Make sure the engine is completely cool before beginning work on the exhaust system.*
1 Disconnect the negative battery cable from the battery. **Caution:** *If the radio in your vehicle is equipped with an anti-theft system, make sure you have the correct activation code before disconnecting the battery.* **Note**: *If, after connecting the battery, the wrong language appears on the instrument panel display, refer to page 0-7 for the language resetting procedure.*
2 On models where the air cleaner is on the exhaust manifold side of the engine, remove the air cleaner housing assembly and/or airflow sensor to provide working area (see Chapter 4, if necessary).
3 Unplug the spark plug wires and set the spark plug wire harness aside (see Chapter 1).
4 Clearly label, then disconnect or remove all wires, hoses, fittings, etc. that are in the way. Be sure to disconnect the oxygen sensor.
5 Raise the vehicle and support it securely on jackstands. Working

from under the vehicle, separate the exhaust pipe from the manifold. Use penetrating oil on the fasteners to ease removal **(see illustration)**.
6 Remove the jackstands and lower the vehicle. Working from the ends of the manifold toward the center, loosen the retaining nuts gradually until they can be removed. Again, penetrating oil may prove helpful.
7 Pull the manifold off of the head, then remove the old gasket.
Note: *Be very careful not to damage the oxygen sensor.*
8 Clean the gasket mating surfaces of the head and manifold and make sure the threads on the exhaust manifold studs are in good condition.
9 Check for corrosion, warping, cracks and other damage. Repair or replace the manifold as necessary.
10 When installing the manifold, use a new gasket and tighten the retaining nuts gradually, starting at the center and working out to the ends, to the torque listed in this Chapter's Specifications.
11 The remaining steps are simply a reversal of the removal procedure.

7 Timing chain covers - removal and installation

Note 1: *This procedure applies to all four-cylinder and 3.2L and 3.4L six-cylinder engines.*
Note 2: *The upper timing chain cover can be removed separately. If you need to remove both the upper and lower covers, special tools are required. Read Steps 6 and 7 before beginning work.*

Removal
Refer to illustrations 7.5, 7.8 and 7.10
1 Disconnect the negative battery cable from the battery. **Caution:** *If the radio in your vehicle is equipped with an anti-theft system, make sure you have the correct activation code before disconnecting the battery.* **Note:** *If, after connecting the battery, the wrong language appears on the instrument panel display, refer to page 0-7 for the language resetting procedure.*
2 If you're removing the lower timing chain cover (the upper cover can be removed separately), remove the cooling fan and fan shroud, the radiator, the fan belt pulley and the water pump (see Chapter 3).
3 On models where the distributor cap is mounted directly to the timing chain cover, remove the cap, rotor and the black plastic cover beneath the rotor (see Chapter 1).
4 Remove the valve cover (see Section 4).
5 If you'll be removing the lower timing chain cover on a six-cylinder model, remove the crankshaft pulley by holding the pulley stationary with a socket on the center bolt and removing the pulley bolts with another socket **(see illustration)**.

7.8 Unscrew the plug from the timing chain cover and remove the tensioner spring and plunger

7.10 From underneath the vehicle, remove the three bolts that connect the cover and the oil pan

6 If you'll be removing the lower timing chain cover, remove the vibration damper by locking the crankshaft in position and loosening the large center bolt. Since the bolt is on very tight, you'll need to use a large breaker bar and socket to break it loose. On six-cylinder engines, BMW recommends using a 3/4-inch drive socket and breaker bar, since the bolt is extremely tight on these engines. To lock the crankshaft in place while the bolt is being loosened, use one of the following BMW special tools (or equivalent):
  a) No. 11 2 100 for four-cylinder engines
  b) No. 11 2 200 for 1985 and earlier six-cylinder engines
  c) No. 11 2 220 for 1986 and later six-cylinder engines

7 On four-cylinder models, if the special tool listed in the previous Step is not available, you may try locking the crankshaft by removing the flywheel/driveplate inspection cover and jamming a large prybar into the ring gear teeth. On six-cylinder models, since the bolt is so extremely tight, we don't recommend substitute methods. Use the correct tool. On four-cylinder models, after the center bolt is removed, it will probably be necessary to use a jaw-type puller to pull the vibration damper off the crankshaft. Position the jaws behind the inner pulley groove and tighten the puller center bolt very slowly, checking the pulley to make sure it does not get bent or otherwise damaged by the puller. If the pulley seems to be sticking on the crankshaft, it may help to spray the hub area with some penetrating oil and gently tap on the hub area with a hammer.

8 Unscrew the plug and remove the timing chain tensioner spring **(see illustration). Caution:** *The spring is under tension and can cause the plug to be ejected from its hole with considerable force. Hold the tensioner plug securely as it's being unscrewed and release the spring tension slowly.* The tensioner plunger may come out with the spring. If not, reach down into the hole where the tensioner spring was and remove the plunger. To check the plunger for proper operation, see Section 8.

9 Remove the bolts and nuts securing the upper timing chain cover to the engine block and remove the cover. Draw a simple diagram showing the location of the bolts so they can be returned to the same holes from which they're removed. Remove the upper timing chain cover. If it sticks to the engine block, tap it gently with a rubber mallet or place a piece of wood against the cover and hit the wood with a hammer.

10 Remove the bolts and nuts attaching the lower timing chain cover to the engine block. Be sure to remove the three bolts from underneath that connect the front of the oil pan to the bottom of the front cover **(see illustration).** Loosen the remaining oil pan bolts.

11 Run a sharp, thin knife, such as an Xacto knife, between the oil pan gasket and lower timing chain cover, cutting the cover free from the gasket. Be very careful not to damage or dirty the gasket, so you can reuse it.

12 Break the lower timing chain cover-to-block gasket seal by tapping the cover with a rubber mallet or block of wood and hammer. Do not pry between the cover and the engine block, as damage to the gasket sealing surfaces will result.

13 Using a scraper, remove all traces of old gasket material from the sealing surfaces of the covers and engine block. **Caution:** *Be very careful not to scratch or gouge the delicate aluminum surfaces. Also, do not damage the oil pan gasket and keep it clean.* Gasket removal solvents are available at auto parts stores and may prove helpful. After all gasket material has been removed, the gasket surfaces can be degreased by wiping them with a rag dampened with lacquer thinner or acetone.

Installation

14 Replace the front oil seals (see Section 11). It's not wise to take a chance on an old seal, since replacement with the covers removed is very easy.

15 Apply a film of RTV-type gasket sealant to the surface of the oil pan gasket that mates with the lower timing chain cover. Apply extra beads of RTV sealant to the edges where the gasket meets the engine block. **Note:** *If the oil pan gasket is damaged, instead of replacing the whole gasket, you might try trimming the front portion of the gasket off at the point where it meets the engine block, then trim off the front portion of a new oil pan gasket so it's exactly the same size. Cover the exposed inside area of the oil pan with a rag, then clean all traces of old gasket material off the area where the gasket was removed. Attach the new gasket piece to the oil pan with contact-cement-type gasket adhesive, then apply RTV-type sealant as described at the beginning of this Step.*

16 Coat both sides of the new gasket with RTV-type gasket sealant, then attach the lower timing chain cover to the front of the engine. Install the bolts and tighten them evenly to the torque listed in this Chapter's Specifications. Work from bolt-to-bolt in a criss-cross pattern to be sure they're tightened evenly. **Note 1:** *Tighten the lower cover-to-block bolts first, then tighten the oil pan-to-cover bolts. If the gasket protrudes above the cover-to-block joint or bunches up at the cover-to-oil pan joint, trim the gasket so it fits correctly.* **Note 2:** *After applying RTV-type sealant, assembly must be completed in about 10 minutes so the RTV won't prematurely set up.*

17 Install the upper timing chain cover in the same way as the lower cover. If the gasket protrudes beyond the top of the cover and the engine block, trim off the excess with a razor blade.

18 Installation is otherwise the reverse of removal. Be sure to apply a little oil to the front oil seal lips.

8 Timing chain and sprockets - removal, inspection and installation

Note: *This procedure applies to all four-cylinder and 3.2L and 3.4L six-cylinder engines.*

Removal

1 Position the number one cylinder at Top Dead Center (TDC) on the compression stroke (see Section 3). **Caution:** *Once the engine is set at TDC, do not rotate the camshaft or crankshaft until the timing chain is reinstalled. If the crankshaft or camshaft is rotated with the timing chain removed, the valves could hit the pistons, causing expensive internal engine damage.*

2 Remove the valve cover (see Section 4). Double-check that the number one cylinder is at TDC on the compression stroke by making sure the number one cylinder rocker arms are loose (not compressing their valve springs).

3 Remove the upper timing chain cover (see Section 7). Note the location of the camshaft timing marks, which should now be aligned. On four-cylinder engines, there's usually a stamped line on the camshaft flange that aligns with a cast mark on the top of the cylinder head; also, the camshaft sprocket dowel pin hole will be at its lowest point. On six-cylinder engines, a line drawn through two of the camshaft sprocket bolts opposite each other would be exactly vertical, while a line drawn through the other two bolts would be horizontal; also, the locating pin should be in the lower left corner (between the 7 and 8 O'clock positions). Be sure you've identified the correct camshaft TDC position before disassembly, because correct valve timing depends on you aligning them exactly on reassembly.

4 Secure the crankshaft with a socket and ratchet on the vibration damper center bolt, then break loose (but don't unscrew completely) the four bolts attaching the camshaft sprocket to the camshaft. Be very careful not to rotate the camshaft or crankshaft. **Note:** *Some earlier models may be equipped with locking tabs for the camshaft sprocket bolts. Bend the tabs down before loosening the bolts. The tabs are no longer available from the manufacturer and do not have to be reinstalled.*

5 Remove the lower timing chain cover (see Section 7).

6 Unscrew and remove the four camshaft sprocket bolts, then disengage the chain from the crankshaft sprocket and carefully remove the chain and camshaft sprocket from the engine. It may be necessary to gently pry the camshaft sprocket loose from the camshaft with a screwdriver.

Inspection

Timing sprockets

7 Examine the teeth on both the crankshaft sprocket and the camshaft sprocket for wear. Each tooth forms an inverted V. If worn, the side of each tooth under tension will be slightly concave in shape when compared with the other side of the tooth (i.e. one side of the inverted V will be concave when compared with the other). If the teeth appear to be worn, the sprockets must be replaced with new ones. **Note:** *The crankshaft sprocket is a press-fit on the crankshaft and can be removed with a jaw-type puller after the Woodruff key and oil pump are removed (see Section 14). However, BMW recommends the new sprocket be pressed onto the crankshaft after being heated to 390-degrees F. For this reason, if the crankshaft sprocket requires replacement, we recommend removing the crankshaft (see Part B of this Chapter) and taking it to an automotive machine shop to have the old sprocket pressed off and a new one pressed on.*

Timing chain

8 The chain should be replaced with a new one if the sprockets are worn or if the chain is loose. It's a good idea to replace the chain if the engine is stripped down for overhaul. The rollers on a very badly worn chain may be slightly grooved. To avoid future problems, if there's any doubt at all about the chain's condition, replace it with a new one.

Chain rail and tensioner

Refer to illustration 8.9

9 Inspect the chain guide rail and tensioner rail for deep grooves caused by chain contact. Replace them if they are excessively worn. The rails can be replaced after removing the circlips with a pointed tool or needle-nose pliers **(see illustration)**.

10 Shake the tensioner plunger and listen for a rattling sound from

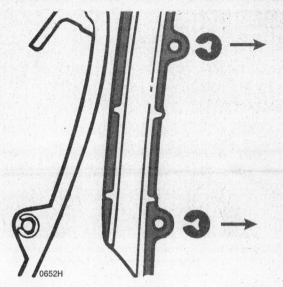

8.9 To remove the tensioner or chain guide rail, remove the circlips with a pointed tool or needle-nose pliers - the circlips tend to fly off when they're released, so make sure you catch them or they'll get lost (or, worst yet, wind up in the engine!)

the check ball. If you can't hear the ball ratting, replace the plunger.

11 To further check the tensioner plunger, blow through it first from the closed end, then from the slotted, guide end. No air should flow through the plunger when you blow through the closed end, and air should flow through it freely when you blow through the slotted end. If the tensioner fails either test, replace it.

Installation

12 Reinstall the tensioner rail and chain guide rail, if removed.

13 Temporarily install the lower timing chain cover and vibration damper so you can check the crankshaft timing marks. Once you've verified the TDC marks are aligned, remove the damper and cover.

14 Loop the timing chain over the crankshaft sprocket, then loop it over the camshaft sprocket and, guiding the chain between the chain guide and tensioner rail, install the camshaft sprocket on the camshaft. Make sure the camshaft timing marks are aligned.

15 The remainder of installation is the reverse of removal. Be sure to tighten the fasteners to the correct torques (see this Chapter's Specifications).

9 Timing belt covers - removal and installation

Refer to illustrations 9.8 and 9.9

Note: *This procedure applies to all 2.5L and 2.7L six-cylinder engines.*

1 Disconnect the cable from the negative battery terminal. **Caution:** *If the radio in your vehicle is equipped with an anti-theft system, make sure you have the correct activation code before disconnecting the battery.* **Note:** *If, after connecting the battery, the wrong language appears on the instrument panel display, refer to page 0-7 for the language resetting procedure.*

2 Remove the fan clutch and fan shroud (see Chapter 3).

3 Remove the radiator (see Chapter 3).

4 Remove the fan belt pulley.

5 If equipped, disconnect the reference sensor wiring harness which runs across the front of the timing belt cover and set it aside.

6 If the distributor cap is mounted directly to the upper timing belt cover, remove the cap, rotor and the black plastic cover beneath the rotor (see Chapter 1).

7 Remove the lower fan belt pulley and vibration damper. Secure the crankshaft pulley center bolt while you loosen the outer pulley/damper bolts **(see illustration 7.5)**.

2A-8   Chapter 2 Part A   Engines

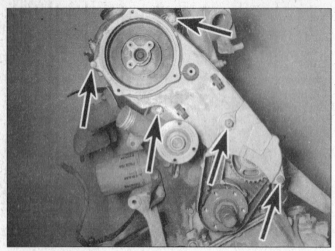

9.8 Remove all the nuts and bolts that attach the upper and lower covers (arrows) (engine removed for clarity)

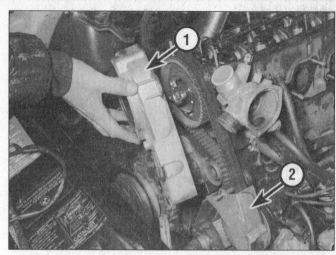

9.9 Remove the upper timing belt cover first, then the lower cover

1 Upper timing belt cover   2 Lower timing belt cover

10.4a Align the groove in the hub on the end of the crankshaft with the notch in the front inner cover and mark them for assembly reference later on

10.4b Align the mark on the camshaft sprocket with the mark on the cylinder head

8 Remove the nuts attaching the timing belt covers to the engine **(see illustration)**.

9 Remove the upper cover first **(see illustration)**, then the lower cover. **Note:** *The upper cover has two alignment sleeves in the top bolt positions. Be sure these are in place upon reassembly.*

10 Installation is the reverse of the removal procedure. Tighten the cover bolts securely.

10 Timing belt and sprockets - removal, inspection and installation

Note: *This procedure applies to all 2.5L and 2.7L six-cylinder engines.*

Removal

Refer to illustrations 10.4a, 10.4b, 10.5 and 10.6

1 Disconnect the negative cable from the battery. **Caution:** *If the radio in your vehicle is equipped with an anti-theft system, make sure you have the correct activation code before disconnecting the battery.* **Note**: *If, after connecting the battery, the wrong language appears on the instrument panel display, refer to page 0-7 for the language resetting procedure.*

2 Remove the timing belt covers (see Section 9).

3 Set the number one piston at TDC (see Section 3). **Caution:** *Once the engine is set at TDC, do not rotate the camshaft or crankshaft until the timing belt is reinstalled. If the crankshaft or camshaft is rotated with the timing belt removed, the valves could hit the pistons, causing expensive internal engine damage.*

4 The crankshaft mark should be aligned with the mark on the inner cover **(see illustration)**. The mark on the camshaft sprocket should be aligned with the stamped line on the cylinder head **(see illustration)**.

5 Loosen the two idler pulley retaining bolts a little and push the idler pulley toward the water pump **(see illustration)**. With the timing belt tension relieved, re-tighten the retaining bolt. If the same belt is to be reinstalled, mark it with an arrow indicating direction of rotation.

6 Remove the timing belt by slipping it off of the tensioner pulley and the other sprockets **(see illustration)**.

7 If it's necessary to remove the camshaft or the intermediate shaft sprocket, remove the sprocket bolt while holding the sprocket from moving. To hold the sprocket, wrap it with a piece of an old timing belt (with the tooth side engaging the sprocket teeth) or a piece of leather, then secure the sprocket with a chain wrench, but don't tighten the chain wrench so tight that you bend the sprocket. If a chain wrench is not available, clamp the ends of the piece of belt or leather tightly together with a pair of Vise Grips. **Caution:** *Do not use the timing belt you're planning to install to hold the sprocket. Also, be sure to hold the camshaft sprocket very steady, because if it moves more than a few*

Chapter 2 Part A Engines

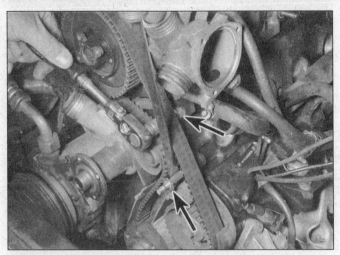

10.5 Loosen the idler pulley bolts to relieve the tension on the timing belt so it can be removed

10.6 When removing the timing belt on models with a two-piece crankshaft hub, it's a tight fit to remove it around the hub, but it's a lot easier than removing the crankshaft hub assembly, which is secured by a very tight bolt

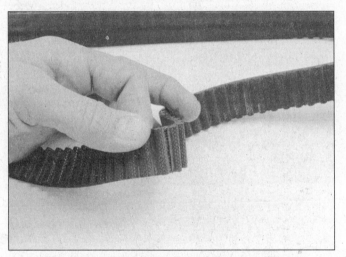

10.9a Inspect the timing belt carefully for cracking, as shown here...

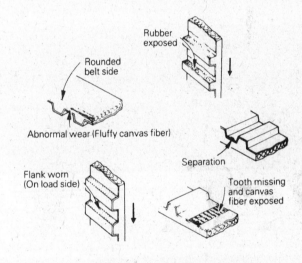

10.9b ...and any other damage

degrees, the valves could hit the pistons.

8 If it's necessary to remove the crankshaft sprocket, remove the crankshaft hub center bolt while holding the crankshaft steady. **Note:** *The removal of the crankshaft hub mounting bolt requires a heavy-duty holding device because of the 302 ft-lbs of torque used to tighten the bolt. BMW has a special tool, numbered 11 2 150, for this purpose, or you can check with an automotive tool dealer or parts store for a tool capable of doing the job. On models with a two-piece hub, after removing the outer hub piece, you'll then need to remove the sprocket with a bolt-type puller (available at most auto parts stores). When using the puller, thread the crankshaft center bolt in approximately three turns to be used as a bottoming point for the puller's center bolt.*

Inspection

Refer to illustration 10.9a and 10.9b

9 Check for a cracked, worn or damaged belt. Replace it if any of these conditions are found. **(see illustrations).** Also look at the sprockets for any signs of irregular wear or the possible need for replacement. **Note:** *If any parts need to be replaced, check with your local BMW dealer parts department to be sure compatible parts are used. Later replacement parts are marked as "Z 127". Replacement of the timing belt will require the use of a the newer version belt tensioner. Replace it at this time if it has not already been updated previously.*

10 Inspect the idler pulley and tension spring. Rotate the tensioner pulley to be sure it rotates freely, with no noise or play. **Note:** *When replacing the timing belt, it is recommended that the tensioner be replaced also.*

Installation

Refer to illustration 10.16

11 Install the idler pulley and spring, pushing the pulley towards the air conditioning compressor side as far as it will go before tightening the bolt.

12 If you are reinstalling the old belt, check to make sure the mark made to indicate belt direction of rotation is pointing the right way (the belt should rotate in a clockwise direction as you face the front of the engine).

13 Install the timing belt, placing the belt under the crankshaft gear first to get by the housing. Then guide the belt around the other gears.

14 Finally place the belt over the idler pulley.

15 Loosen the tensioner bolts and allow the spring tension to be applied to the belt.

16 Lightly apply pressure behind the tensioner to be sure spring pressure is being applied to the belt **(see illustration).** But don't tighten the bolts while prying. Lightly tighten the bolts only after the prybar has been released.

17 Check to make sure the camshaft and crankshaft timing marks

10.16 After the belt has been installed correctly around all sprockets and the tensioner pulley, lightly apply pressure to the tensioner to be sure the tensioner isn't stuck and has full movement against the timing belt

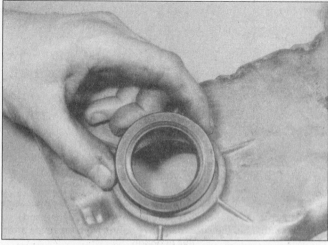

11.7 The crankshaft front oil seal is pressed into the front of the lower timing chain cover (cover removed from the engine for clarity)

are still aligned **(see illustrations 10.4a and 10.4b)**.
18 Turn the crankshaft clockwise through two complete revolutions. Loosen, then tighten, the idler pulley bolts to the torque listed in this Chapter's Specifications. **Caution:** *This is absolutely necessary to stretch the new belt. If not done, the belt tensioner will be too loose and damage could result.*
19 Verify the timing marks are still perfectly aligned. If not, remove and reinstall the timing belt.
20 The remainder of installation is the reverse of removal.

11 Front oil seals - replacement

Refer to illustration 11.3

Timing chain engines

Front camshaft seal (533i and 535i models with Motronic ignition system only)

1 Remove the upper timing chain cover only (see Section 7).
2 Support the cover on two blocks of wood and drive out the seal front behind with a hammer and screwdriver. Be very careful not to damage the seal bore in the process.
3 Coat the outside diameter and lip of the new seal with multi-purpose grease and drive the seal into the cover with a hammer and a socket slightly smaller in diameter than the outside diameter of the seal.
4 The remainder of installation is the reverse of removal.

Front crankshaft seal

5 Remove the drivebelts, crankshaft pulley and vibration damper (see Section 7).
6 Carefully pry the old seal out of the cover with a large screwdriver. Be very careful not to damage the seal bore or the crankshaft with the tool. Wrap the tip of the screwdriver with a piece of tape to prevent damage.
7 Clean the bore in the cover and coat the outer edge of the new seal with engine oil or multi-purpose grease. Also lubricate the lips of the seal with multi-purpose grease. Using a socket with an outside diameter slightly smaller than the outside diameter of the seal, carefully drive the new seal into place with a hammer **(see illustration)**. If a socket isn't available, a short section of large-diameter pipe will work. Check the seal after installation to be sure the spring around the inside of the seal lip didn't pop out of place.
8 The remainder of installation is the reverse of removal.

Timing belt engines

Front camshaft seal

9 Remove the timing belt and camshaft sprocket (see Section 10).
10 Remove the two screws and separate the camshaft seal housing from the cylinder head by pulling it as you rotate it back and forth.
11 Support the housing on two blocks of wood and drive the seal out of the housing from behind using a hammer and screwdriver. Be very careful not to damage the seal housing.
12 Coat the lip and outside diameter of the new seal with multi-purpose grease.
13 Drive the new seal into the housing using a hammer and socket with a diameter slightly smaller than the outside diameter of the seal.
14 Replace the O-ring on the back of the seal housing and work the lip of the seal over the end of the camshaft. Install the screws and tighten them securely. The remainder of installation is the reverse of removal.

Front crankshaft and intermediate shaft seals

15 Remove the timing belt and crankshaft and intermediate shaft pulleys (see Section 10). **Note:** *We recommend the timing belt be replaced any time it is removed.*
16 Remove the bolts and nuts attaching the front cover to the engine block. Be sure to remove the three bolts from underneath that connect the front of the oil pan to the bottom of the front cover **(see illustration 7.10)**.
17 Run a sharp, thin knife, such as an Xacto knife, between the oil pan gasket and the front cover, cutting the cover free from the gasket. Be very careful not to damage or dirty the gasket, so you can reuse it.
18 Break the front cover-to-block gasket seal by tapping the cover with a rubber mallet or block of wood and hammer. Do not pry between the cover and the engine block, as damage to the gasket sealing surfaces will result.
19 Using a scraper, remove all traces of old gasket material from the sealing surfaces of the covers and engine block. **Caution:** *Be very careful not to scratch or gouge the delicate aluminum surfaces. Also, do not damage the oil pan gasket and keep it clean.* Gasket removal solvents are available at auto parts stores and may prove helpful. After all gasket material has been removed, the gasket surfaces can be degreased by wiping them with a rag dampened with lacquer thinner or acetone.
20 Support the cover on two blocks of wood and drive out the seals from behind with a hammer and screwdriver. Be very careful not to damage the seal bores in the process.
21 Coat the outside diameters and lips of the new seals with multi-purpose grease and drive the seals into the cover with a hammer and a socket slightly smaller in diameter than the outside diameter of the seal.

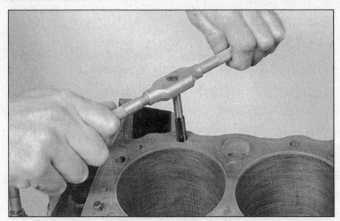

12.22 The cylinder head bolt holes should be cleaned and restored with a tap (be sure to remove debris from the holes after this is done)

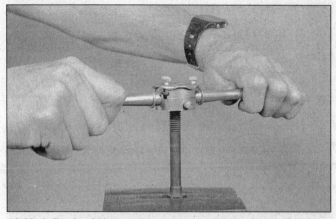

12.23 A die should be used to remove sealant and corrosion from the head bolt threads prior to installation

22 Apply a film of RTV-type gasket sealant to the surface of the oil pan gasket that mates with the front cover. Apply extra beads of RTV sealant to the edges where the gasket meets the engine block. **Note:** *If the oil pan gasket is damaged, instead of replacing the whole gasket, you might try trimming the front portion of the gasket off at the point where it meets the engine block, then trim off the front portion of a new oil pan gasket so it's exactly the same size. Cover the exposed inside area of the oil pan with a rag, then clean all traces of old gasket material off the area where the gasket was removed. Attach the new gasket piece to the oil pan with contact-cement-type gasket adhesive, then apply RTV-type sealant as described at the beginning of this Step.*

23 Coat both sides of the new gasket with RTV-type gasket sealant, then attach the front cover to the front of the engine, carefully working the seals over the crankshaft and intermediate shaft. Install the bolts and tighten them evenly to the torque listed in this Chapter's Specifications. Work from bolt-to-bolt in a criss-cross pattern to be sure they're tightened evenly. **Note 1:** *Tighten the front cover-to-block bolts first, then tighten the oil pan-to-cover bolts.* **Note 2:** *After applying RTV-type sealant, assembly must be completed in about 10 minutes so the RTV won't prematurely set up.*

24 The remainder of installation is the reverse of removal.

12 Cylinder head - removal and installation

Removal

1 Relieve the fuel pressure (see Chapter 4).
2 Remove the battery (see Chapter 5). **Caution:** *If the radio in your vehicle is equipped with an anti-theft system, make sure you have the correct activation code before disconnecting the battery.* **Note:** *If, after connecting the battery, the wrong language appears on the instrument panel display, refer to page 0-7 for the language resetting procedure.*
3 Remove the air cleaner assembly (see Chapter 4).
4 Disconnect the primary lead from the distributor and the high tension lead from the coil (see Chapter 5).
5 Disconnect the lead from the coolant temperature sending unit (see Chapter 3).
6 Disconnect the fuel lines from the fuel rail (see Chapter 4).
7 Clearly label, then disconnect, all other hoses from the throttle body, intake manifold and/or cylinder head.
8 Disconnect the throttle cable from the throttle linkage (see Chapter 4).
9 Disconnect the exhaust manifold from the cylinder head (see Section 6). It is not usually necessary to disconnect the manifold from the exhaust pipe.
10 Remove and disconnect any remaining hoses or lines from the intake manifold, including the ignition advance vacuum line(s), and the coolant and heater hoses.
11 Remove the intake manifold (see Section 5). Do not disassemble any fuel injection system components unless it is absolutely necessary.
12 Remove the fan belt and fan (see Chapter 3).
13 Remove the valve cover (see Section 4). Remove the semi-circular rubber seal from the front of the cylinder head.
14 Set the number one piston at Top Dead Center on the compression stroke (see Section 3).
15 Remove the timing chain or belt (see Section 8 or 10). **Note:** *If you want to save time by not removing and installing the timing belt or chain and re-timing the engine, you can unfasten the camshaft sprocket and suspend it out of the way - with the belt or chain still attached - by a piece of rope. Be sure the rope keeps firm tension on the belt or chain so it won't become disengaged from any of the sprockets.*
16 Loosen the cylinder head bolts 1/4-turn at a time each, in the reverse of the sequence shown **(see illustrations 12.26a, 12.26b or 12.26c)**, but **Do not** disassemble the rocker arm assembly at this time.
17 Remove the cylinder head by lifting it straight up and off the engine block. Do not pry between the cylinder head and the engine block as damage to the gasket sealing surfaces may result. Instead, use a blunt prybar positioned in an intake port to gently pry the head loose.
18 Remove any remaining external components from the head to allow for thorough cleaning and inspection. See Chapter 2B for cylinder head servicing procedures.

Installation

Refer to illustrations 12.22, 12.23, 12.26a, 12.26b and 12.26c

19 The mating surfaces of the cylinder head and block must be perfectly clean when the head is installed.
20 Use a gasket scraper to remove all traces of carbon and old gasket material, then clean the mating surfaces with lacquer thinner or acetone. If there's oil on the mating surfaces when the head is installed, the gasket may not seal correctly and leaks could develop. When working on the block, stuff the cylinders with clean shop rags to keep out debris. Use a vacuum cleaner to remove material that falls into the cylinders.
21 Check the block and head mating surfaces for nicks, deep scratches and other damage. If damage is slight, it can be removed with a file; if it's excessive, machining may be the only alternative.
22 Use a tap of the correct size to chase the threads in the head bolt holes, then clean the holes with compressed air - make sure that nothing remains in the holes **(see illustration)**.
23 Mount each bolt in a vise and run a die down the threads to remove corrosion and restore the threads. Dirt, corrosion, sealant and damaged threads will affect torque readings **(see illustration)**.
24 Install any components removed from the head prior to cleaning and inspection.
25 Make sure the gasket sealing surfaces of the engine block and cylinder head are clean and oil-free, then lay the head gasket in place on the block with the manufacturer's stamped mark facing up (it usually says "UP," "OBEN" or something similar). Use the dowel pins in the top of the block to properly locate the gasket.
26 Carefully set the cylinder head in place on the block. Use the

2A-12   Chapter 2 Part A   Engines

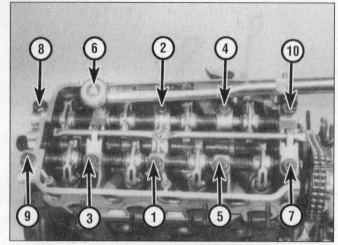

12.26a Cylinder head bolt TIGHTENING sequence for four-cylinder engines

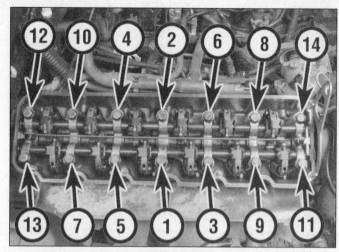

12.26b Cylinder head bolt TIGHTENING sequence for all six-cylinder engines, except those used in 533i and 535i models

12.26c Cylinder head bolt TIGHTENING sequence for six-cylinder engines used in 533i and 535i models

13.4 If equipped, remove the nut attaching the power steering lines to the oil pan and move the lines off to one side to allow you to get at the oil pan bolts

dowel pins to properly align it. Tighten the cylinder head bolts, in the sequence shown, to the torque listed in this Chapter's Specifications **(see illustrations)**.

27 The remainder of installation is the reverse of removal. Set the valve clearances (see Chapter 1) before installing the valve cover (check them again after the engine is warmed up). Run the engine and check for leaks.

13 Oil pan - removal and installation

Refer to illustrations 13.4, 13.5, 13.6a, 13.6b and 13.7

1 Drain the engine oil (see Chapter 1).
2 Raise the front of the vehicle and place it securely on jackstands.
3 Remove the splash shields from under the engine.
4 On 5-series models, disconnect the hoses attached to the oil pan and pull them to one side **(see illustration)**.
5 On 5-series models, disconnect the oil level sensor electrical connector **(see illustration)**.
6 On 5-series models, remove the cast-aluminum inspection cover that covers the rear of the oil pan **(see illustrations)**.
7 Remove the bolts securing the oil pan to the engine block and front cover **(see illustration)**.
8 Tap on the pan with a soft-faced hammer to break the gasket seal, and lower the oil pan from the engine.

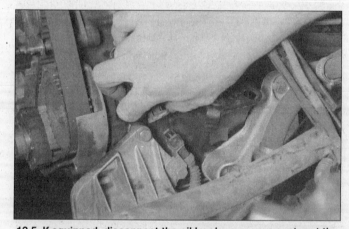

13.5 If equipped, disconnect the oil level sensor connector at the left side of the engine, down near the power steering pump mounting bracket

9 Using a gasket scraper, scrape off all traces of the old gasket from the engine block, the timing chain cover, the rear main oil seal housing and the oil pan. Be especially careful not to nick or gouge the gasket sealing surfaces of the timing chain cover and the oil seal hous-

Chapter 2 Part A Engines   2A-13

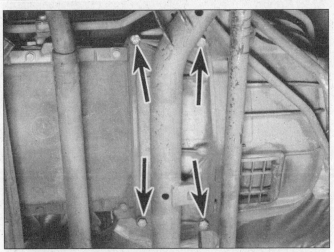

13.6a Remove the four bolts attaching the inspection cover and . . .

13.6b . . . remove the cover to get to all the oil pan bolts

13.7 Remove the bolts holding the oil pan to the engine block and front cover, as shown here on a six-cylinder engine

14.2 On all four-cylinder models and 533i and 535i models, remove the three bolts that hold the driven gear to the oil pump and remove the gear

14.3a On six-cylinder engines with a timing belt, the oil pump is bolted across the engine block from side to side, towards the front of the engine

ing (they are made of aluminum and are quite soft).

10 Clean the oil pan with solvent and dry it thoroughly. Check the gasket sealing surfaces for distortion. Clean any residue from the gasket sealing surfaces on the pan and engine with a rag dampened with lacquer thinner or acetone.

11 Before installing the oil pan, apply a thin coat of RTV-type gasket sealant to the oil pan gasket sealing surfaces. Lay a new oil pan gasket in place and carefully apply a coat of gasket sealant to the exposed side of the gasket.

12 Gently lay the oil pan in place (do not disturb the gasket) and install the bolts. Start with the bolts closest to the center of the pan and tighten them to the torque listed in this Chapter's Specifications using a criss-cross pattern. Do not overtighten them or leakage may occur.

13 The remainder of installation is the reverse of removal.

14 Oil pump and driveshaft - removal, inspection and installation

Removal

Refer to illustrations 14.3a, 14.3b and 14.4

1 Remove the oil pan (see Section 13).

2 On timing chain engines, remove the three bolts that attach the gear to the front of the pump (see illustration). **Note**: *Some models have a single center nut attaching the gear to the oil pump*

3 Unbolt the oil pump from the engine block (see illustrations) and remove it.

Chapter 2 Part A Engines

14.3b On all four-cylinder models and 533i and 535i models, the oil pump is bolted to the block at the front and center of the engine block

14.4 If necessary, remove the plug and oil pump driveshaft from the engine and inspect the driveshaft gear, as well as this intermediate shaft gear in the engine block (arrow)

15.3 Using a socket and a large ratchet or breaker bar, remove the eight bolts that hold the flywheel/driveplate to the crankshaft flange - prevent the flywheel/driveplate from turning by locking the ring gear with a prybar

4   If the engine has a timing belt, the intermediate shaft drives the oil pump driveshaft, which drives the oil pump. To remove the driveshaft, remove the hold-down plate from the block and lift out the plug. Check the O-ring and replace it, if necessary. Lift the gear out and check both gears for wear, replacing them, if necessary **(see illustration)**.
5   If the gear on the intermediate shaft is worn, or the intermediate shaft bearing is worn or damaged, the intermediate shaft must be removed. Remove the engine (see Chapter 2B), then remove the timing belt, crankshaft and intermediate shaft sprockets (see Section 10) and the engine front cover (see Section 11). The intermediate shaft can be slid out the front of the engine.

Inspection

Note: Considering that a malfunctioning oil pump can easily cause major engine damage, we recommend routinely replacing the oil pump unless it's in like-new condition.

6   Remove the cover and check the pump body, gears or rotors and cover for cracks and wear (especially in the gear or rotor contact areas).
7   Check the strainer to make sure it is not clogged or damaged.
8   Lubricate the gears with clean engine oil, then attach the pump cover to the body and tighten the bolts evenly and securely.
9   Before installing the pump - new, rebuilt or original - on the engine, check it for proper operation. Fill a clean drain pan to a depth of one inch with new, clean engine oil of the recommended viscosity.
10   Immerse the oil pump inlet in the oil and turn the driveshaft counterclockwise by hand. As the shaft is turned, oil should be discharged from the pump outlet.

Installation

11   Make sure the mounting surfaces are clean, then insert the pump into the engine block recess. Install the bolts and tighten them to the torque specifications at the beginning of this Chapter.
12   Installation is the reverse of removal.

15   Flywheel/driveplate - removal and installation

Refer to illustration 15.3

1   Remove the transmission (on vehicles with a manual transmission, see Chapter 7A; on vehicles with an automatic transmission, see Chapter 7B).
2   On vehicles with a manual transmission, remove the clutch (see Chapter 8).
3   Mark the relationship of the flywheel/driveplate to the crankshaft so it can be reinstalled the same way.
4   The flywheel/driveplate is attached to the rear of the crankshaft with eight bolts. Loosen and remove the bolts, then separate it from the crankshaft flange **(see illustration)**. Be careful. The flywheel is heavy.
5   To install the flywheel/driveplate on the crankshaft, use a liquid thread locking compound on the bolts and tighten them gradually, using a criss-cross pattern, to the torque listed in this Chapter's Specifications.
6   The remainder of installation is the reverse of removal.

16   Rear main oil seal - replacement

Refer to illustrations 16.2, 16.6 and 16.7

1   Remove the flywheel or driveplate (see Section 15).
2   Remove the bolts and nuts attaching the seal retainer to the engine block. Be sure to remove the two bolts from underneath that connect the rear of the oil pan to the bottom of the seal retainer **(see illustration)**.
3   Run a sharp, thin knife, such as an Xacto knife, between the oil

Chapter 2 Part A Engines

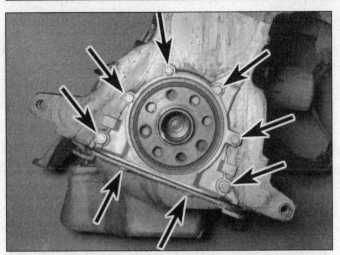

16.2 Remove the six bolts from the rear of the block and the two from underneath at the oil pan

16.6 After removing the retainer from the block, support it on two wood blocks and drive out the old seal with a punch and hammer

16.7 Drive the new seal into the retainer with a wood block or a section of pipe, if you have one large enough - make sure you don't cock the seal in the retainer bore

pan gasket and the seal retainer, cutting the retainer free from the gasket. Be very careful not to damage or dirty the gasket, so you can reuse it.

4   Break the seal retainer-to-block gasket seal by tapping the retainer with a plastic mallet or block of wood and hammer. Do not pry between the retainer and the engine block, as damage to the gasket sealing surfaces will result.

5   Using a scraper, remove all traces of old gasket material from the sealing surfaces of the retainer and engine block. **Caution:** *Be very careful not to scratch or gouge the delicate aluminum surfaces. Also, do not damage the oil pan gasket and keep it clean.* Gasket removal solvents are available at auto parts stores and may prove helpful. After all gasket material has been removed, the gasket surfaces can be degreased by wiping them with a rag dampened with lacquer thinner or acetone.

6   Support the retainer on two blocks of wood and drive out the seal from behind with a hammer and screwdriver **(see illustration)**. Be very careful not to damage the seal bore in the process.

7   Coat the outside diameter and lip of the new seal with multi-purpose grease and drive the seal into the retainer with a hammer and a block of wood **(see illustration)**.

8   Apply a film of RTV-type gasket sealant to the surface of the oil pan gasket that mates with the seal retainer. Apply extra beads of RTV sealant to the edges where the gasket meets the engine block. **Note:** *If the oil pan gasket is damaged, instead of replacing the whole gasket, you might try trimming the rear portion of the gasket off at the point where it meets the engine block, then trim off the rear portion of a new oil pan gasket so it's exactly the same size. Cover the exposed inside area of the oil pan with a rag, then clean all traces of old gasket material off the area where the gasket was removed. Attach the new gasket piece to the oil pan with contact-cement-type gasket adhesive, then apply RTV-type sealant as described at the beginning of this Step.*

9   Coat both sides of the new retainer gasket with RTV-type gasket sealant, then attach the gasket to the seal retainer and the seal retainer to the rear of the engine. Install the bolts and tighten them evenly to the torque listed in this Chapter's Specifications. Work from bolt-to-bolt in a criss-cross pattern to be sure they're tightened evenly. **Note 1:** *Tighten the retainer-to-block bolts first, then tighten the oil pan-to-retainer bolts.* **Note 2:** *After applying RTV-type sealant, assembly must be completed in about 10 minutes so the RTV won't prematurely set up.*

10 Install the flywheel/driveplate (see Section 15).

11 Install the transmission (on vehicles with a manual transmission, see Chapter 7A; on vehicles with an automatic transmission, see Chapter 7B).

17 Engine mounts - check and replacement

Refer to illustrations 17.4, 17.5 and 17.8

1   Engine mounts seldom require attention, but broken or deteriorated mounts should be replaced immediately or the added strain placed on the driveline components may cause damage or wear.

Check

2   During the check, the engine must be raised slightly to remove the weight from the mounts.

3   Raise the vehicle and support it securely on jackstands, then position a jack under the engine oil pan. Place a large block of wood between the jack head and the oil pan, then carefully raise the engine just enough to take the weight off the mounts. **Warning:** *DO NOT place any part of your body under the engine when it's supported only by a jack!*

4   Check the mounts to see if the rubber is cracked **(see illustration)**, hardened or separated from the metal plates. Sometimes the rubber will split right down the center.

5   Check for relative movement between the mount plates and the engine or frame (use a large screwdriver or pry bar to attempt to move the mounts). If movement is noted, lower the engine and tighten the mount fasteners **(see illustration)**.

17.4 As engine mounts wear or age, they should be inspected for cracking or separating from their metal plates

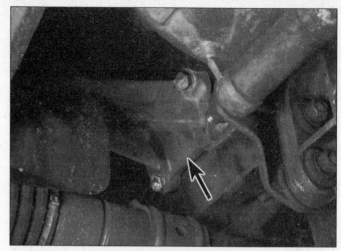

17.5 Pry gently between the block and where the engine mount attaches (arrow) - if there is movement, tighten the bolts

6 Rubber preservative should be applied to the mounts to slow deterioration.

Replacement

7 Disconnect the negative battery cable from the battery, then raise the vehicle and support it securely on jackstands if you haven't already done so. Support the engine as described in Step 3. **Caution:** *If the radio in your vehicle is equipped with an anti-theft system, make sure you have the correct activation code before disconnecting the battery.* **Note:** *If, after connecting the battery, the wrong language appears on the instrument panel display, refer to page 0-7 for the language resetting procedure.*

8 Remove the large bracket-to-mount nut **(see illustration)**. Raise the engine slightly, then remove the two mount-to-frame bolts and nuts from either side of the mount and detach the mount.

9 Installation is the reverse of removal. Use thread locking compound on the mount bolts/nuts and be sure to tighten them securely.

17.8 To remove an engine mount, first remove the stud nut (arrow) - 535i model shown, others similar

Chapter 2 Part B
General engine overhaul procedures

Contents

Compression check	3
Crankshaft - inspection	19
Crankshaft - installation and main bearing oil clearance check	24
Crankshaft - removal	13
Cylinder head, camshaft and rocker arms - cleaning and inspection	9
Cylinder head - disassembly	8
Cylinder head - reassembly	11
Cylinder honing	17
Engine block - cleaning	15
Engine block - inspection	16
Engine overhaul - disassembly sequence	7
Engine overhaul - general information	2
Engine overhaul - reassembly sequence	21
Engine rebuilding alternatives	6
Engine - removal and installation	5
Engine removal - methods and precautions	4
General information	1
Initial start-up and break-in after overhaul	27
Intermediate shaft - removal and inspection	14
Intermediate shaft - installation	23
Main and connecting rod bearings - inspection	20
Pistons/connecting rods - inspection	18
Pistons/connecting rods - installation and rod bearing oil clearance check	26
Pistons/connecting rods - removal	12
Piston rings - installation	22
Rear main oil seal - installation	25
Valves - servicing	10

Specifications

General
Cylinder compression pressure (all engines)	142 to 156 psi
Cylinder head warpage limit	0.004 inch
Minimum cylinder head thickness (do not resurface the head to a thickness less than listed)	
All four-cylinder engines and 533i and 535i six-cylinder engines	5.063 inches
325i, 525i and 528e six-cylinder engines	4.909 inches

Camshaft and rocker arms
Camshaft bearing oil clearance	0.0013 to 0.0030 inch
Camshaft endplay	
3-series body style	0.0012 to 0.0071 inch
5-series body style	
525i and 528e	0.008 inch
533i and 535i	0.0012 to 0.0071 inch
Rocker arm radial clearance	0.0006 to 0.0020 inch

Valves
Valve stem diameter
   318i (1984 and 1985) and 533i and 535i (1983 through 1988)
      Standard .... 0.315 inch
      First oversize .... 0.319 inch
      Second oversize .... 0.323 inch
   325 and 325i, 325e and 325es (1984 through 1988); 525i and 528e (1982 through 1988)
      Standard .... 0.275 inch
      First oversize .... 0.279 inch
      Second oversize .... 0.283 inch
Minimum valve margin width
   Intake .... 3/64 inch
   Exhaust .... 5/64 inch
Valve stem-to-guide clearance .... 0.031 inch
Valve face angle
   Intake 45-degrees
   Exhaust 45-degrees

(handwritten note: Guide OD > 522 thou, 523 ish, ID .275)

Crankshaft
Runout
   3-series body style
      Four-cylinder engine .... 0.004 inch
      Six cylinder engine .... 0.006 inch
   5-series body style
      525i .... 0.006 inch
      528e .... 0.0031 to 0.0064 inch
      533i and 535i .... 0.0033 to 0.0069 inch
Endplay .... 0.0031 to 0.0069 inch
Main bearing journal diameter (standard)
   Four-cylinder engine .... 2.1654 inches
   Six-cylinder engines .... 2.3622 inches
Main bearing oil clearance .... 0.0008 to 0.0028 inch
Connecting rod journal diameter (standard)
   Four-cylinder engines .... 1.8912 inches
   Six-cylinder engines, except 528e .... 2.364 inches
   528e .... 1.773 inches

Connecting rods
Connecting rod side play (all engines) .... 0.0016 inch
Connecting rod bearing oil clearance .... 0.0008 to 0.0028 inch

Engine block
Cylinder bore - diameter (standard)
   Four-cylinder engines .... 3.5039 to 3.5043 inches
   325i, 525i and 528e six-cylinder engines .... 3.3071 to 3.3075 inches
   533i six-cylinder engine .... 3.5309 to 3.5313 inches
Cylinder out-of-round limit (maximum)
   All four-cylinder engines and 533i and 535i
      six-cylinder engines .... 0.0004 inch
   325i, 525i and 528e six-cylinder engines .... 0.0012 inch
Cylinder taper
   All four-cylinder engines and 533i and 535i
      six-cylinder engines .... 0.0004 inch
   325i, 525i and 528e six-cylinder engines .... 0.0008 inch

Pistons and piston rings
Piston diameter
   All four-cylinder engines and 533i six-cylinder engine ...... 3.5027 inches
   325i, 525i and 528e six-cylinder engines .... 3.3063 inches
   535i six-cylinder engine .... 3.6209 inches
Piston-to-cylinder wall clearance
   All four-cylinder engines and 533i and 535i six-cylinder engines
      New .... 0.0008 to 0.0020 inch
      Service limit .... 0.0060 inch
   325i, 525i and 528e six-cylinder engines
      New .... 0.0004 to 0.0016 inch
      Service limit .... 0.0047 inch
Piston ring end gap
   3-series body style
      Four-cylinder engine
         Top compression ring .... 0.012 to 0.028 inch
         Second compression ring .... 0.008 to 0.016 inch
         Oil ring .... 0.010 to 0.020 inch

Chapter 2 Part B General engine overhaul procedures

  Six-cylinder engine
    Compression rings .... 0.012 to 0.020 inch
    Oil ring .... 0.010 to 0.020 inch
  5-series body style
    525i
      Compression rings .... 0.008 to 0.020 inch
      Oil ring .... 0.008 to 0.020 inch
    528e
      Compression rings .... 0.012 to 0.020 inch
      Oil ring .... 0.010 to 0.020 inch
    533i and 535i
      Top compression ring .... 0.012 to 0.020 inch
      Second compression ring .... 0.008 to 0.016 inch
      Oil ring .... 0.010 to 0.020 inch
Piston ring side clearance
  3-series body style
    Four-cylinder engine
      Top compression ring .... 0.0024 to 0.0035 inch
      Second compression ring .... 0.0012 to 0.0028 inch
      Oil ring .... 0.0008 to 0.0024 inch
    Six-cylinder engine
      First compression ring .... 0.0016 to 0.0028 inch
      Second compression ring .... 0.0012 to 0.0024 inch
      Oil ring .... 0.0008 to 0.0017 inch
  5-series body style
    525i
      Top compression ring .... 0.0016 to 0.0031 inch
      Second compression ring .... 0.0012 to 0.0027 inch
      Oil ring .... 0.0008 to 0.0020 inch
    528e
      Top compression ring .... 0.0016 to 0.0028 inch
      Second compression ring .... 0.0012 to 0.0024 inch
      Oil ring .... 0.0008 to 0.0017 inch
    533i and 535i
      Top compression ring .... 0.0020 to 0.0032 inch
      Second compression ring .... 0.0016 to 0.0028 inch
      Oil ring .... 0.0008 to 0.0020 inch

Torque specifications   Ft-lbs (unless otherwise noted)
Main bearing cap-to-engine block bolts .... 43 to 46
Connecting rod cap nuts
  All four-cylinder engines and 533i and 535i
    six-cylinder engines .... 38 to 42
  325i, 525i and 528e six-cylinder engines
    Step 1 15
    Step 2 .... turn an additional 70-degrees
Intermediate shaft sprocket-to-shaft bolt... 44
Oil supply tube-to-cylinder head bolt(s)
  3-series body style
    Four-cylinder engine .... 96 to 120 in-lbs
    six-cylinder engine .... 48 to 72 in-lbs
  5-series body style
    528e .... 53 to 71 in-lbs
    533i and 535i .... 97 to 115 in-lbs

1 General information

Included in this portion of Chapter 2 are the general overhaul procedures for the cylinder head(s) and internal engine components.

The information ranges from advice concerning preparation for an overhaul and the purchase of replacement parts to detailed, step-by-step procedures covering removal and installation of internal engine components and the inspection of parts.

The following Sections have been written based on the assumption that the engine has been removed from the vehicle. For information concerning in-vehicle engine repair, as well as removal and installation of the external components necessary for the overhaul, see Chapter 2A and Section 7 of this Part.

The Specifications included in this Part are only those necessary for the inspection and overhaul procedures which follow. Refer to Part A for additional Specifications.

2 Engine overhaul - general information

It's not always easy to determine when, or if, an engine should be completely overhauled, as a number of factors must be considered.

High mileage is not necessarily an indication that an overhaul is needed, while low mileage doesn't preclude the need for an overhaul. Frequency of servicing is probably the most important consideration. An engine that's had regular and frequent oil and filter changes, as well as other required maintenance, will most likely give many thousands of miles of reliable service. Conversely, a neglected engine may require an overhaul very early in its life.

Chapter 2 Part B General engine overhaul procedures

3.5 As a safety precaution, before performing a compression check, remove the cover and the main relay from the left side of the engine compartment to disable the fuel and ignition systems (525i model shown, other models similar)

Excessive oil consumption is an indication that piston rings, valve seals and/or valve guides are in need of attention. Make sure that oil leaks aren't responsible before deciding that the rings and/or guides are bad. Perform a cylinder compression check to determine the extent of the work required (see Section 3).

Check the oil pressure: Unscrew the oil pressure sending unit and hook up an oil pressure gauge in its place. Measure the oil pressure with the engine at its normal operating temperature. Compare your measurements to the oil pressure listed in this Chapter's Specifications. If it's extremely low, the bearings and/or oil pump are probably worn out. On timing chain engines, the oil pressure sending unit is located high on the left rear of the cylinder head. On timing belt engines, the sending unit is threaded into the side of the engine block, below the oil filter.

Loss of power, rough running, knocking or metallic engine noises, excessive valve train noise and high fuel consumption rates may also point to the need for an overhaul, especially if they're all present at the same time. If a complete tune-up doesn't remedy the situation, major mechanical work is the only solution.

An engine overhaul involves restoring the internal parts to the specifications of a new engine. During an overhaul, the piston rings are replaced and the cylinder walls are reconditioned (rebored and/or honed). If a rebore is done by an automotive machine shop, new oversize pistons will also be installed. The main bearings and connecting rod bearings and are generally replaced with new ones and, if necessary, the crankshaft may be reground to restore the journals. Generally, the valves are serviced as well, since they're usually in less-than-perfect condition at this point. While the engine is being overhauled, other components, such as the distributor, starter and alternator, can be rebuilt as well. The end result should be a like-new engine that will give many thousands of trouble-free miles. **Note:** *Critical cooling system components such as the hoses, drivebelts, thermostat and water pump MUST be replaced with new parts when an engine is overhauled. The radiator should be checked carefully to ensure that it isn't clogged or leaking (see Chapter 1). Also, we don't recommend overhauling the oil pump - always install a new one when an engine is rebuilt.*

Before beginning the engine overhaul, read through the entire procedure to familiarize yourself with the scope and requirements of the job. Overhauling an engine isn't difficult if you follow all of the instructions carefully, have the necessary tools and equipment and pay close attention to all specifications; however, it is time consuming. Plan on the vehicle being tied up for a minimum of two weeks, especially if parts must be taken to an automotive machine shop for repair or reconditioning. Check on availability of parts and make sure that any necessary special tools and equipment are obtained in advance. Most work can be done with typical hand tools, although a number of precision measuring tools are required for inspecting parts to determine if they must be replaced. Often an automotive machine shop will handle the inspection of parts and offer advice concerning reconditioning and replacement. **Note:** *Always wait until the engine has been completely disassembled and all components, especially the engine block, have been inspected before deciding what service and repair operations must be performed by an automotive machine shop. Since the block's condition will be the major factor to consider when determining whether to overhaul the original engine or buy a rebuilt one, never purchase parts or have machine work done on other components until the block has been thoroughly inspected. As a general rule, time is the primary cost of an overhaul, so it doesn't pay to install worn or substandard parts.*

As a final note, to ensure maximum life and minimum trouble from a rebuilt engine, everything must be assembled with care in a spotlessly clean environment.

3 Compression check

Refer to illustration 3.5

1 A compression check will tell you what mechanical condition the upper end (pistons, rings, valves, head gaskets) of your engine is in. Specifically, it can tell you if the compression is down due to leakage caused by worn piston rings, defective valves and seats or a blown head gasket. **Note:** *The engine must be at normal operating temperature and the battery must be fully charged for this check.*

2 Begin by cleaning the area around the spark plugs before you remove them (compressed air should be used, if available, otherwise a small brush or even a bicycle tire pump will work). The idea is to prevent dirt from getting into the cylinders as the compression check is being done.

3 Remove all of the spark plugs from the engine (see Chapter 1).

4 Block the throttle wide open.

5 Detach the coil wire from the center of the distributor cap and ground it on the engine block. Use a jumper wire with alligator clips on each end to ensure a good ground. The fuel pump and ignition circuit should also be disabled by removing the main relay **(see illustration)** to avoid the possibility of the engine starting or causing a fire from fuel being sprayed in the engine compartment. The location of the main relay is generally near the fuse panel area on the drivers side of the vehicle under the hood, but refer to Chapter 12 for the specific location on your model.

6 Install the compression gauge in the number one spark plug hole.

7 Crank the engine over at least seven compression strokes and watch the gauge. The compression should build up quickly in a healthy engine. Low compression on the first stroke, followed by gradually increasing pressure on successive strokes, indicates worn piston rings. A low compression reading on the first stroke, which doesn't build up during successive strokes, indicates leaking valves or a blown head gasket (a cracked head could also be the cause). Deposits on the undersides of the valve heads can also cause low compression. Record the highest gauge reading obtained.

8 Repeat the procedure for the remaining cylinders and compare the results to the compression listed in this Chapter's Specifications.

9 Add some engine oil (about three squirts from a plunger-type oil can) to each cylinder, through the spark plug hole, and repeat the test.

10 If the compression increases after the oil is added, the piston rings are definitely worn. If the compression doesn't increase significantly, the leakage is occurring at the valves or head gasket. Leakage past the valves may be caused by burned valve seats and/or faces or warped, cracked or bent valves.

11 If two adjacent cylinders have equally low compression, there's a strong possibility that the head gasket between them is blown. The appearance of coolant in the combustion chambers or the crankcase would verify this condition.

12 If one cylinder is 20 percent lower than the others, and the engine has a slightly rough idle, a worn exhaust lobe on the camshaft could be the cause.

13 If the compression is unusually high, the combustion chambers are probably coated with carbon deposits. If that's the case, the cylinder head(s) should be removed and decarbonized.

Chapter 2 Part B General engine overhaul procedures

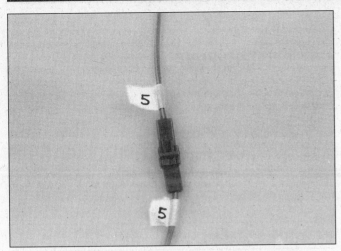

5.5 Label each wire before unplugging the connector

14 If compression is way down or varies greatly between cylinders, it would be a good idea to have a leak-down test performed by an automotive repair shop. This test will pinpoint exactly where the leakage is occurring and how severe it is.

4 Engine removal - methods and precautions

If you've decided that an engine must be removed for overhaul or major repair work, several preliminary steps should be taken.

Locating a suitable place to work is extremely important. Adequate work space, along with storage space for the vehicle, will be needed. If a shop or garage isn't available, at the very least a flat, level, clean work surface made of concrete or asphalt is required.

Cleaning the engine compartment and engine before beginning the removal procedure will help keep tools clean and organized.

An engine hoist or A-frame will also be necessary. Make sure the equipment is rated in excess of the combined weight of the engine and accessories. Safety is of primary importance, considering the potential hazards involved in lifting the engine out of the vehicle.

If the engine is being removed by a novice, a helper should be available. Advice and aid from someone more experienced would also be helpful. There are many instances when one person cannot simultaneously perform all of the operations required when lifting the engine out of the vehicle.

Plan the operation ahead of time. Arrange for or obtain all of the tools and equipment you'll need prior to beginning the job. Some of the equipment necessary to perform engine removal and installation safely and with relative ease are (in addition to an engine hoist) a heavy duty floor jack, complete sets of wrenches and sockets as described in the front of this manual, wooden blocks and plenty of rags and cleaning solvent for mopping up spilled oil, coolant and gasoline. If the hoist must be rented, make sure that you arrange for it in advance and perform all of the operations possible without it beforehand. This will save you money and time.

Plan for the vehicle to be out of use for quite a while. A machine shop will be required to perform some of the work which the do-it-yourselfer can't accomplish without special equipment. These shops often have a busy schedule, so it would be a good idea to consult them before removing the engine in order to accurately estimate the amount of time required to rebuild or repair components that may need work.

Always be extremely careful when removing and installing the engine. Serious injury can result from careless actions. Plan ahead, take your time and a job of this nature, although major, can be accomplished successfully. **Warning:** *The air conditioning system is under high pressure. Do not loosen any fittings or remove any components until after the system has been discharged. Air conditioning refrigerant should be properly discharged into an EPA-approved container at a dealer service department or an automotive air conditioning repair facility. Always wear eye protection when disconnecting air conditioning system fittings.*

5 Engine - removal and installation

Removal
Refer to illustration 5.5

1 Relieve the fuel system pressure (see Chapter 4), then disconnect the negative cable from the battery. **Caution:** *If the radio in your vehicle is equipped with an anti-theft system, make sure you have the correct activation code before disconnecting the battery.* **Note**: *If, after connecting the battery, the wrong language appears on the instrument panel display, refer to page -0-7 for the language resetting procedure.*
2 Cover the fenders and cowl and remove the hood (see Chapter 11). Special pads are available to protect the fenders, but an old bedspread or blanket will also work.
3 Remove the air cleaner housing and intake ducts (see Chapter 4).
4 Drain the cooling system (see Chapter 1).
5 Label the vacuum lines, emissions system hoses, wiring connectors, ground straps and fuel lines, to ensure correct reinstallation, then detach them. Pieces of masking tape with numbers or letters written on them work well **(see illustration)**. If there's any possibility of confusion, make a sketch of the engine compartment and clearly label the lines, hoses and wires.
6 Label and detach all coolant hoses from the engine (see Chapter 3).
7 Remove the cooling fan, shroud and radiator (see Chapter 3).
8 Remove the drivebelts (see Chapter 1).
9 Disconnect the fuel lines from the fuel rail (see Chapter 4). **Warning:** *Gasoline is extremely flammable, so take extra precautions when you work on any part of the fuel system. Don't smoke or allow open flames or bare light bulbs near the work area, and don't work in a garage where a natural gas-type appliance (such as a water heater or clothes dryer) with a pilot light is present. If you spill any fuel on your skin, rinse it off immediately with soap and water. When you perform any kind of work on the fuel system, wear safety glasses and have a Class B type fire extinguisher on hand.*
10 Disconnect the accelerator cable (see Chapter 4) and TV linkage/speed control cable (see Chapter 7), if equipped, from the engine.
11 Unbolt the power steering pump (see Chapter 10). Leave the lines/hoses attached and make sure the pump is kept in an upright position in the engine compartment (use wire or rope to restrain it out of the way).
12 On air-conditioned models, unbolt the compressor (see Chapter 3) and set it aside and tie it up out of the way. Do not disconnect the hoses.
13 Drain the engine oil (see Chapter 1) and remove the filter.
14 Remove the starter motor (see Chapter 5).
15 Remove the alternator (see Chapter 5).
16 Unbolt the exhaust system from the engine (see Chapter 4).
17 If you're working on a vehicle with an automatic transmission, remove the torque converter-to-driveplate fasteners (see Chapter 7B).
18 Support the transmission with a jack. Position a block of wood between them to prevent damage to the transmission. Special transmission jacks with safety chains are available - use one if possible.
19 Attach an engine sling or a length of chain to the lifting brackets on the engine or, if the brackets have been removed, the chain can be bolted directly to the intake manifold studs, but place a flat washer against the chain before the nut and run the nut all the way up tight to the chain to avoid the possibility of the studs bending.
20 Roll the hoist into position and connect the sling to it. Take up the slack in the sling or chain, but don't lift the engine. **Warning:** *DO NOT place any part of your body under the engine when it's supported only by a hoist or other lifting device.*
21 Remove the transmission rear crossmember and slightly lower the rear of the transmission.

22 Remove the transmission-to-engine block bolts using a Torx socket. **Note:** *The bolts holding the bellhousing to the engine block will require a swivel at the socket and a very long extension going back towards the transmission.*
23 Remove the engine mount-to-frame bracket nuts.
24 Recheck to be sure nothing is still connecting the engine to the transmission or vehicle. Disconnect anything still remaining.
25 Raise the engine slightly. Carefully work it forward to separate it from the transmission. If you're working on a vehicle with an automatic transmission, you may find the torque converter comes forward with the engine. If it stays with the transmission, leave it, but you may find it easier to let it come forward until it can be grasped easier and be pulled from the crankshaft **Note:** *When reinstalling the torque converter into the transmission before engine reinstallation be sure to replace the transmission front pump seal, which was probably damaged when the converter came out with the engine.* Either method is acceptable but be prepared for some fluid to leak from the torque converter if it comes out of the transmission. If you're working on a vehicle with a manual transmission, the input shaft must be completely disengaged from the clutch. Slowly raise the engine out of the engine compartment. Check carefully to make sure nothing is hanging up.
26 Remove the flywheel/driveplate and mount the engine on an engine stand.

Installation

27 Check the engine and transmission mounts. If they're worn or damaged, replace them.
28 If you're working on a manual-transmission-equipped vehicle, install the clutch and pressure plate (see Chapter 7A). Now is a good time to install a new clutch.
29 If the torque converter came out with the engine during removal carefully reinstall the converter into the transmission before the engine is lowered into the vehicle.
30 Carefully lower the engine into the engine compartment - make sure the engine mounts line up.
31 If you're working on an automatic transmission equipped vehicle, guide the torque converter into the crankshaft following the procedure outlined in Chapter 7B.
32 If you're working on a manual transmission equipped vehicle, apply a dab of high-temperature grease to the input shaft and guide it into the crankshaft pilot bearing until the bellhousing is flush with the engine block.
33 Install the transmission-to-engine bolts and tighten them securely. **Caution:** *DO NOT use the bolts to force the transmission and engine together.*
34 Reinstall the remaining components in the reverse order of removal.
35 Add coolant, oil, power steering and transmission fluid as needed.
36 Run the engine and check for leaks and proper operation of all accessories, then install the hood and test drive the vehicle.
37 Have the air conditioning system recharged and leak tested.

6 Engine rebuilding alternatives

The do-it-yourselfer is faced with a number of options when performing an engine overhaul. The decision to replace the engine block, piston/connecting rod assemblies and crankshaft depends on a number of factors, with the number one consideration being the condition of the block. Other considerations are cost, access to machine shop facilities, parts availability, time required to complete the project and the extent of prior mechanical experience on the part of the do-it-yourselfer.

Some of the rebuilding alternatives include:

Individual parts - If the inspection procedures reveal that the engine block and most engine components are in reusable condition, purchasing individual parts may be the most economical alternative. The block, crankshaft and piston/connecting rod assemblies should all be inspected carefully. Even if the block shows little wear, the cylinder bores should be surface honed.

Crankshaft kit - A crankshaft kit consists of a re-ground crankshaft with matched undersize new main and connecting rod bearings. Sometimes, reconditioned connecting rods and new pistons and rings are included with the kit (such a kit is sometimes called an "engine kit"). If the block is in good condition, but the crankshaft journals are scored or worn, a crankshaft kit and other individual parts may be the most economical alternative.

Short block - A short block consists of an engine block with a crankshaft and piston/connecting rod assemblies already installed. All new bearings are incorporated and all clearances will be correct. The existing camshaft, valve train components, cylinder head(s) and external parts can be bolted to the short block with little or no machine shop work necessary.

Long block - A long block consists of a short block plus an oil pump, oil pan, cylinder head, valve cover, camshaft and valve train components, timing sprockets and chain or gears and timing cover. All components are installed with new bearings, seals and gaskets incorporated throughout. The installation of manifolds and external parts is all that's necessary.

Give careful thought to which alternative is best for you and discuss the situation with local automotive machine shops, auto parts dealers and experienced rebuilders before ordering or purchasing replacement parts.

7 Engine overhaul - disassembly sequence

1 It's much easier to disassemble and work on the engine if it's mounted on a portable engine stand. A stand can often be rented quite cheaply from an equipment rental yard. Before the engine is mounted on a stand, the flywheel/driveplate should be removed from the engine.
2 If a stand isn't available, it's possible to disassemble the engine with it blocked up on the floor. Be extra careful not to tip or drop the engine when working without a stand.
3 If you're going to obtain a rebuilt engine, all external components must come off first, to be transferred to the replacement engine, just as they will if you're doing a complete engine overhaul yourself. These include:

Alternator and brackets
Emissions control components
Distributor, spark plug wires and spark plugs
Thermostat and housing cover
Water pump
Fuel injection and fuel system components
Intake and exhaust manifolds
Oil filter and oil pressure sending unit
Engine mounts
Clutch and flywheel/driveplate
Engine rear plate

Note: *When removing the external components from the engine, pay close attention to details that may be helpful or important during installation. Note the installed position of gaskets, seals, spacers, pins, brackets, washers, bolts and other small items.*

4 If you're obtaining a short block, which consists of the engine block, crankshaft, pistons and connecting rods all assembled, then the cylinder head, oil pan and oil pump will have to be removed as well. See *Engine rebuilding alternatives* for additional information regarding the different possibilities to be considered.
5 If you're planning a complete overhaul, the engine must be disassembled and the internal components removed in the following general order:

Valve cover
Intake and exhaust manifolds
Timing belt or chain covers
Timing chain/belt
Water pump
Cylinder head
Oil pan
Oil pump
Piston/connecting rod assemblies
Crankshaft and main bearings
Camshaft

Chapter 2 Part B General engine overhaul procedures

8.2 Remove the oiling tube from the top of the cylinder head. Be sure to note the placement of all gaskets and washers for later reassembly

8.4 To check camshaft endplay, mount a dial indicator so that its stem is in-line with the camshaft and just touching the camshaft at the front

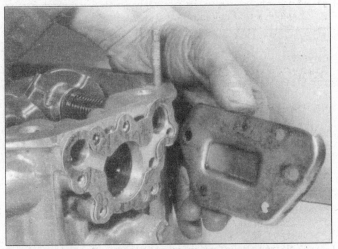

8.6 Remove the cover from the rear of the cylinder head - be sure to note the locations of any washers, gaskets and seals while you are removing the cover

Rocker shafts and rocker arms
Valve spring retainers and springs
Valves

6   Before beginning the disassembly and overhaul procedures, make sure the following items are available. Also, refer to *Engine overhaul - reassembly sequence* for a list of tools and materials needed for engine reassembly.
Common hand tools
Small cardboard boxes or plastic bags for storing parts
Gasket scraper
Ridge reamer
Vibration damper puller
Micrometers
Telescoping gauges
Dial indicator set
Valve spring compressor
Cylinder surfacing hone
Piston ring groove cleaning tool
Electric drill motor
Tap and die set
Wire brushes
Oil gallery brushes
Cleaning solvent

8.7 Remove the retaining clips from sides of the rocker arms - the wire-type clip is shown here

8   Cylinder head - disassembly

Refer to illustrations 8.2, 8.4, 8.6, 8.7, 8.8, 8.13, 8.15, 8.16 and 8.17

1   Remove the cylinder head (see Chapter 2A)
2   Remove the oil supply tube from its mounting on top of the cylinder head **(see illustration)**. **Note:** *It's important to replace the seals that the tube has under the mounting bolts.*
3   Adjust all valves to their maximum clearance by rotating the eccentric on the valve end of the rocker arm towards the center of the head (see Chapter 1, if necessary).
4   Before removing the thrust plate, measure the camshaft endplay by mounting a dial indicator to the front end of the cylinder head, with the probe resting on the camshaft **(see illustration)**. Pry the camshaft back-and-forth in the cylinder head. The reading is the camshaft endplay. Compare the reading to this Chapter's Specifications.
5   Unbolt and remove the camshaft thrust plate. **Note:** *There are two different locations for thrust plates. On timing chain engines, it is attached on the front of the cylinder head, behind the timing gear flange. On timing belt engines, the plate is located inside the head, by the rocker shafts, at the forward end of the cylinder head.*
6   Remove the rear cover plate from the back of the cylinder head **(see illustration)**.
7   Remove the retaining clips from each of the rocker arms. **Note:** *There is more than one style of clip. There are wire type* **(see illustration)** *with one on each side of the rocker arm or there is a spring-steel type which goes over the rocker arm and clips on either side of the rocker arm.*

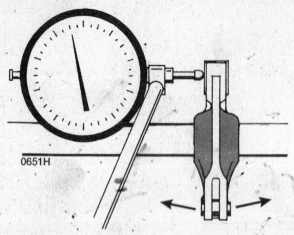

8.8 Check the rocker arm-to-shaft radial clearance by setting up a dial indicator as shown and trying to rotate the rocker arm against the shaft - DO NOT slide the rocker arm along the shaft

8.13 Removing a rocker arm shaft from the front of the cylinder head - the shaft must be either driven out from the rear of the head with a hardwood dowel or, on models where the rocker shaft is threaded at the front, pulled out from the front with a slide-hammer-type puller

8.15 A small plastic bag, with an appropriate label, can be used to store the valve train components so they can be kept together and reinstalled in the original position

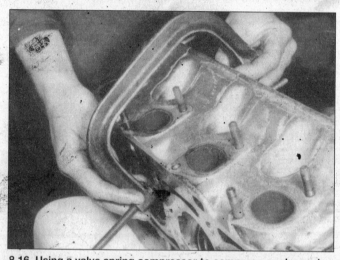

8.16 Using a valve spring compressor to compress a valve spring

8 Before removing the rocker arm shafts, measure the rocker arm radial clearance, using a dial indicator, and compare your measurement to the Specifications at the beginning of this Chapter. Without sliding the rocker arm along the shaft, try to rotate the rocker arm against the shaft in each direction (see illustration). The total movement measured at the camshaft end of the rocker arm is the radial clearance. If the clearance is excessive, either the rocker arm bushing, rocker arm shaft or both will need replacing.

9 Remove the rubber retaining plugs or the threaded plugs at the front of the cylinder head, whichever applies to your engine. There is a plug in front of each rocker shaft. Caution: If your engine has welded-in retaining plugs at the front of the rocker shafts, take the cylinder head to an automotive machine shop for removal to avoid possible damage to the cylinder head or the rocker arm shafts.

10 Rotate the camshaft until the most rocker arms possible are loose (not compressing their associated valve springs).

11 For the remaining rocker arms that are still compressing their valve springs, BMW recommends using a special forked tool to compress the rocker arms against the valve springs (and therefore take the valve spring tension off the camshaft lobe). If the tool is not available, insert a standard screwdriver into the gap above the adjuster eccentric at the valve-end tip of each rocker arm. Using the screwdrivers, pry the rocker arms against the valve springs and hold them in place as the camshaft is removed (see the next Step). At least one assistant will be necessary for this operation, since three or four valve springs usually need compressing. If assistants are not available, you may try securing the screwdrivers that are compressing the valve springs to the bench with lengths sturdy of wire. Warning: Be sure the wire is securely attached to the bench and screwdrivers or the screwdrivers could fly off the cylinder head, possibly causing injury.

12 When all the rocker arms are not contacting the camshaft lobes, slowly and carefully pull the camshaft out the front of the cylinder head. It may be necessary to rotate the camshaft as it is removed. Caution: Be very careful not to scratch the camshaft bearing journals in the cylinder head as the camshaft is withdrawn.

13 After removing the camshaft, carefully remove the rocker arm shafts. On models without threaded holes at the front of the shafts, drive them out from the rear of the cylinder head with a hammer and hardwood dowel that is slightly smaller in diameter than the rocker arm shaft (see illustration). For rocker shafts with a threaded front hole, screw in a slide hammer to pull the shaft from the head.

14 As each rocker arm shaft is slid out of the cylinder head, the rocker arms will be released, one by one. Drop each rocker arm into a labeled bag so they can be returned to their original locations on reassembly. While you're removing the rocker arm shafts, note their orientation. The guide plate notches and the small oil holes face in; the large oil holes face down, toward the valve guides. Also, label the rocker shafts so they can be returned to their original locations in the cylinder head.

15 Before the valves are removed, arrange to label and store them, along with their related components, so they can be kept separate and reinstalled in the same valve guides from which they're removed (see illustration).

16 Compress the springs on the first valve with a spring compressor and remove the keepers (see illustration). Carefully release the valve spring compressor and remove the retainer, the spring and the spring seat (if used).

Chapter 2 Part B General engine overhaul procedures

8.17 If the valve won't pull through the guide, deburr the edge of the stem end and the area around the top of the keeper groove with a file or whetstone

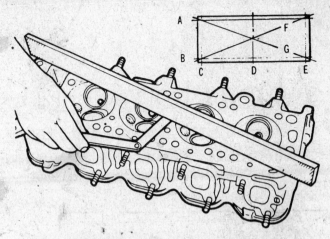

9.10 Check the cylinder head gasket surface for warpage by trying to slip a feeler gauge under the straightedge (see this Chapter's Specifications for the maximum warpage allowed and use a feeler gauge of that thickness)

17 Pull the valve out of the head, then remove the oil seal from the guide. If the valve binds in the guide (won't pull through), push it back into the head and deburr the area around the keeper groove with a fine file or whetstone **(see illustration)**.

18 Repeat the procedure for the remaining valves. Remember to keep all the parts for each valve together so they can be reinstalled in the same locations.

19 Once the valves and related components have been removed and stored in an organized manner, the head should be thoroughly cleaned and inspected. If a complete engine overhaul is being done, finish the engine disassembly procedures before beginning the cylinder head cleaning and inspection process.

9 Cylinder head, camshaft and rocker arms - cleaning and inspection

1 Thorough cleaning of the cylinder head(s) and related valve train components, followed by a detailed inspection, will enable you to decide how much valve service work must be done during the engine overhaul. **Note:** *If the engine was severely overheated, the cylinder head is probably warped (see Step 10).*

Cleaning

2 Scrape all traces of old gasket material and sealing compound off the cylinder head, intake manifold and exhaust manifold sealing surfaces. Be very careful not to gouge the cylinder head. Special gasket removal solvents that soften gaskets and make removal much easier are available at auto parts stores.

3 Remove all built-up scale from the coolant passages.

4 Run a stiff bristled brush through the various holes to remove deposits that may have formed in them.

5 Run an appropriate size tap into each of the threaded holes to remove corrosion and thread sealant that may be present. If compressed air is available, use it to clear the holes of debris produced by this operation. **Warning:** *Wear eye protection when using compressed air!*

6 Clean the cylinder head with solvent and dry it thoroughly. Compressed air will speed the drying process and ensure that all holes and recessed areas are clean. **Note:** *Decarbonizing chemicals are available and may prove very useful when cleaning cylinder heads and valve train components. They are very caustic and should be used with caution. Be sure to follow the instructions on the container.*

7 Clean all the rocker shafts, rocker arms, springs, valve springs, spring seats, keepers and retainers with solvent and dry them thoroughly. Do the components from one valve at a time to avoid mixing up the parts.

8 Scrape off any heavy deposits that may have formed on the valves, then use a motorized wire brush to remove deposits from the valve heads and stems. Again, make sure the valves don't get mixed up.

Inspection

Note: *Be sure to perform all of the following inspection procedures before concluding that machine shop work is required. Make a list of the items that need attention.*

Cylinder head

Refer to illustrations 9.10 and 9.12

9 Inspect the head very carefully for cracks, evidence of coolant leakage and other damage. If cracks are found, check with an automotive machine shop concerning repair. If repair isn't possible, a new cylinder head should be obtained.

10 Using a straightedge and feeler gauge, check the head gasket mating surface for warpage **(see illustration)**. If the warpage exceeds the limit listed in this Chapter's Specifications, it can be resurfaced at an automotive machine shop.

11 Examine the valve seats in each of the combustion chambers. If they're pitted, cracked or burned, the head will require valve service that's beyond the scope of the home mechanic.

12 Check the valve stem-to-guide clearance by measuring the lateral movement of the valve stem with a dial indicator **(see illustration)**. The valve must be in the guide and approximately 1/16-inch off the seat.

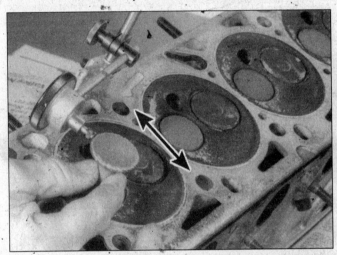

9.12 A dial indicator can be used to determine the valve stem-to-guide clearance (move the valve as indicated by the arrows)

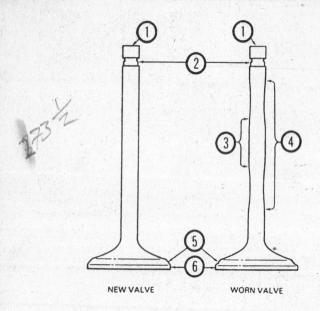

9.13 Check for valve wear at the points shown here

1 Valve tip
2 Keeper groove
3 Stem (least worn area)
4 Stem (most worn area)
5 Valve face
6 Margin

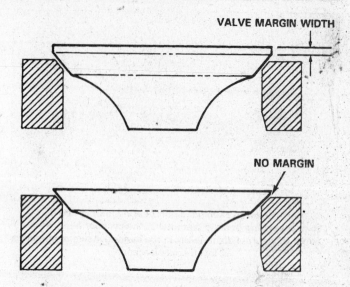

9.14 The margin width on each valve must be as specified (if no margin exists, the valve cannot be reused)

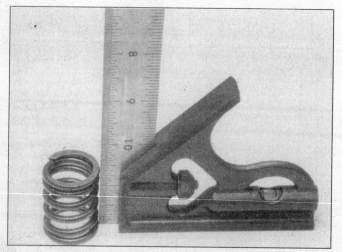

9.16 Check each valve spring for squareness

9.18 Look for signs of pitting, discoloration or excessive wear on the ends of the rocker arms where they contact the camshaft and the valve stem tip

The total valve stem movement indicated by the gauge needle must be divided by two to obtain the actual clearance. After this is done, if there's still some doubt regarding the condition of the valve guides they should be checked by an automotive machine shop (the cost should be minimal).

Valves

Refer to illustrations 9.13 and 9.14

13 Carefully inspect each valve face for uneven wear, deformation, cracks, pits and burned areas **(see illustration)**. Check the valve stem for scuffing and galling and the neck for cracks. Rotate the valve and check for any obvious indication that it's bent. Look for pits and excessive wear on the end of the stem. The presence of any of these conditions indicates the need for valve service by an automotive machine shop.

14 Measure the margin width on each valve **(see illustration)**. Any valve with a margin narrower than specified will have to be replaced with a new one.

Valve components

Refer to illustration 9.16

15 Check each valve spring for wear (on the ends) and pits. The tension of all springs should be checked with a special fixture before deciding that they're suitable for use in a rebuilt engine (take the springs to an automotive machine shop for this check).
16 Stand each spring on a flat surface and check it for squareness **(see illustration)**. If any of the springs are distorted or sagged, or possibly have a broken coil, replace all of them with new parts.
17 Check the spring retainers and keepers for obvious wear and cracks. Any questionable parts should be replaced with new ones, as extensive damage will occur if they fail during engine operation.

Rocker arms

Refer to illustration 9.18

Note: *The rocker arms for the exhaust valves are the most subject to wear and should be checked with particular care.*

18 Inspect all the rocker arms for excessive wear on the tips that contact the valve stem and camshaft **(see illustration)**.
19 Check the rocker arm radial clearance (see Section 8). If it's excessive, either the rocker arm bushing or the shaft (or both) is exces-

Chapter 2 Part B General engine overhaul procedures

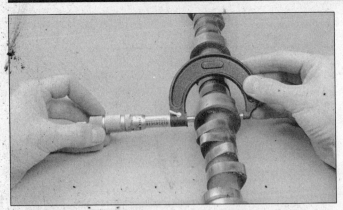

9.23 Measure each camshaft bearing journal and its corresponding bearing diameter in the cylinder head, then subtract the journal diameter from the bearing inside diameter to obtain the oil clearance

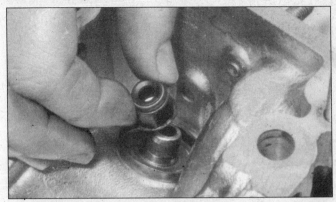

11.4a Lubricate the end of the valve stem and place the seal over the stem . . .

11.4b . . .then lightly drive the seal to the spring seat with a socket or piece of tubing

sively worn. To determine which is more worn, slide the rocker arm onto an unworn portion of the rocker arm shaft and check the radial clearance again. If it's now within specifications, the shaft is probably the most-worn component. If it's not within specifications, the rocker arm bushings should be replaced.

Rocker arm shafts

20 Check the shafts for scoring, excessive wear and other damage. The areas where the rocker arms contact the shafts should be smooth. If there is a visible ridge at the edge of where the rocker arm rides, the shaft is probably worn excessively.

Camshaft

Refer to illustration 9.23

21 Inspect the camshaft journals (the round bearing areas) and lobes for scoring, pitting, flaking and excessive wear. Using a micrometer, measure the height of each exhaust and intake lobe. Compare the heights of all the exhaust lobes and intake lobes. If the readings among the exhaust valve lobes or intake valve lobes vary more than about 0.003 inch, or if the camshaft is exhibiting any signs of wear, replace the camshaft.

22 Inspect the camshaft bearing surfaces in the cylinder head for scoring and other damage. If the bearing surfaces are scored or damaged, you'll normally have to replace the cylinder head, since the bearings are simply a machined surface in the cylinder head. **Note:** *An automotive machine shop (particularly one that specializes in BMW's) or dealer service department may be able to provide an alternative to replacing the cylinder head, if the only problem with the head is mildly scored camshaft bearing surfaces.*

23 Using a micrometer, measure the journals on the camshaft and record the measurements **(see illustration)**. Using a telescoping gauge or inside micrometer, measure the camshaft bearing diameters in the cylinder head. Subtract the camshaft journal measurement from its corresponding bearing inside diameter to obtain the oil clearance. Compare the oil clearance to what's listed in this Chapter's specifications. If it's not within specifications, the camshaft and/or cylinder head will have to be replaced. **Note:** *Before replacing the cylinder head, check with an automotive machine shop that specializes in BMW's. They may be able to repair the head.*

10 Valves - servicing

1 Because of the complex nature of the job and the special tools and equipment needed, servicing of the valves, the valve seats and the valve guides, commonly known as a valve job, should be done by a professional.

2 The home mechanic can remove and disassemble the head, do the initial cleaning and inspection, then reassemble and deliver it to a dealer service department or an automotive machine shop for the actual service work. Doing the inspection will enable you to see what condition the head and valvetrain components are in and will ensure that you know what work and new parts are required when dealing with an automotive machine shop.

3 The dealer service department, or automotive machine shop, will remove the valves and springs, recondition or replace the valves and valve seats, recondition the valve guides, check and replace the valve springs, spring retainers or rotators and keepers (as necessary), replace the valve seals with new ones, reassemble the valve components and make sure the installed spring height is correct. The cylinder head gasket surface will also be resurfaced if it's warped.

4 After the valve job has been performed by a professional, the head will be in like-new condition. When the head is returned, be sure to clean it again before installation on the engine to remove any metal particles and abrasive grit that may still be present from the valve service or head resurfacing operations. Use compressed air, if available, to blow out all the oil holes and passages.

11 Cylinder head - reassembly

Refer to illustrations 11.4a, 11.4b, 11.6a, 11.6b and 11.9

1 Regardless of whether or not the head was sent to an automotive repair shop for valve servicing, make sure it's clean before beginning reassembly.

2 If the head was sent out for valve servicing, the valves and related components will already be in place. Begin the reassembly procedure with Step 8.

3 Starting at one end of the head, apply moly-base grease or clean engine oil to each valve stem and install the valve.

4 Lubricate the lip of each valve guide seal, carefully slide it over the tip of the valve, then slide it all the way down the stem to the guide. Using a hammer and a deep socket or seal installation tool, gently tap the seal into place until it's completely seated on the guide (see illustra-

Chapter 2 Part B General engine overhaul procedures

11.6a With the retainer installed, compress the valve spring and install the keepers as shown

11.6b Apply a small dab of grease to each keeper as shown here before installation - it'll hold them in place on the valve stem as the spring is released

tions). Don't twist or cock a seal during installation or it won't seal properly against the valve stem. **Note:** *The seals for intake and exhaust valves are different - don't mix them up.*

5   Drop the spring seat or shim(s) over the valve guide and set the valve spring and retainer in place.

6   Compress the spring with a valve spring compressor and carefully install the keepers in the upper groove, then slowly release the compressor and make sure the keepers seat properly. Apply a small dab of grease to each keeper to hold it in place, if necessary **(see illustrations)**.

7   Repeat Steps 3 through 6 for each of the valves. Be sure to return the components to their original locations - don't mix them up!

8   Install the rocker arms and shafts by reversing the disassembly sequence. Be sure to install the rocker shafts in the correct orientation. The guide plate notches and the small oil holes face in; the large oil holes face down, toward the valve guides.

9   Lubricate the camshaft journals and lobes **(see illustration)**, then carefully insert it into the cylinder head, rotating it as you go so the camshaft lobes will clear the rocker arms. It will also be necessary to compress rocker arms, as described in Section 8, so they'll clear the camshaft lobes. Be very careful not to scratch or gouge the camshaft bearing surfaces in the cylinder head.

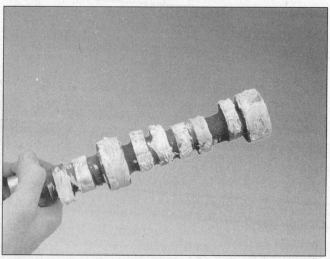

11.9 Lubricate the camshaft bearing journals and lobes with engine assembly lube or moly-based grease

12 Pistons/connecting rods - removal

Refer to illustrations 12.1, 12.3, 12.4 and 2.6

Note: *Prior to removing the piston/connecting rod assemblies, remove the cylinder head(s), the oil pan and the oil pump by referring to the appropriate Sections in Chapter 2A.*

1   Use your fingernail to feel if a ridge has formed at the upper limit of ring travel (about 1/4-inch down from the top of each cylinder). If carbon deposits or cylinder wear have produced ridges, they must be completely removed with a special tool **(see illustration)**. Follow the manufacturer's instructions provided with the tool. Failure to remove the ridges before attempting to remove the piston/connecting rod assemblies may result in piston breakage.

2   After the cylinder ridges have been removed, turn the engine upside-down so the crankshaft is facing up.

3   Before the connecting rods are removed, check the endplay with feeler gauges. Slide them between the first connecting rod and the crankshaft throw until the play is removed **(see illustration)**. The endplay is equal to the thickness of the feeler gauge(s). If the endplay exceeds the service limit, new connecting rods will be required. If new rods (or a new crankshaft) are installed, the endplay may fall under the specified minimum (if it does, the rods will have to be machined to restore it - consult an automotive machine shop for advice if necessary). Repeat the procedure for the remaining connecting rods.

12.1 A ridge reamer is required to remove the ridge from the top of each cylinder - do this before removing the pistons!

Chapter 2 Part B General engine overhaul procedures

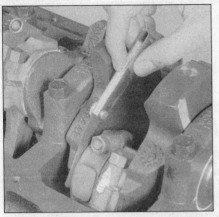

12.3 Check the connecting rod side clearance with a feeler gauge as shown

12.4 Mark the rod bearing caps in order from the front of the engine to the rear (one mark for the front cap, two for the second one and so on)

12.6 To prevent damage to the crankshaft journals and cylinder walls, slip sections of rubber or plastic hose over the rod bolts before removing the pistons

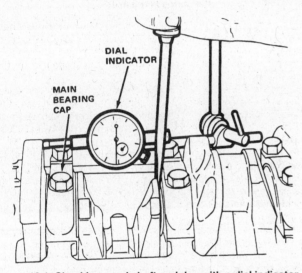

13.1 Checking crankshaft endplay with a dial indicator

13.4 Use a center-punch or number stamping dies to mark the main bearing caps to ensure installation in their original locations on the block (make the punch marks near one of the bolt heads)

4   Check the connecting rods and caps for identification marks. If they aren't plainly marked, use a small center-punch to make the appropriate number of indentations **(see illustration)** on each rod and cap (1, 2, 3, etc., depending on the engine type and cylinder they're associated with).

5   Loosen each of the connecting rod cap nuts 1/2-turn at a time until they can be removed by hand. Remove the number one connecting rod cap and bearing insert. Don't drop the bearing insert out of the cap.

6   Slip a short length of plastic or rubber hose over each connecting rod cap bolt to protect the crankshaft journal and cylinder wall as the piston is removed **(see illustration)**.

7   Remove the bearing insert and push the connecting rod/piston assembly out through the top of the engine. Use a wooden hammer handle to push on the upper bearing surface in the connecting rod. If resistance is felt, double-check to make sure that all of the ridge was removed from the cylinder.

8   Repeat the procedure for the remaining cylinders.

9   After removal, reassemble the connecting rod caps and bearing inserts in their respective connecting rods and install the cap nuts finger tight. Leaving the old bearing inserts in place until reassembly will help prevent the connecting rod bearing surfaces from being accidentally nicked or gouged.

10 Don't separate the pistons from the connecting rods (see Section 18).

13 Crankshaft - removal

Refer to illustrations 13.1, 13.4 and 13.6

Note: *The crankshaft can be removed only after the engine has been removed from the vehicle. It's assumed that the flywheel or driveplate, vibration damper, timing chain or belt, oil pan, oil pump and piston/connecting rod assemblies have already been removed. The rear main oil seal housing must be unbolted and separated from the block before proceeding with crankshaft removal.*

1   Before the crankshaft is removed, check the endplay. Mount a dial indicator with the stem in line with the crankshaft and touching the nose or a cheek of the crank **(see illustration)**.

2   Push the crankshaft all the way to the rear and zero the dial indicator. Next, pry the crankshaft to the front as far as possible and check the reading on the dial indicator. The distance that it moves is the endplay. If it's greater than the maximum endplay listed in this Chapter's Specifications, check the crankshaft thrust surfaces for wear. If no wear is evident, new main bearings should correct the endplay.

3   If a dial indicator isn't available, feeler gauges can be used. Gently pry or push the crankshaft all the way to the front of the engine. Slip feeler gauges between the crankshaft and the front face of the thrust main bearing to determine the clearance.

4   Check the main bearing caps to see if they're marked to indicate their locations. They should be numbered consecutively from the front of the engine to the rear. If they aren't, mark them with number stamping dies or a center-punch **(see illustration)**. Main bearing caps generally have a cast-in arrow, which points to the front of the engine. Loosen the main bearing cap bolts 1/4-turn at a time each, working

13.6 Remove the crankshaft by lifting straight up. Be very careful when removing the crankshaft - it is very heavy

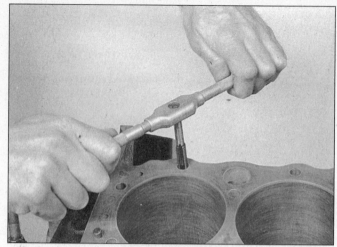

15.8 All bolt holes in the block - particularly the main bearing cap and head bolt holes - should be cleaned and restored with a tap (be sure to remove debris from the holes after this is done)

from the outer ends towards the center, until they can be removed by hand. Note if any stud bolts are used and make sure they're returned to their original locations when the crankshaft is reinstalled.

5   Gently tap the caps with a soft-face hammer, then separate them from the engine block. If necessary, use the bolts as levers to remove the caps. Try not to drop the bearing inserts if they come out with the caps.

6   Carefully lift the crankshaft out of the engine. It may be a good idea to have an assistant available, since the crankshaft is quite heavy **(see illustration)**. With the bearing inserts in place in the engine block and main bearing caps, return the caps to their respective locations on the engine block and tighten the bolts finger tight.

14 Intermediate shaft - removal and inspection

Note: *The intermediate shaft is used on timing-belt engines only. The shaft rotates in the engine block parallel to the crankshaft. It is driven by the timing belt and its only purpose is to drive the oil pump.*

1   Remove the timing belt (see Chapter 2A).
2   With the belt removed, unbolt the gear from the intermediate shaft.
3   Remove the oil pump driveshaft (see Chapter 2A).
4   The shaft is held in the cylinder block by a retaining plate with two bolts. Remove the bolts and pull the shaft forward and out of the block.
5   Look for any signs of abnormal wear on the bearing surfaces or the gear at the back end of the shaft, which directs motion down to the oil pump shaft. If the bearing surfaces in the engine block show wear, they'll have to be replaced by an automotive machine shop.

15 Engine block - cleaning

Refer to illustrations 15.8 and 15.10

Caution: *The core plugs (also known as freeze plugs or soft plugs) may be difficult or impossible to retrieve if they're driven into the block coolant passages.*

1   Remove the core plugs from the engine block. To do this, knock one side of the plugs into the block with a hammer and punch, then grasp them with large pliers and pull them out.
2   Using a gasket scraper, remove all traces of gasket material from the engine block. Be very careful not to nick or gouge the gasket sealing surfaces.
3   Remove the main bearing caps and separate the bearing inserts from the caps and the engine block. Tag the bearings, indicating which cylinder they were removed from and whether they were in the cap or the block, then set them aside.

4   Remove all of the threaded oil gallery plugs from the block. The plugs are usually very tight - they may have to be drilled out and the holes retapped. Use new plugs when the engine is reassembled.
5   If the engine is extremely dirty it should be taken to an automotive machine shop to be steam cleaned or hot tanked.
6   After the block is returned, clean all oil holes and oil galleries one more time. Brushes specifically designed for this purpose are available at most auto parts stores. Flush the passages with warm water until the water runs clear, dry the block thoroughly and wipe all machined surfaces with a light, rust preventive oil. If you have access to compressed air, use it to speed the drying process and to blow out all the oil holes and galleries. **Warning:** *Wear eye protection when using compressed air!*
7   If the block isn't extremely dirty or sludged up, you can do an adequate cleaning job with hot soapy water and a stiff brush. Take plenty of time and do a thorough job. Regardless of the cleaning method used, be sure to clean all oil holes and galleries very thoroughly, dry the block completely and coat all machined surfaces with light oil.
8   The threaded holes in the block must be clean to ensure accurate torque readings during reassembly. Run the proper size tap into each of the holes to remove rust, corrosion, thread sealant or sludge and restore damaged threads **(see illustration)**. If possible, use compressed air to clear the holes of debris produced by this operation. Now is a good time to clean the threads on the head bolts and the main bearing cap bolts as well.
9   Reinstall the main bearing caps and tighten the bolts finger tight.
10   After coating the sealing surfaces of the new core plugs with Permatex no. 2 sealant, install them in the engine block **(see illustration)**. Make sure they're driven in straight and seated properly or leakage could result. Special tools are available for this purpose, but a large socket, with an outside diameter that will just slip into the core plug, a 1/2-inch drive extension and a hammer will work just as well.
11   Apply non-hardening sealant (such as Permatex no. 2 or Teflon pipe sealant) to the new oil gallery plugs and thread them into the holes in the block. Make sure they're tightened securely.
12   If the engine isn't going to be reassembled right away, cover it with a large plastic trash bag to keep it clean.

16 Engine block - inspection

Refer to illustrations 16.4a, 16.4b and 16.4c

1   Before the block is inspected, it should be cleaned (see Section 15).
2   Visually check the block for cracks, rust and corrosion. Look for stripped threads in the threaded holes. It's also a good idea to have the block checked for hidden cracks by an automotive machine shop

Chapter 2 Part B General engine overhaul procedures

15.10 A large socket on an extension can be used to drive the new core plugs into the bores

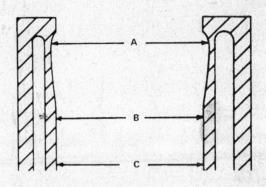

16.4a Measure the diameter of each cylinder just under the wear ride (A), at the center (B) and at the bottom (C)

16.4b The ability to "feel" when the telescoping gauge is at the correct point will be developed over time, so work slowly and repeat the check until you're satisfied the bore measurement is accurate

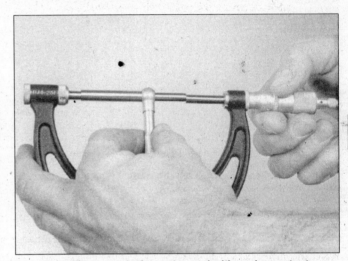

16.4c The gauge is then measured with a micrometer to determine the bore size

that has the special equipment to do this type of work. If defects are found, have the block repaired, if possible, or replaced.
3   Check the cylinder bores for scuffing and scoring.
4   Measure the diameter of each cylinder at the top (just under the ridge area), center and bottom of the cylinder bore, parallel to the crankshaft axis **(see illustrations)**.
5   Next, measure each cylinder's diameter at the same three locations across the crankshaft axis. Compare the results to this Chapter's Specifications.
6   If the required precision measuring tools aren't available, the piston-to-cylinder clearances can be obtained, though not quite as accurately, using feeler gauge stock. Feeler gauge stock comes in 12-inch lengths and various thicknesses and is generally available at auto parts stores.
7   To check the clearance, select a feeler gauge and slip it into the cylinder along with the matching piston. The piston must be positioned exactly as it normally would be. The feeler gauge must be between the piston and cylinder on one of the thrust faces (90-degrees to the piston pin bore).
8   The piston should slip through the cylinder (with the feeler gauge in place) with moderate pressure.
9   If it falls through or slides through easily, the clearance is excessive and a new piston will be required. If the piston binds at the lower end of the cylinder and is loose toward the top, the cylinder is tapered. If tight spots are encountered as the piston/feeler gauge is rotated in the cylinder, the cylinder is out-of-round.
10   Repeat the procedure for the remaining pistons and cylinders.
11   If the cylinder walls are badly scuffed or scored, or if they're out-of-round or tapered beyond the limits given in this Chapter's Specifications, have the engine block rebored and honed at an automotive machine shop. If a rebore is done, oversize pistons and rings will be required.
12   If the cylinders are in reasonably good condition and not worn to the outside of the limits, and if the piston-to-cylinder clearances can be maintained properly, then they don't have to be rebored. Honing is all that's necessary (see Section 17).

17 Cylinder honing

Refer to illustrations 17.3a, and 17.3b

1   Prior to engine reassembly, the cylinder bores must be honed so the new piston rings will seat correctly and provide the best possible combustion chamber seal. **Note:** *If you don't have the tools or don't want to tackle the honing operation, most automotive machine shops will do it for a reasonable fee.*
2   Before honing the cylinders, install the main bearing caps and tighten the bolts to the torque listed in this Chapter's Specifications.
3   Two types of cylinder hones are commonly available - the flex hone or "bottle brush" type and the more traditional surfacing hone with spring-loaded stones. Both will do the job, but for the less experienced mechanic the "bottle brush" hone will probably be easier to use.

17.3a A "bottle brush" hone will produce better results if you've never honed cylinders before

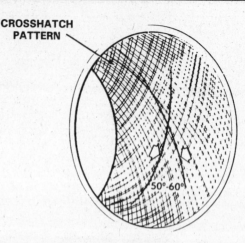

17.3b The cylinder hone should leave a smooth, crosshatch pattern with the lines intersecting at approximately a 60-degree angle

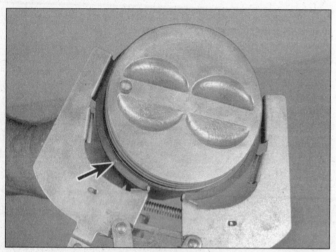

18.2 Installing the compression rings with a ring expander - the mark (arrow) must face up

You'll also need some kerosene or honing oil, rags and an electric drill motor. Proceed as follows:
  a) Mount the hone in the drill motor, compress the stones and slip it into the first cylinder (see illustration). Be sure to wear safety goggles or a face shield!
  b) Lubricate the cylinder with plenty of honing oil, turn on the drill and move the hone up-and-down in the cylinder at a pace that will produce a fine crosshatch pattern on the cylinder walls. Ideally, the crosshatch lines should intersect at approximately a 60-degree angle (see illustration). Be sure to use plenty of lubricant and don't take off any more material than is absolutely necessary to produce the desired finish. Note: *Piston ring manufacturers may specify a smaller crosshatch angle than the traditional 60-degrees - read and follow any instructions included with the new rings.*
  c) Don't withdraw the hone from the cylinder while it's running. Instead, shut off the drill and continue moving the hone up-and-down in the cylinder until it comes to a complete stop, then compress the stones and withdraw the hone. If you're using a "bottle brush" type hone, stop the drill motor, then turn the chuck in the normal direction of rotation while withdrawing the hone from the cylinder.
  d) Wipe the oil out of the cylinder and repeat the procedure for the remaining cylinders.

4 After the honing job is complete, chamfer the top edges of the cylinder bores with a small file so the rings won't catch when the pistons are installed. Be very careful not to nick the cylinder walls with the end of the file.
5 The entire engine block must be washed again very thoroughly with warm, soapy water to remove all traces of the abrasive grit produced during the honing operation. Note: *The bores can be considered clean when a lint-free white cloth - dampened with clean engine oil - used to wipe them out doesn't pick up any more honing residue, which will show up as gray areas on the cloth. Be sure to run a brush through all oil holes and galleries and flush them with running water.*
6 After rinsing, dry the block and apply a coat of light rust preventive oil to all machined surfaces. Wrap the block in a plastic trash bag to keep it clean and set it aside until reassembly.

18 Pistons/connecting rods - inspection

Refer to illustrations 18.2, 18.4a, 18.4b, 18.10 and 18.11

1 Before the inspection process can be carried out, the piston/connecting rod assemblies must be cleaned and the original piston rings removed from the pistons. Note: *Always use new piston rings when the engine is reassembled.*
2 Using a piston ring installation tool, carefully remove the rings from the pistons. Be careful not to nick or gouge the pistons in the process (see illustration).
3 Scrape all traces of carbon from the top of the piston. A handheld wire brush or a piece of fine emery cloth can be used once the majority of the deposits have been scraped away. Do not, under any circumstances, use a wire brush mounted in a drill motor to remove deposits from the pistons. The piston material is soft and may be eroded away by the wire brush.
4 Use a piston ring groove cleaning tool to remove carbon deposits from the ring grooves. If a tool isn't available, a piece broken off the old ring will do the job. Be very careful to remove only the carbon deposits - don't remove any metal and do not nick or scratch the sides of the ring grooves (see illustrations).
5 Once the deposits have been removed, clean the piston/rod assemblies with solvent and dry them with compressed air (if available). Make sure the oil return holes in the back sides of the ring grooves are clear.
6 If the pistons and cylinder walls aren't damaged or worn excessively, and if the engine block is not rebored, new pistons won't be necessary. Normal piston wear appears as even vertical wear on the piston thrust surfaces and slight looseness of the top ring in its groove. New piston rings, however, should always be used when an engine is rebuilt.

Chapter 2 Part B General engine overhaul procedures

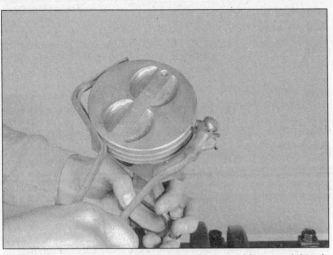

18.4a The piston ring grooves can be cleaned with a special tool, as shown here, . . .

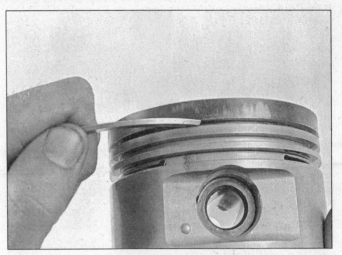

18.4b . . . or a section of a broken ring

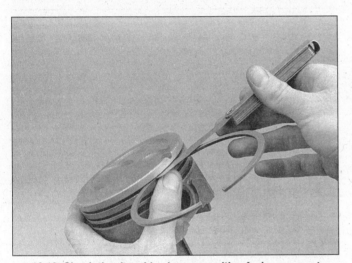

18.10 Check the ring side clearance with a feeler gauge at several points around the groove

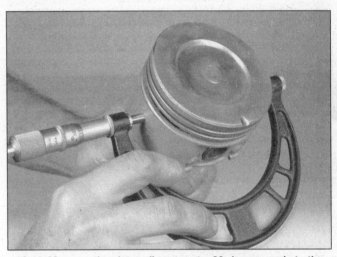

18.11 Measure the piston diameter at a 90-degree angle to the piston pin and in line with it

7 Carefully inspect each piston for cracks around the skirt, at the pin bosses and at the ring lands.

8 Look for scoring and scuffing on the thrust faces of the skirt, holes in the piston crown and burned areas at the edge of the crown. If the skirt is scored or scuffed, the engine may have been suffering from overheating and/or abnormal combustion, which caused excessively high operating temperatures. The cooling and lubrication systems should be checked thoroughly. A hole in the piston crown is an indication that abnormal combustion (pre-ignition) was occurring. Burned areas at the edge of the piston crown are usually evidence of spark knock (detonation). If any of the above problems exist, the causes must be corrected or the damage will occur again. The causes may include intake air leaks, incorrect fuel/air mixture, incorrect ignition timing and EGR system malfunctions.

9 Corrosion of the piston, in the form of small pits, indicates that coolant is leaking into the combustion chamber and/or the crankcase. Again, the cause must be corrected or the problem may persist in the rebuilt engine.

10 Measure the piston ring side clearance by laying a new piston ring in each ring groove and slipping a feeler gauge in beside it **(see illustration)**. Check the clearance at three or four locations around each groove. Be sure to use the correct ring for each groove - they are different. If the side clearance is greater than the figure listed in this Chapter's Specifications, new pistons will have to be used.

11 Check the piston-to-bore clearance by measuring the bore (see Section 16) and the piston diameter. Make sure the pistons and bores are correctly matched. Measure the piston across the skirt, at a 90-degree angle to and in line with the piston pin **(see illustration)**. Subtract the piston diameter from the bore diameter to obtain the clearance. If it's greater than specified, the block will have to be rebored and new pistons and rings installed.

12 Check the piston-to-rod clearance by twisting the piston and rod in opposite directions. Any noticeable play indicates excessive wear, which must be corrected. The piston/connecting rod assemblies should be taken to an automotive machine shop to have the pistons and rods resized and new pins installed.

13 If the pistons must be removed from the connecting rods for any reason, they should be taken to an automotive machine shop. While they are there have the connecting rods checked for bend and twist, since automotive machine shops have special equipment for this purpose. **Note:** *Unless new pistons and/or connecting rods must be installed, do not disassemble the pistons and connecting rods.*

14 Check the connecting rods for cracks and other damage. Temporarily remove the rod caps, lift out the old bearing inserts, wipe the rod and cap bearing surfaces clean and inspect them for nicks, gouges and scratches. After checking the rods, replace the old bearings, slip the caps into place and tighten the nuts finger tight. **Note:** *If the engine is being rebuilt because of a connecting rod knock, be sure to install new rods.*

19.1 The oil holes should be chamfered so sharp edges don't gouge or scratch the new bearings

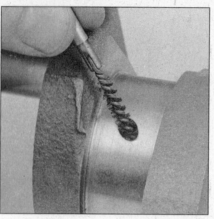

19.2 Use a wire or stiff plastic bristle brush to clean the oil passages in the crankshaft

19.4 Rubbing a penny lengthwise on each journal will reveal its condition - if copper rubs off and is embedded in the crankshaft, the journals should be reground

19.6 Measure the diameter of each crankshaft journal at several points to detect taper and out-of-round conditions

19.8 If the seals have worn grooves in the crankshaft journals, or if the seal contact surfaces are nicked or scratched, the new seals will leak

19 Crankshaft - inspection

Refer to illustrations 19.1, 19.2, 19.4, 19.6 and 19.8

1 Remove all burrs from the crankshaft oil holes with a stone, file or scraper **(see illustration)**.
2 Clean the crankshaft with solvent and dry it with compressed air (if available). Be sure to clean the oil holes with a stiff brush **(see illustration)** and flush them with solvent.
3 Check the main and connecting rod bearing journals for uneven wear, scoring, pits and cracks.
4 Rub a penny across each journal several times **(see illustration)**. If a journal picks up copper from the penny, it's too rough and must be reground.
5 Check the rest of the crankshaft for cracks and other damage. It should be magnafluxed to reveal hidden cracks - an automotive machine shop will handle the procedure.
6 Using a micrometer, measure the diameter of the main and connecting rod journals and compare the results to this Chapter's Specifications **(see illustration)**. By measuring the diameter at a number of points around each journal's circumference, you'll be able to determine whether or not the journal is out-of-round. Take the measurement at each end of the journal, near the crank throws, to determine if the journal is tapered.
7 If the crankshaft journals are damaged, tapered, out-of-round or worn beyond the limits given in the Specifications, have the crankshaft reground by an automotive machine shop. Be sure to use the correct size bearing inserts if the crankshaft is reconditioned.
8 Check the oil seal journals at each end of the crankshaft for wear and damage. If the seal has worn a groove in the journal, or if it's nicked or scratched **(see illustration)**, the new seal may leak when the engine is reassembled. In some cases, an automotive machine shop may be able to repair the journal by pressing on a thin sleeve. If repair isn't feasible, a new or different crankshaft should be installed.
9 Examine the main and rod bearing inserts (see Section 20).

20 Main and connecting rod bearings - inspection

Refer to illustration 20.1

1 Even though the main and connecting rod bearings should be replaced with new ones during the engine overhaul, the old bearings should be retained for close examination, as they may reveal valuable information about the condition of the engine.
2 Bearing failure occurs because of lack of lubrication, the presence of dirt or other foreign particles, overloading the engine and corrosion. Regardless of the cause of bearing failure, it must be corrected before the engine is reassembled to prevent it from happening again.
3 When examining the bearings, remove them from the engine block, the main bearing caps, the connecting rods and the rod caps and lay them out on a clean surface in the same general position as their location in the engine. This will enable you to match any bearing problems with the corresponding crankshaft journal.

Chapter 2 Part B General engine overhaul procedures

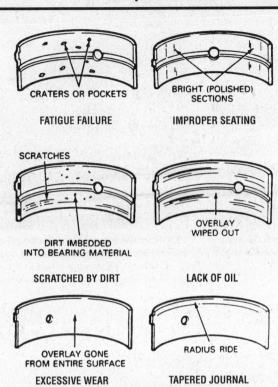

20.1 Typical bearing failures

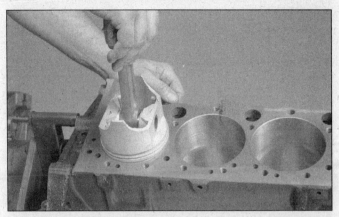

22.3 When checking piston ring end gap, the ring must be square in the cylinder bore (this is done by pushing the ring down with the top of a piston as shown)

4   Dirt and other foreign particles get into the engine in a variety of ways. It may be left in the engine during assembly, or it may pass through filters or the PCV system. It may get into the oil, and from there into the bearings. Metal chips from machining operations and normal engine wear are often present. Abrasives are sometimes left in engine components after reconditioning, especially when parts are not thoroughly cleaned using the proper cleaning methods. Whatever the source, these foreign objects often end up embedded in the soft bearing material and are easily recognized. Large particles will not embed in the bearing and will score or gouge the bearing and journal. The best prevention for this cause of bearing failure is to clean all parts thoroughly and keep everything spotlessly clean during engine assembly. Frequent and regular engine oil and filter changes are also recommended.

5   Lack of lubrication (or lubrication breakdown) has a number of interrelated causes. Excessive heat (which thins the oil), overloading (which squeezes the oil from the bearing face) and oil leakage or throw off (from excessive bearing clearances, worn oil pump or high engine speeds) all contribute to lubrication breakdown. Blocked oil passages, which usually are the result of misaligned oil holes in a bearing shell, will also oil starve a bearing and destroy it. When lack of lubrication is the cause of bearing failure, the bearing material is wiped or extruded from the steel backing of the bearing. Temperatures may increase to the point where the steel backing turns blue from overheating.

6   Driving habits can have a definite effect on bearing life. Full throttle, low speed operation (lugging the engine) puts very high loads on bearings, which tends to squeeze out the oil film. These loads cause the bearings to flex, which produces fine cracks in the bearing face (fatigue failure). Eventually the bearing material will loosen in pieces and tear away from the steel backing. Short trip driving leads to corrosion of bearings because insufficient engine heat is produced to drive off the condensed water and corrosive gases. These products collect in the engine oil, forming acid and sludge. As the oil is carried to the engine bearings, the acid attacks and corrodes the bearing material.

7   Incorrect bearing installation during engine assembly will lead to bearing failure as well. Tight-fitting bearings leave insufficient bearing oil clearance and will result in oil starvation. Dirt or foreign particles trapped behind a bearing insert result in high spots on the bearing which lead to failure.

21   Engine overhaul - reassembly sequence

1   Before beginning engine reassembly, make sure you have all the necessary new parts, gaskets and seals as well as the following items on hand:

   Common hand tools
   A torque wrench
   Piston ring installation tool
   Piston ring compressor
   Vibration damper installation tool
   Short lengths of rubber or plastic hose to fit over connecting rod bolts
   Plastigage
   Feeler gauges
   A fine-tooth file
   New engine oil
   Engine assembly lube or moly-base grease
   Gasket sealant
   Thread locking compound

2   In order to save time and avoid problems, engine reassembly must be done in the following general order:

   Piston rings
   Crankshaft and main bearings
   Piston/connecting rod assemblies
   Oil pump
   Oil pan
   Cylinder head assembly
   Timing belt or chain and tensioner assemblies
   Water pump
   Timing belt or chain covers
   Intake and exhaust manifolds
   Valve cover
   Engine rear plate
   Flywheel/driveplate

22   Piston rings - installation

Refer to illustrations 22.3, 22.4, 22.5, 22.9a and 22.9b

1   Before installing the new piston rings, the ring end gaps must be checked. It's assumed that the piston ring side clearance has been checked and verified correct (see Section 18).

2   Lay out the piston/connecting rod assemblies and the new ring sets so the ring sets will be matched with the same piston and cylinder during the end gap measurement and engine assembly.

3   Insert the top (number one) ring into the first cylinder and square it up with the cylinder walls by pushing it in with the top of the piston **(see illustration)**. The ring should be near the bottom of the cylinder, at the lower limit of ring travel.

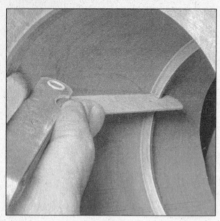

22.4 With the ring square in the cylinder, measure the end gap with a feeler gauge

22.5 If the end gap is too small, clamp a file in a vise and file the ring ends (from the outside in only) to enlarge the gap slightly

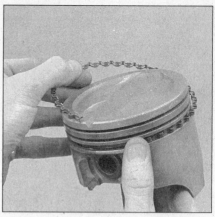

22.9a Installing the spacer/expander in the oil control ring groove

22.9b DO NOT use a piston ring installation tool when installing the oil ring side rails

4   To measure the end gap, slip feeler gauges between the ends of the ring until a gauge equal to the gap width is found (see illustration). The feeler gauge should slide between the ring ends with a slight amount of drag. Compare the measurement to this Chapter's Specifications. If the gap is larger or smaller than specified, double-check to make sure you have the correct rings before proceeding.

5   If the gap is too small, it must be enlarged or the ring ends may come in contact with each other during engine operation, which can cause serious damage to the engine. The end gap can be increased by filing the ring ends very carefully with a fine file. Mount the file in a vise equipped with soft jaws, slip the ring over the file with the ends contacting the file face and slowly move the ring to remove material from the ends. When performing this operation, file only from the outside in (see illustration).

6   Excess end gap isn't critical unless it's greater than 0.040-inch. Again, double-check to make sure you have the correct rings for your engine.

7   Repeat the procedure for each ring that will be installed in the first cylinder and for each ring in the remaining cylinders. Remember to keep rings, pistons and cylinders matched up.

8   Once the ring end gaps have been checked/corrected, the rings can be installed on the pistons.

9   The oil control ring (lowest one on the piston) is usually installed first. It's composed of three separate components. Slip the spacer/expander into the groove (see illustration). If an anti-rotation tang is used, make sure it's inserted into the drilled hole in the ring groove. Next, install the lower side rail. Don't use a piston ring installation tool on the oil ring side rails, as they may be damaged. Instead, place one end of the side rail into the groove between the spacer/expander and the ring land, hold it firmly in place and slide a finger around the piston while pushing the rail into the groove (see illustration). Next, install the upper side rail in the same manner.

10   After the three oil ring components have been installed, check to make sure that both the upper and lower side rails can be turned smoothly in the ring groove.

11   The number two (middle) ring is installed next. It's usually stamped with a mark which must face up, toward the top of the piston. **Note:** *Always follow the instructions printed on the ring package or box - different manufacturers may require different approaches. Do not mix up the top and middle rings, as they have different cross sections.*

12   Use a piston ring installation tool and make sure the identification mark is facing the top of the piston, then slip the ring into the middle groove on the piston (see illustration 18.2). Don't expand the ring any more than necessary to slide it over the piston.

13   Install the number one (top) ring in the same manner. Make sure the mark is facing up. Be careful not to confuse the number one and number two rings.

14   Repeat the procedure for the remaining pistons and rings.

23   Intermediate shaft - installation

1   Clean the intermediate shaft bearing surfaces and the pressed-in bearing sleeves in the cylinder block.
2   Lubricate the shaft and slide it into the block.
3   Reinstall the two bolts that hold the retaining plate to the block.
4   Bolt the gear onto the shaft.
5   The remainder of the parts are installed in the reverse order of removal.

24   Crankshaft - installation and main bearing oil clearance check

1   Crankshaft installation is the first step in engine reassembly. It's assumed at this point that the engine block and crankshaft have been cleaned, inspected and repaired or reconditioned.
2   Position the engine with the bottom facing up.
3   Remove the main bearing cap bolts and lift out the caps. Lay them out in the proper order to ensure correct installation.
4   If they're still in place, remove the original bearing inserts from the block and the main bearing caps. Wipe the bearing surfaces of the block and caps with a clean, lint-free cloth. They must be kept spotlessly clean.

Chapter 2 Part B General engine overhaul procedures

24.6 Installing a thrust bearing in the engine block bearing saddle - the thrust bearings must be installed in the No. 3 cap and saddle on four-cylinder engines and the No. 6 cap and saddle on six-cylinder engines

24.11 Lay the Plastigage strips (arrow) on the main bearing journals, parallel to the crankshaft centerline

24.15 Compare the width of the crushed Plastigage to the scale on the envelope to determine the main bearing oil clearance (always take the measurement at the widest point of the Plastigage); be sure to use the correct scale - standard and metric ones are included

Main bearing oil clearance check

Refer to illustrations 24.6, 24.11, and 24.15

5   Clean the back sides of the new main bearing inserts and lay one in each main bearing saddle in the block. If one of the bearing inserts from each set has a large groove in it, make sure the grooved insert is installed in the block. Lay the other bearing from each set in the corresponding main bearing cap. Make sure the tab on the bearing insert fits into the recess in the block or cap. **Caution:** *The oil holes in the block must line up with the oil holes in the bearing insert. Do not hammer the bearing into place and don't nick or gouge the bearing faces. No lubrication should be used at this time.*

6   The flanged thrust bearing must be installed in the No. 3 bearing cap and saddle in the four-cylinder engine and in the No. 6 bearing cap and saddle in the six-cylinder-engine **(see illustration)**.

7   Clean the faces of the bearings in the block and the crankshaft main bearing journals with a clean, lint-free cloth.

8   Check or clean the oil holes in the crankshaft, as any dirt here can go only one way - straight through the new bearings.

9   Once you're certain the crankshaft is clean, carefully lay it in position in the main bearings.

10 Before the crankshaft can be permanently installed, the main bearing oil clearance must be checked.

11 Cut several pieces of the appropriate size Plastigage (they must be slightly shorter than the width of the main bearings) and place one piece on each crankshaft main bearing journal, parallel with the journal axis **(see illustration)**.

12 Clean the faces of the bearings in the caps and install the caps in their respective positions (don't mix them up) with the arrows pointing toward the front of the engine. Don't disturb the Plastigage.

13 Starting with the center main and working out toward the ends, tighten the main bearing cap bolts, in three steps, to the torque listed in this Chapter's Specifications. Don't rotate the crankshaft at any time during this operation.

14 Remove the bolts and carefully lift off the main bearing caps. Keep them in order. Don't disturb the Plastigage or rotate the crankshaft. If any of the main bearing caps are difficult to remove, tap them gently from side-to-side with a soft-face hammer to loosen them.

15 Compare the width of the crushed Plastigage on each journal to the scale printed on the Plastigage envelope to obtain the main bearing oil clearance **(see illustration)**. Check the Specifications to make sure it's correct.

16 If the clearance is not as specified, the bearing inserts may be the wrong size (which means different ones will be required). Before deciding that different inserts are needed, make sure that no dirt or oil was between the bearing inserts and the caps or block when the clearance was measured. If the Plastigage was wider at one end than the other, the journal may be tapered (see Section 19).

17 Carefully scrape all traces of the Plastigage material off the main bearing journals and/or the bearing faces. Use your fingernail or the edge of a credit card - don't nick or scratch the bearing faces.

Final crankshaft installation

18 Carefully lift the crankshaft out of the engine.

19 Clean the bearing faces in the block, then apply a thin, uniform layer of moly-base grease or engine assembly lube to each of the bearing surfaces. Be sure to coat the thrust faces as well as the journal face of the thrust bearing.

20 Make sure the crankshaft journals are clean, then lay the crankshaft back in place in the block.

21 Clean the faces of the bearings in the caps, then apply lubricant to them.

22 Install the caps in their respective positions with the arrows pointing toward the front of the engine.

23 Install the bolts.

24 Tighten all except the thrust bearing cap bolts to the specified torque (work from the center out and approach the final torque in three steps).

25 Tighten the thrust bearing cap bolts to the torque listed in this Chapter's Specifications.

26 Tap the ends of the crankshaft forward and backward with a lead or brass hammer to line up the main bearing and crankshaft thrust surfaces.

27 Retighten all main bearing cap bolts to the specified torque, starting with the center main and working out toward the ends.

28 On manual transmission equipped models, install a new pilot bearing in the end of the crankshaft (see Chapter 8).

29 Rotate the crankshaft a number of times by hand to check for any obvious binding.

30 The final step is to check the crankshaft endplay with a feeler gauge or a dial indicator as described in Section 13. The endplay should be correct if the crankshaft thrust faces aren't worn or damaged and new bearings have been installed.

31 Install the new seal, then bolt the housing to the block (see Section 25).

25 Rear main oil seal installation

Refer to illustration 25.3, 25.4 and 25.5

1   The crankshaft must be installed first and the main bearing caps bolted in place, then the new seal should be installed in the retainer and the retainer bolted to the block.

25.3 After removing the retainer from the block, support it on two wood blocks and drive out the old seal with a punch and hammer

25.4 Drive the new seal into the retainer with a wood block or a section of pipe, if you have one large enough - make sure you don't cock the seal in the retainer bore

25.5 Lubricate the lip of the seal and bolt the retainer to the rear of the engine block

2 Before installing the crankshaft, check the seal contact surface very carefully for scratches and nicks that could damage the new seal lip and cause oil leaks. If the crankshaft is damaged, the only alternative is a new or different crankshaft.

3 The old seal can be removed from the housing with a hammer and punch by driving it out from the back side **(see illustration)**. Be sure to note how far it's recessed into the housing bore before removing it; the new seal will have to be recessed an equal amount. Be very careful not to scratch or otherwise damage the bore in the housing or oil leaks could develop.

4 Make sure the retainer is clean, then apply a thin coat of engine oil to the outer edge of the new seal. The seal must be pressed squarely into the housing bore, so hammering it into place is not recommended. If you don't have access to a press, sandwich the housing and seal between two smooth pieces of wood and press the seal into place with the jaws of a large vise. The pieces of wood must be thick enough to distribute the force evenly around the entire circumference of the seal **(see illustration)**. Work slowly and make sure the seal enters the bore squarely.

5 The seal lips must be lubricated with multi-purpose grease or clean engine oil before the seal/retainer is slipped over the crankshaft and bolted to the block **(see illustration)**. Use a new gasket - no sealant is required - and make sure the dowel pins are in place before installing the retainer.

6 Tighten the retainer nuts/screws a little at a time until they're all snug, then tighten them to the torque listed in this Chapter's Specifications.

26 Pistons/connecting rods - installation and rod bearing oil clearance check

Refer to illustrations 26.11 and 26.14

1 Before installing the piston/connecting rod assemblies, the cylinder walls must be perfectly clean, the top edge of each cylinder must be chamfered, and the crankshaft must be in place.

2 Remove the cap from the end of the number one connecting rod (refer to the marks made during removal). Remove the original bearing inserts and wipe the bearing surfaces of the connecting rod and cap with a clean, lint-free cloth. They must be kept spotlessly clean.

Connecting rod bearing oil clearance check

3 Clean the back side of the new upper bearing insert, then lay it in place in the connecting rod. Make sure the tab on the bearing fits into the recess in the rod. Don't hammer the bearing insert into place and be very careful not to nick or gouge the bearing face. Don't lubricate the bearing at this time.

4 Clean the back side of the other bearing insert and install it in the rod cap. Again, make sure the tab on the bearing fits into the recess in the cap, and don't apply any lubricant. It's critically important that the mating surfaces of the bearing and connecting rod are perfectly clean and oil free when they're assembled.

5 Position the piston ring gaps so they're staggered 120-degrees from each other.

6 Slip a section of plastic or rubber hose over each connecting rod cap bolt.

7 Lubricate the piston and rings with clean engine oil and attach a piston ring compressor to the piston. Leave the skirt protruding about 1/4-inch to guide the piston into the cylinder. The rings must be compressed until they're flush with the piston.

8 Rotate the crankshaft until the number one connecting rod journal is at BDC (bottom dead center) and apply a coat of engine oil to the cylinder walls.

9 With the mark or notch on top of the piston facing the front of the engine, gently insert the piston/connecting rod assembly into the number one cylinder bore and rest the bottom edge of the ring compressor on the engine block.

10 Tap the top edge of the ring compressor to make sure it's contacting the block around its entire circumference.

Chapter 2 Part B General engine overhaul procedures

26.11 Drive the piston gently into the cylinder bore with the end of a wooden or plastic hammer handle

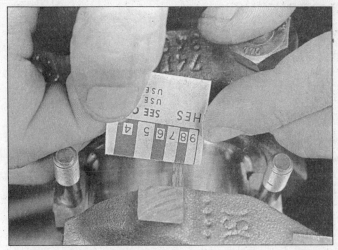

26.17 Measuring the width of the crushed Plastigage to determine the rod bearing oil clearance (be sure to use the correct scale - standard and metric ones are included)

11   Gently tap on the top of the piston with the end of a wooden hammer handle **(see illustration)** while guiding the end of the connecting rod into place on the crankshaft journal. The piston rings may try to pop out of the ring compressor just before entering the cylinder bore, so keep some downward pressure on the ring compressor. Work slowly, and if any resistance is felt as the piston enters the cylinder, stop immediately. Find out what's hanging up and fix it before proceeding. Do not, for any reason, force the piston into the cylinder - you might break a ring and/or the piston.

12   Once the piston/connecting rod assembly is installed, the connecting rod bearing oil clearance must be checked before the rod cap is permanently bolted in place.

13   Cut a piece of the appropriate size Plastigage slightly shorter than the width of the connecting rod bearing and lay it in place on the number one connecting rod journal, parallel with the journal axis.

14   Clean the connecting rod cap bearing face, remove the protective hoses from the connecting rod bolts and install the rod cap. Make sure the mating mark on the cap is on the same side as the mark on the connecting rod.

15   Install the nuts and tighten them to the torque listed in this Chapter's Specifications, working up to it in three steps. **Note:** *Use a thin-wall socket to avoid erroneous torque readings that can result if the socket is wedged between the rod cap and nut. If the socket tends to wedge itself between the nut and the cap, lift up on it slightly until it no longer contacts the cap.* Do not rotate the crankshaft at any time during this operation.

16   Remove the nuts and detach the rod cap, being very careful not to disturb the Plastigage.

17   Compare the width of the crushed Plastigage to the scale printed on the Plastigage envelope to obtain the oil clearance **(see illustration)**. Compare it to the Specifications to make sure the clearance is correct.

18   If the clearance is not as specified, the bearing inserts may be the wrong size (which means different ones will be required). Before deciding that different inserts are needed, make sure that no dirt or oil was between the bearing inserts and the connecting rod or cap when the clearance was measured. Also, recheck the journal diameter. If the Plastigage was wider at one end than the other, the journal may be tapered (see Section 19).

Final connecting rod installation

19   Carefully scrape all traces of the Plastigage material off the rod journal and/or bearing face. Be very careful not to scratch the bearing - use your fingernail or the edge of a credit card.

20   Make sure the bearing faces are perfectly clean, then apply a uniform layer of clean moly-base grease or engine assembly lube to both of them. You'll have to push the piston into the cylinder to expose the face of the bearing insert in the connecting rod - be sure to slip the protective hoses over the rod bolts first.

21   Slide the connecting rod back into place on the journal, remove the protective hoses from the rod cap bolts, install the rod cap and tighten the nuts to the specified torque. Again, work up to the torque in three steps.

22   Repeat the entire procedure for the remaining pistons/connecting rods.

23   The important points to remember are:
 a) Keep the back sides of the bearing inserts and the insides of the connecting rods and caps perfectly clean when assembling them.
 b) Make sure you have the correct piston/rod assembly for each cylinder.
 c) The notch or mark on the piston must face the front of the engine.
 d) Lubricate the cylinder walls with clean oil.
 e) Lubricate the bearing faces when installing the rod caps after the oil clearance has been checked.

24   After all the piston/connecting rod assemblies have been properly installed, rotate the crankshaft a number of times by hand to check for any obvious binding.

25   Check the connecting rod endplay (see Section 13).

26   Compare the measured endplay to the Specifications to make sure it's correct. If it was correct before disassembly and the original crankshaft and rods were reinstalled, it should still be right. If new rods or a new crankshaft were installed, the endplay may be inadequate. If so, the rods will have to be removed and taken to an automotive machine shop for resizing.

27   Initial start-up and break-in after overhaul

Warning: *Have a fire extinguisher handy when starting the engine for the first time.*

1   Once the engine has been installed in the vehicle, double-check the engine oil and coolant levels.

2   With the spark plugs out of the engine and the ignition system disabled (see Section 3), crank the engine until oil pressure registers on the gauge or the light goes out.

3   Install the spark plugs, hook up the plug wires and restore the ignition system functions (see Section 3).

4   Start the engine. It may take a few moments for the fuel system to build up pressure, but the engine should start without a great deal of effort. **Note:** *If backfiring occurs through the throttle body, recheck the valve timing, firing order and ignition timing.*

5   After the engine starts, it should be allowed to warm up to normal operating temperature. While the engine is warming up, make a thorough check for fuel, oil and coolant leaks.

6   Shut the engine off and recheck the engine oil and coolant levels.
7   Drive the vehicle to an area with minimum traffic, accelerate at full throttle from 30 to 50 mph, then allow the vehicle to slow to 30 mph with the throttle closed. Repeat the procedure 10 or 12 times. This will load the piston rings and cause them to seat properly against the cylinder walls. Check again for oil and coolant leaks.
8   Drive the vehicle gently for the first 500 miles (no sustained high speeds) and keep a constant check on the oil level. It is not unusual for an engine to use oil during the break-in period.
9   At approximately 500 to 600 miles, change the oil and filter.
10 For the next few hundred miles, drive the vehicle normally. Do not pamper it or abuse it.
11 After 2000 miles, change the oil and filter again and consider the engine broken in.

Chapter 3 Cooling, heating and air conditioning systems

Contents

Air conditioner receiver-drier - removal and installation	16
Air conditioning and heating system - check and maintenance	12
Air conditioning blower motor (5-Series models through mid-year 1988) - removal and installation	14
Air conditioning compressor - removal and installation	13
Air conditioning condenser - removal and installation	15
Antifreeze - general information	2
Coolant level check	See Chapter 1
Coolant temperature sending unit - check and replacement	8
Cooling system check	See Chapter 1
Cooling system servicing (draining, flushing and refilling)	See Chapter 1
Engine cooling fan(s) and clutch - check and replacement	5
Evaporator core - removal and installation	17
General information	1
Heater and air conditioner control assembly - removal and installation	10
Heater and air conditioning blower motor - removal and installation	9
Heater core - removal and installation	11
Radiator - removal and installation	4
Thermostat - check and replacement	3
Water pump - check	6
Water pump - removal and installation	7

Specifications

General

Coolant capacity	See Chapter 1
Thermostat rating	
Opening temperature	176 degrees
Fully open at	212 degrees
Cooling fan thermo-switch - switching temperatures	
Low speed	196 degrees
High speed	210 degrees
Refrigerant capacity	
3-Series models	2.1 lbs
5-Series models	2.9 lbs
Refrigerant oil capacity	
3-Series models	
Swash plate-type compressor	10.2 ounces
Valve-type compressor	6.80 ounces
5-Series models	5.78 ounces
Oil to add when replacing air conditioning components	
Receiver-drier	0.35 ounces
Evaporator	1.4 ounces
Condenser	0.70 ounces
Hose assemblies	0.35 ounces

Chapter 3 Cooling, heating and air conditioning systems

Torque specifications

Ft-lbs (unless otherwise indicated)

Mechanical cooling fan clutch-to-water pump securing nut	29
Mechanical cooling fan-to-clutch bolts	80 in-lbs
Water pump bolts	
3-Series models	
Small bolts (M6)	80 in-lbs
Large bolts (M8)	16
535i and 533i	80 in-lbs
528e	16
Thermostat housing bolts	80 in-lbs
Water pump pulley bolts	13

1 General information

Engine cooling system

All vehicles covered by this manual employ a pressurized engine cooling system with thermostatically controlled coolant circulation.

An impeller-type water pump mounted on the front of the block pumps coolant through the engine. The coolant flows around each cylinder and toward the rear of the engine. Cast-in coolant passages direct coolant around the intake and exhaust ports, near the spark plug areas and in close proximity to the exhaust valve guides.

A wax-pellet-type thermostat is located in a housing near the front of the engine (six-cylinder engines) or in-line in the radiator hose (four-cylinder engines). During warm up, the closed thermostat prevents coolant from circulating through the radiator. As the engine nears normal operating temperature, the thermostat opens and allows hot coolant to travel through the radiator, where it's cooled before returning to the engine.

The cooling system is sealed by a pressure-type cap, which raises the boiling point of the coolant and increases the cooling efficiency of the radiator. If the system pressure exceeds the cap pressure relief value, the excess pressure in the system forces the spring-loaded valve inside the cap off its seat and allows the coolant to escape through the overflow tube.

The pressure cap on four-cylinder models is on the top of the radiator; on six-cylinder models, it's on top of a translucent plastic expansion tank. The cap pressure rating is molded into the top of the cap. The pressure rating is either 1.0 Bar (14 psi) or 1.2 Bar (17 psi). **Warning:** *Do not remove the pressure cap from the radiator or expansion tank until the engine has cooled completely and there's no pressure remaining in the cooling system.*

Heating system

The heating system consists of a blower fan and heater core located in the heater box, the hoses connecting the heater core to the engine cooling system and the heater/air conditioning control head on the dashboard. Hot engine coolant is circulated through the heater core. When the heater mode is activated, a flap door opens to expose the heater box to the passenger compartment. A fan switch on the control head activates the blower motor, which forces air through the core, heating the air.

Air conditioning system

The air conditioning system consists of a condenser mounted in front of the radiator, an evaporator mounted adjacent to the heater core, a compressor mounted on the engine, a filter-drier (receiver-drier) which contains a high pressure relief valve and the plumbing connecting all of the above components.

A blower fan forces the warmer air of the passenger compartment through the evaporator core (sort of a radiator-in-reverse), transferring the heat from the air to the refrigerant. The liquid refrigerant boils off into low pressure vapor, taking the heat with it when it leaves the evaporator.

2 Antifreeze - general information

Warning: *Do not allow antifreeze to come in contact with your skin or painted surfaces of the vehicle. Rinse off spills immediately with plenty of water. If consumed, antifreeze can be fatal; children and pets are attracted by its sweet taste, so wipe up garage floor and drip pan coolant spills immediately. Keep antifreeze containers covered and repair leaks in your cooling system as soon as they are noticed.*

The cooling system should be filled with a water/ethylene glycol based antifreeze solution, which will prevent freezing down to at least -20-degrees F, or lower if local climate requires it. It also provides protection against corrosion and increases the coolant boiling point.

The cooling system should be drained, flushed and refilled at the specified intervals (see Chapter 1). Old or contaminated antifreeze solutions are likely to cause damage and encourage the formation of rust and scale in the system. Use distilled water with the antifreeze.

Before adding antifreeze, check all hose connections, because antifreeze tends to search out and leak through very minute openings. Engines don't normally consume coolant, so if the level goes down, find the cause and correct it.

The exact mixture of antifreeze-to-water which you should use depends on the relative weather conditions. The mixture should contain at least 50 percent antifreeze, but should never contain more than 70 percent antifreeze. Consult the mixture ratio chart on the antifreeze container before adding coolant. Hydrometers are available at most auto parts stores to test the coolant. Use antifreeze which meets the vehicle manufacturer's specifications.

3 Thermostat - check and replacement

Warning: *Do not remove the radiator cap, drain the coolant or replace the thermostat until the engine has cooled completely.*

Check

1   Before assuming the thermostat is to blame for a cooling system problem, check the coolant level, drivebelt tension (see Chapter 1) and temperature gauge (or light) operation.

2   If the engine seems to be taking a long time to warm up (based on heater output or temperature gauge operation), the thermostat is probably stuck open. Replace the thermostat with a new one.

3   If the engine runs hot, use your hand to check the temperature of the upper radiator hose. If the hose isn't hot, but the engine is, the thermostat is probably stuck closed, preventing the coolant inside the engine from escaping to the radiator. Replace the thermostat. **Caution:** *Don't drive the vehicle without a thermostat. The computer may stay in open loop and emissions and fuel economy will suffer.*

4   If the upper radiator hose is hot, it means that the coolant is flowing and the thermostat is open. Consult the *Troubleshooting* Section at the front of this manual for cooling system diagnosis.

Replacement

Refer to illustrations 3.8a, 3.8b and 3.12

All models

5   Disconnect the negative battery cable from the battery. **Caution:** *If the radio in your vehicle is equipped with an anti-theft system, make sure you have the correct activation code before disconnecting the battery.* **Note:** *If, after connecting the battery, the wrong language appears on the instrument panel display, refer to page 0-7 for the language resetting procedure.*

Chapter 3 Cooling, heating and air conditioning systems

3-3

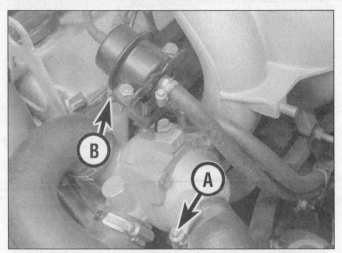

3.8a On six-cylinder-engine models, loosen the hose clamp (A) and disconnect the hose from the thermostat housing cover - note that the coolant temperature sending unit (barely visable behind the fuel pressure regulator) is located at the top of the thermostat housing (B)

3.8b On four-cylinder engine models, the thermostat (arrow) is connected in-line in the radiator hose

3.12 On six-cylinder-engine models, after the housing and thermostat have been removed, take the gasket out of the housing and clean the recess in the housing to be sure of a good seal upon reassembly

6   Drain the cooling system (see Chapter 1). If the coolant is relatively new or in good condition save it and reuse it.
7   Follow the radiator hose to the engine to locate the thermostat housing.
8   Loosen the hose clamp, then detach the hose(s) from the fitting **(see illustrations)**. If the hose clamps are the crimped type, grasp the clamp near the end with a pair of Channelock pliers and twist it to break the seal, then pull it off. If the clamps are the screw type, remove them by using a socket or a nut driver. If the hose is old or deteriorated, cut it off and install a new one. On four-cylinder models, lift the thermostat assembly out of the engine compartment.

Six-cylinder models only

9   If the outer surface of the fitting that mates with the hose is deteriorated (corroded, pitted, etc.), it may be damaged further by hose removal. If it is, the thermostat housing cover will have to be replaced.
10 Remove the bolts and detach the housing cover. If the cover is stuck, tap it with a soft-face hammer to jar it loose. Be prepared for some coolant to spill as the gasket seal is broken.
11 Note how it's installed (which end is facing out), then remove the thermostat.

12 Stuff a rag into the engine opening, then remove all traces of old gasket material, (if the gasket is the paper type material) or remove the rubber O-ring type gasket **(see illustration)** and sealant from the housing and cover with a gasket scraper. Remove the rag from the opening and clean the gasket mating surfaces with lacquer thinner or acetone.
13 Install the new thermostat and gasket in the housing. Make sure the correct end faces out - the spring end is normally directed toward the engine.
14 Install the cover and bolts. Tighten the bolts to the torque listed in this Chapter's Specifications.

All models

15 Reattach the hose(s) to the thermostat assembly or housing and tighten the hose clamp(s) securely.
16 Refill the cooling system (see Chapter 1).
17 Start the engine and allow it to reach normal operating temperature, then check for leaks and proper thermostat operation (as described in Steps 2 through 4).

4   Radiator - removal and installation

Removal

Refer to illustrations 4.8, 4.9 and 4.13
Warning: *Wait until the engine is completely cool before beginning this procedure.*
1   Disconnect the negative battery cable from the battery. **Caution:** *If the radio in your vehicle is equipped with an anti-theft system, make sure you have the correct activation code before disconnecting the battery.* **Note**: *If, after connecting the battery, the wrong language appears on the instrument panel display, refer to page 0-7 for the language resetting procedure.*
2   Drain the cooling system (see Chapter 1). If the coolant is relatively new or in good condition, save it and reuse it.
3   Loosen the hose clamps, then detach the radiator hoses from the fittings. If they're stuck, grasp each hose near the end with a pair of Channelock pliers and twist it to break the seal, then pull it off - be careful not to distort the radiator fittings! If the hoses are old or deteriorated, cut them off and install new ones.
4   Disconnect the reservoir hose from the radiator filler neck.
5   Remove the screws that attach the shroud to the radiator and slide the shroud toward the engine.
6   If the vehicle is equipped with an automatic transmission, disconnect the cooler lines from the radiator. Use a drip pan to catch spilled fluid.

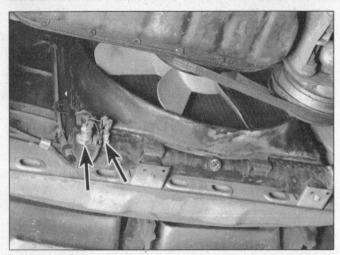

4.8 Sensors that control the high and low speed operation of the auxiliary cooling fan are located in various places in the radiator tanks - sensors for a 318i model are shown here

4.9 The radiator is bolted to the front core support at either the sides or the top of the radiator - the mounting positions of the 318i radiator are shown here

4.13 When the radiator is out, the radiator mounts can be seen - check for signs of deterioration and replace them, if needed

7   Plug the lines and fittings.
8   Disconnect the electrical connections of the coolant sensors located on the radiator **(see illustration)**. The thermostatically controlled switches for high and low-speed operation of the auxiliary fan are located in different spots in the radiators tanks, depending on the engine or model of the vehicle. The single switch on the 325i model is located on the upper left hand radiator tank, while dual switches on the four-cylinder models are located on the left hand side of the lower radiator tank. The dual switches for the auxiliary fan on the 533i and 535i are on the upper right hand radiator tank and on the left hand tank for the 528e models.
9   Remove the radiator mounting bolts. On four-cylinder models, the mounting bolts are located at the top of the radiator, and on six-cylinder models, the mounting bolts are located on the sides of the radiator **(see illustration)**.
10 Carefully lift out the radiator. Don't spill coolant on the vehicle or scratch the paint.
11 With the radiator removed, it can be inspected for leaks and damage. If it needs repair, have a radiator shop or dealer service department perform the work as special techniques are required.
12 Bugs and dirt can be removed from the radiator with compressed air and a soft brush. Don't bend the cooling fins as this is done.
13 Check the radiator mounts for deterioration and make sure there's nothing in them **(see illustration)** when the radiator is installed.

Installation

14 Installation is the reverse of the removal procedure.
15 After installation, fill the cooling system with the proper mixture of antifreeze and water. Refer to Chapter 1 if necessary.
16 Start the engine and check for leaks. Allow the engine to reach normal operating temperature, indicated by the upper radiator hose becoming hot. Recheck the coolant level and add more if required.
17 If you're working on an automatic transmission equipped vehicle, check and add fluid as needed.

5   Engine cooling fan(s) and clutch - check and replacement

Warning: *To avoid possible injury or damage, DO NOT operate the engine with a damaged fan. Do not attempt to repair fan blades - replace a damaged fan with a new one. Also, the electric auxiliary fan in front of the radiator or air conditioning condenser can come on without the engine running or ignition being on. It is controlled by the coolant temperature of the thermo-switches located in the radiator.*

Check

Electric auxiliary fan

Note: *This fan on most models is controlled by two thermo-switches placed in the radiator: one for low speed/low temperature operation and one for high speed/high temperature operation. Each switch comes on at a different coolant temperature (see the specifications at the beginning of this Chapter).*
1   The thermostatically controlled switches for high and low speed operation of the auxiliary fan are located in different spots in the radiators tanks **(see illustration 4.8)**, depending on the engine or model of the vehicle. The single switch on the 325i model is located on the upper left radiator tank, while dual switches on the four-cylinder models are located on the left side of the lower radiator tank. The dual switches for the auxiliary fan on the 533i and 535i are on the upper right hand radiator tank and on the left hand tank for the 528e models.
2   Insert a small screwdriver into the connector to lift the lock tab and unplug the fan wire harness.
3   As a fan motor test, unplug the electrical connector at the motor and use jumper wires to connect the fan directly to the battery. If the fan still doesn't work, replace the motor. If the fan works, there's a good chance the switch is malfunctioning. To more accurately diagnose the problem, follow the steps that apply to your model. **Note**: *Spin the auxiliary fan motor by hand to be sure the motor or fan isn't binding.*
4   To test the low-speed circuit, disconnect the electrical connector from the low-speed fan switch and jumper the terminals of the switch's

Chapter 3 Cooling, heating and air conditioning systems

5.27a The cooling fan on the water pump is attached to the shaft by a left-hand-threaded nut located directly behind the fan . . .

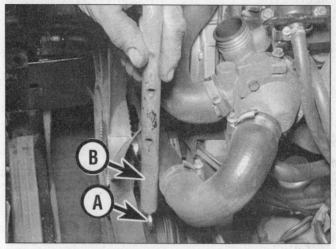

5.27b . . . to loosen the nut, place a 32 mm open-ended wrench on the nut and sharply strike the wrench (A) with a metal drift (B) and hammer; this will loosen the nut and allow it to be turned easily so the fan can be removed

electrical connector indicated for your model. When jumpered, the fan should run at low speed.
   a) On 325i models, jumper the green/black wire to the black wire.
   b) On 318i, 325 and 325e models, jumper the wires on the white terminal.
   c) On 528e, 533i and 535i models, jumper the green/brown wire with the black wire on the white-bodied switch.
5   To test the high-speed circuit, disconnect the electrical connector from the low-speed fan switch and jumper the terminals of the switch's electrical connector indicated for your model. When jumpered, with the ignition switch on, the fan should run at high speed.
   a) On 528e, 533i and 535i models, jumper the green/brown wire with the green/yellow wire on the red-bodied switch connector.
   b) On 318i, 325 and 325e models, jumper the wires on the red terminals.
   c) On 325i models, jumper the green/black wire to the black/brown wire.
6   If the tests don't give the indicated results, replace the switch. To remove the switch, drain the coolant below the level of the switch (see Chapter 1), then unscrew the switch and screw in the new one. Refill the system with coolant.
7   If the motor does not operate with the wires jumpered, the problem lies in the fuse, relay, wiring which connects the components or the fan motor itself. Carefully check the fuse, relay and all wiring and connections. If no obvious problems are found, further diagnosis should be done by a dealer service department or repair shop.

Mechanical fan with viscous clutch

8   Disconnect the negative battery cable and rock the fan back and forth by hand to check for excessive bearing play. **Caution:** *If the radio in your vehicle is equipped with an anti-theft system, make sure you have the correct activation code before disconnecting the battery.* **Note**: *If, after connecting the battery, the wrong language appears on the instrument panel display, refer to page 0-7 for the language resetting procedure.*
9   With the engine cold, turn the fan blades by hand. The fan should turn with slight resistance.
10   Visually inspect for substantial fluid leakage from the clutch assembly. If problems are noted, replace the clutch assembly.
11   With the engine completely warmed up, turn off the ignition switch and disconnect the negative battery cable from the battery. Turn the fan by hand. Heavier resistance should be evident. If the fan turns easily, replace the fan clutch.

Removal and installation

Electric auxiliary fan

12   Disconnect the negative battery cable from the battery.
13   To remove the auxiliary fan follow the procedure that applies to your vehicle

3-Series models
14   Remove the grill assembly (see Chapter 11)
15   Unbolt and remove the fan bracket and shroud assembly from the radiator (see Section 5)
16   Remove the radiator (see Section 4)
17   Unbolt the air conditioning condenser mounting bolts. Do not remove the condenser or disconnect any refrigerant lines from the condenser.
18   Carefully pull the condenser back towards the engine, slightly, to gain access to lift the auxiliary fan.
19   Disconnect the fan motor electrical connection.
20   Remove the auxiliary fan. Installation is the reverse of removal.
5-Series models
21   Remove the screws and trim panel in front of the radiator.
22   Unbolt the fan assembly from the condenser mounting points.
23   Disconnect the fan electrical connector.
24   Remove the fan and housing from the car, being careful not to damage the air conditioning condenser while removing the fan.
25   Installation is the reverse of removal.

Mechanical fan with viscous clutch

Refer to illustrations 5.27a and 5.27b
26   Disconnect the negative battery cable. Remove the fan shroud mounting screws and detach the shroud (see Section 4).
27   Use a 32mm open-ended wrench to remove the fan/clutch assembly. Place the wrench on the large nut ahead of the pulley **(see illustrations)** and tap the end of the wrench to loosen the nut. **Caution:** *The nut has left-handed threads, so it loosens by being turned clockwise, as viewed from the front of the vehicle.*
28   Lift the fan/clutch assembly (and shroud, if necessary) out of the engine compartment.
29   If necessary, remove the four bolts attaching the pulley to the water pump hub **(see illustration 5.27a)**.
30   Carefully inspect the fan blades for damage and defects. Replace it if necessary.
31   At this point, the fan may be unbolted from the clutch, if necessary. If the fan clutch is stored, position it with the radiator side facing down.
32   Installation is the reverse of removal.

6   Water pump - check

1   A failure in the water pump can cause serious engine damage due to overheating.

Chapter 3 Cooling, heating and air conditioning systems

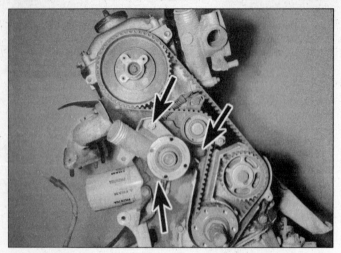

7.14 Water pump bolt locations - timing-belt engines

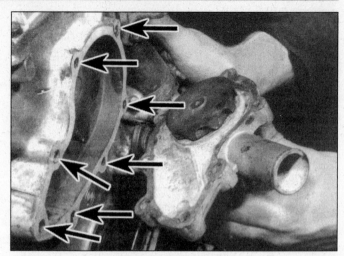

7.17 On four-cylinder engines, there are seven bolts that hold the water pump to the block

2   There are two ways to check the operation of the water pump while it's installed on the engine. If the pump is defective, it should be replaced with a new or rebuilt unit.
3   Water pumps are equipped with weep or vent holes. If a failure occurs in the pump seal, coolant will leak from the hole. In most cases you'll need a flashlight to find the hole on the water pump from underneath to check for leaks.
4   If the water pump shaft bearings fail there may be a howling sound at the front of the engine while it's running. Shaft wear can be felt if the water pump pulley is rocked up and down. Don't mistake drivebelt slippage, which causes a squealing sound, for water pump bearing failure.

7   Water pump - removal and installation

Removal

Warning: *Wait until the engine is completely cool before beginning this procedure.*

1   Disconnect the negative battery cable from the battery. **Caution**: *If the radio in your vehicle is equipped with an anti-theft system, make sure you have the correct activation code before disconnecting the battery.* **Note**: *If, after connecting the battery, the wrong language appears on the instrument panel display, refer to page 0-7 for the language resetting procedure.*
2   Drain the cooling system (see Chapter 1). If the coolant is relatively new or in good condition, save it and reuse it.
3   Remove the cooling fan and shroud (see Section 5).
4   Remove the drivebelts (see Chapter 1) and the pulley at the end of the water pump shaft.
5   Loosen the clamps and detach the hoses from the water pump. If they're stuck, grasp each hose near the end with a pair of Channelock pliers and twist it to break the seal, then pull it off. If the hoses are deteriorated, cut them off and install new ones.
6   Remove the fan and clutch assembly. (see Section 5)
7   To remove the water pump, follow the specific steps that apply to your vehicle.

535i and 533i

8   Remove the bolts that mount the water pump to the engine block.
9   Remove the engine lifting bracket.
10 Remove the water pump.

528e and 325i, and 325e

Refer to illustration 7.14

11 Remove the distributor cap, wires, ignition rotor and dust shield (see Chapter 1).

12 On 528e and 325i models built after March 1987, remove the timing sensor (see Chapter 5).
13 Remove the camshaft drivebelt cover (see Chapter 2A).
14 Loosen all 3 water pump mounting bolts and remove the top and right side bolts, but DON'T remove the lower bolt **(see illustration)**.
15 Rotate the pump down and remove the drivebelt tensioner spring and pin.
16 Remove the final water pump bolt and remove the pump. **Caution**: *Leave the tensioner bolt tight. Be careful to not move the camshaft gear. Damage can occur if the valves are moved.*

318i

Refer to illustration 7.17

17 Remove the seven mounting bolts and remove the water pump **(see illustration)**.

Installation

18 Clean the bolt threads and the threaded holes in the engine to remove corrosion and sealant.
19 Compare the new pump to the old one to make sure they're identical.
20 Remove all traces of old gasket material from the engine with a gasket scraper.
21 Clean the engine and new water pump mating surfaces with lacquer thinner or acetone.
22 Apply a thin coat of RTV sealant to the engine side of the new gasket.
23 Apply a thin layer of RTV sealant to the gasket mating surface of the new pump, then carefully mate the gasket and the pump. Slip a couple of bolts through the pump mounting holes to hold the gasket in place.
24 Carefully attach the pump and gasket to the engine and thread the bolts into the holes finger tight. **Note**: *On 528e, 325i and 325e models, install the lower bolt finger-tight, then rotate the water pump into position with the drivebelt tensioner spring and pin in position.*
25 Install the remaining bolts (if they also hold an accessory bracket in place, be sure to reposition the bracket at this time). Tighten them to the torque listed in this Chapter's Specifications in 1/4-turn increments. Don't over tighten them or the pump may be distorted.
26 Reinstall all parts removed for access to the pump.
27 Refill the cooling system and check the drivebelt tension (see Chapter 1). Run the engine and check for leaks.

8   Coolant temperature sending unit - check and replacement

Warning: *Wait until the engine is completely cool before beginning this procedure.*

Chapter 3 Cooling, heating and air conditioning systems

9.2a To get to the heater blower motor, the access panel must be removed - a fastener from the front and . . .

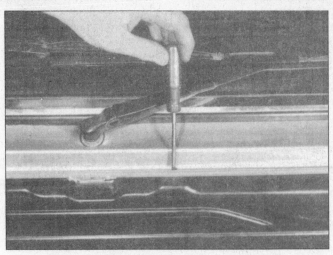

9.2b . . . a fastener from the top secure the panel

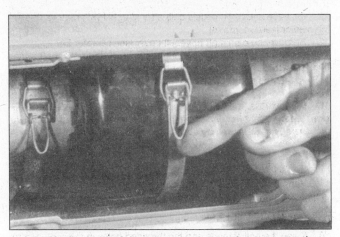

9.5a Unclip the retaining straps to remove the covers on the fan cages . . .

9.5b . . . and unclip the center strap that secures the blower motor assembly

1 The coolant temperature indicator system is composed of a temperature gauge mounted in the instrument panel and a coolant temperature sending unit that's normally mounted on the thermostat housing **(see illustration 3.8a)**. Some vehicles have more than one sending unit, but only one is used for the indicator system.
2 If an overheating indication occurs, check the coolant level in the system and then make sure the wiring between the gauge and the sending unit is secure and all fuses are intact.
3 Test the circuit by grounding the wire connected to the sending unit while the ignition is on (engine not running for safety). **Caution:** *Do not ground the wire for more than a second or two or damage to the gauge could occur.* If the gauge deflects full scale, replace the sending unit. If the gauge does not respond or does not indicate on overheating condition, the gauge or wiring to the gauge is faulty.
4 If the sending unit must be replaced, simply unscrew it from the engine and install the replacement. Use sealant on the threads. Make sure the engine is cool before removing the defective sending unit. There will be some coolant loss as the unit is removed, so be prepared to catch it. Check the level after the replacement has been installed.

9 Heater and air conditioning blower motor - removal and installation

Refer to illustrations 9.2a, 9.2b, 9.5a, 9.5b and 9.6
Note: *3-Series models have always used a single blower motor for vent, heat and air conditioning, while 5-Series models through mid-year 1988 use two separate blower motors: one for vent and heat and another for air conditioning. Mid-year 1988 and later 5-Series models have a single blower motor, like the 3-Series motor. The removal and installation of the single blower motor and early 5-Series vent/heat motor is described below. The removal and installation of the 5-Series air conditioning blower motor is described in Section 14 of this Chapter.*
1 Disconnect the negative battery cable from the battery. **Caution:** *If the radio in your vehicle is equipped with an anti-theft system, make sure you have the correct activation code before disconnecting the battery.* **Note**: *If, after connecting the battery, the wrong language appears on the instrument panel display, refer to page 0-7 for the language resetting procedure.*
2 The blower motor is located behind the firewall, under a trim cover. Remove the screws attaching the panel **(see illustrations)**.
3 Disconnect or cut the plastic ties holding the wiring harness to the cover and move the wiring out of the way.
4 Remove the cover.
5 Unclip the blower housing retaining clip and the clip securing the blower motor **(see illustrations)**.
6 Disconnect the blower wiring from the motor **(see illustration)**.
7 Remove the blower motor. You can test the blower motor by applying battery voltage to the blower motor's terminals with fused jumper wires (be sure the fan cages won't hit anything when they rotate). If the blower motor spins the fan cages rapidly (this test simulates high-speed operation), the blower motor is OK. If the blower motor does not operate, or operates slowly or noisily, replace it.

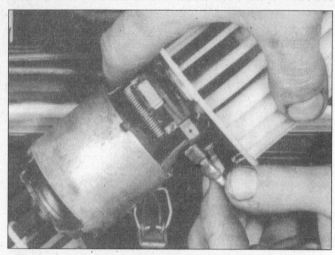

9.6 Lift out the assembly and disconnect the electrical connection from the blower motor

Note: *If the fan cages need to be removed, mark their relationship to the shaft. The cages are balanced during assembly and can cause excessive noise or shortened bearing life if the blower motor is not reinstalled in exactly the same position in relation to the shaft.*

Installation

8   Installation is the reverse of removal. **Note**: *The blower motor may have to be rotated to allow the retaining clip to line up correctly.*

10   Heater and air conditioner control assembly - removal and installation

Removal

1   Disconnect the negative battery cable from the battery. **Caution:** *If the radio in your vehicle is equipped with an anti-theft system, make sure you have the correct activation code before disconnecting the battery.* **Note**: *If, after connecting the battery, the wrong language appears on the instrument panel display, refer to page 0-7 for the language resetting procedure.*

3-Series models

2   Remove the center console and side trim pieces (see Chapter 11).
3   Remove the radio (see Chapter 12), then pull the knobs off the heater/air conditioning control levers.
4   Remove the heater trim panel to get access to the control cables.
5   Disconnect the cables, marking them for accurate reinstallation.
6   Disconnect the electrical connector.
7   Remove the lever assembly.
8   Each lever assembly can be removed separately.

5-Series models

9   Remove the center console (see Chapter 11).
10   Remove the radio (see Chapter 12), then pull the knobs off the heater/air conditioning control levers.
11   Remove the trim bezel and pull the control unit from the dash. This will allow you to disconnect the control cables from the lever assembly.
12   Disconnect the cables from the clips securing them to the lever assembly. **Note**: *Makes notes on cable colors and attaching positions for later reassembly.*
13   Disconnect the electrical connection from the control assembly.
14   Remove the screws attaching the bezel to the control assembly and remove the control assembly.

Installation

15   Installation is the reverse of the removal procedure.

11   Heater core - removal and installation

1   Disconnect the cable from the negative terminal of the battery. **Caution:** *If the radio in your vehicle is equipped with an anti-theft system, make sure you have the correct activation code before disconnecting the battery.* **Note**: *If, after connecting the battery, the wrong language appears on the instrument panel display, refer to page 0-7 for the language resetting procedure.*
2   Drain the cooling system (see Chapter 1).
3   Remove the center console (see Chapter 11). Spread an old blanket on the front carpeting in the center and on the passenger's side; this will prevent carpet stains if any residual coolant spills.

Removal

3-Series models

4   Remove the driver's side heater duct work and set it aside.
5   Remove the heater valve clamp.
6   Remove the screws and detach the flange where the two coolant lines connect the the heater core case. Be careful; some coolant may spill.
7   Remove the two screws holding the heater core case to the heater main assembly.
8   Slide the heater core out of the mounting. Be careful not to spill any of the remaining coolant in the heater core when removing it.

5-Series models

9   Disconnect the temperature sensor electrical connectors.
10   Disconnect the straps holding the wiring to the case and set the wiring out of the way.
11   Unfasten the cover fasteners.
12   Remove the screws holding the cover in place, then remove the cover.
13   Disconnect all heater pipe connections attached to the heater core. Be careful; some coolant may spill.
14   Lifting on the right side of the heater core first, remove the heater core.

Installation

Note: *Always use new O-rings when attaching the coolant lines to the heater core.*
15   Installation is the reverse of removal. Refill the cooling system (see Chapter 1), run the engine with the heater on and check for leaks.

12   Air conditioning and heating system - check and maintenance

Warning: *The air conditioning system is under high pressure. DO NOT loosen any hose or line fittings or remove any components until after the system has been discharged. Air conditioning refrigerant should be properly discharged into an EPA-approved container at a dealer service department or an automotive air conditioning repair facility. Always wear eye protection when disconnecting air conditioning system fittings.*

Check

1   The following maintenance checks should be performed on a regular basis to ensure the air conditioner continues to operate at peak efficiency.
  a) Check the drive-belt. If it's worn or deteriorated, replace it (see Chapter 1).
  b) Check the system hoses. Look for cracks, bubbles, hard spots and deterioration. Inspect the hoses and all fittings for oil bubbles and seepage. If there's any evidence of wear, damage or leaks, replace the hose(s).
  c) Inspect the condenser fins for leaves, bugs and other debris. Use a "fin comb" or compressed air to clean the condenser.
  d) Make sure the system has the correct refrigerant charge.
2   It's a good idea to operate the system for about 10 minutes at

least once a month, particularly during the winter. Long term non-use can cause hardening, and subsequent failure, of the seals.
3 Because of the complexity of the air conditioning system and the special equipment necessary to service it, in-depth troubleshooting and repairs are not included in this manual (refer to the *Haynes Automotive Heating & Air Conditioning Manual*). However, simple checks and component replacement procedures are provided in this Chapter.
4 The most common cause of poor cooling is simply a low system refrigerant charge. If a noticeable loss of cool air output occurs, the following quick check may help you determine if the refrigerant level is low.
5 Warm the engine up to normal operating temperature.
6 Place the air conditioning temperature selector at the coldest setting and put the blower at the highest setting. Open the doors (to make sure the air conditioning system doesn't cycle off as soon as it cools the passenger compartment).
7 With the compressor engaged - the compressor clutch will make an audible click and the center of the clutch will rotate - feel the orifice tube located adjacent to the right front frame rail, near the radiator.
8 If a significant temperature drop is noticed, the refrigerant level is probably okay. Refer to Chapter 6 for information on the air conditioning clutch control system. Further inspection of the system is beyond the scope of the home mechanic and should be left to a professional.
9 If the inlet line has frost accumulation or feels cooler than the receiver-drier surface, the refrigerant charge is low. Add refrigerant.

Adding refrigerant

Note: *Because of recent environmental legislation, 12-ounce cans of refrigerant may not be available in your area. If this is the case, you'll have to take the vehicle to a dealer service department or licensed air conditioning repair shop to have the air conditioning system recharged.*
10 Obtain a 12-ounce can of refrigerant, a tap valve and a short section of hose that can be attached between the tap valve and the system low-side service valve. Because one can of refrigerant may not be sufficient to bring the system charge up to the proper level, it's a good idea to obtain an additional can. **Warning:** *Never add more than two cans of refrigerant to the system.*
11 Use the tap valve and hose to connect the can to the system's low side service port. **Warning:** *DO NOT connect the charging kit hose to the system high side! If you have any doubts about how to hook up the system, obtain a copy of the* Haynes Automotive Heating and Air Conditioning Manual.
12 Warm up the engine and turn on the air conditioner. Keep the charging kit hose away from the fan and other moving parts.
13 Add refrigerant to the low side of the system until both the receiver-drier surface and the evaporator inlet line feel about the same temperature. Allow stabilization time between each addition.
14 Once the receiver-drier surface and the evaporator inlet line feel about the same temperature, add the contents remaining in the can.

13 Air conditioning compressor - removal and installation

Refer to illustration 13.6
Warning: *The air conditioning system is under high pressure. DO NOT disassemble any part of the system (hoses, compressor, line fittings, etc.) until after the system has been de-pressurized by a dealer service department or licensed air conditioning repair shop.*
Note: *The receiver-drier (see Section 16) should be replaced whenever the compressor is replaced.*
1 Have the air conditioning system discharged (see Warning above).
2 Disconnect the negative battery cable from the battery. **Caution:** *If the radio in your vehicle is equipped with an anti-theft system, make sure you have the correct activation code before disconnecting the battery.* **Note:** *If, after connecting the battery, the wrong language appears on the instrument panel display, refer to page 0-7 for the language resetting procedure.*
3 Disconnect the compressor clutch wiring harness.
4 Remove the drivebelt (see Chapter 1).

13.6 From under the vehicle, using a ratchet and socket and a box-end wrench, remove the bolt from the lower air conditioning compressor mount

5 Disconnect the refrigerant lines from the rear of the compressor. Plug the open fittings to prevent entry of dirt and moisture.
6 Unbolt the compressor from the mounting brackets and lift it up and out of the vehicle **(see illustration)**.
7 If a new compressor is being installed, follow the directions with the compressor regarding the draining of excess oil prior to installation.
8 The clutch may have to be transferred from the original to the new compressor.
9 Installation is the reverse of removal. Replace all O-rings with new ones specifically made for air conditioning system use and lubricate them with refrigerant oil.
10 Have the system evacuated, recharged and leak tested by the shop that discharged it.

14 Air conditioning blower motor (5-Series models through mid-year 1988) - removal and installation

Note: *After mid-year 1988, 5-Series models no longer use a separate blower motor for the air conditioning system. The blower motor located under the firewall cowl (described in Section 9) handles all heating, vent and air conditioning functions.*

Removal

1 Disconnect the negative battery cable from the battery. **Caution:** *If the radio in your vehicle is equipped with an anti-theft system, make sure you have the correct activation code before disconnecting the battery.* **Note:** *If, after connecting the battery, the wrong language appears on the instrument panel display, refer to page 0-7 for the language resetting procedure.*
2 Remove the center console (see Chapter 11).
3 Disconnect the blower motor electrical connector from the motor.
4 Unbolt the attaching hardware that holds the blower motor assembly to the main case.
5 Remove the air conditioning blower motor assembly. You can check the motor by following the procedure described in Section 9.

Installation

6 Installation is the reverse of removal.
Note: *If the fan cages need to be removed, mark them so they can be reinstalled in their original positions. They are balanced during assembly and can cause excessive noise or shortened bearing life if they are not installed in exactly the same relationship to the motor shaft.*

Chapter 3 Cooling, heating and air conditioning systems

15 Air conditioning condenser - removal and installation

Warning: *The air conditioning system is under high pressure. DO NOT disassemble any part of the system (hoses, compressor, line fittings, etc.) until after the system has been de-pressurized by a dealer service department or licensed air conditioning service facility.*
Note: *The receiver-drier (see Section 16) should be replaced whenever the condenser is replaced.*

1 Have the air conditioning system discharged (see Warning above).
2 Remove the radiator (see Section 4).
3 Remove the grill (see Chapter 11).
4 Unbolt the auxiliary fan from the air conditioning condenser mounting brackets.
5 Disconnect the refrigerant lines from the condenser.
6 Remove the mounting bolts from the condenser brackets.
7 Lift the condenser out of the vehicle and plug the lines to keep dirt and moisture out.

Installation

8 If the original condenser will be reinstalled, store it with the line fittings on top to prevent oil from draining out.
9 If a new condenser is being installed, add new refrigerant oil to system according to the specifications at the beginning of this Chapter.
10 Reinstall the components in the reverse order of removal. Be sure the rubber pads are in place under the condenser.
11 Have the system evacuated, recharged and leak tested by the shop that discharged it.

16 Air conditioner receiver-drier - removal and installation

Refer to illustration 16.4
Warning: *The air conditioning system is under high pressure. DO NOT loosen any hose or line fittings or remove any components until after the system has been de-pressurized by a dealer service department or licensed air conditioning repair shop. Always wear eye protection when disconnecting air conditioning system fittings.*

Removal

1 Have the system discharged (see Warning above).
2 Disconnect the negative battery cable from the battery. **Caution:** *If the radio in your vehicle is equipped with an anti-theft system, make sure you have the correct activation code before disconnecting the battery.* **Note:** *If, after connecting the battery, the wrong language appears on the instrument panel display, refer to page 0-7 for the language resetting procedure.*
3 Remove the windshield washer fluid tank.
4 Disconnect the electrical connector **(see illustration).**
5 Disconnect the refrigerant lines from the receiver-drier.
6 Plug the open fittings to prevent the entry of dirt and moisture.
7 Remove the mounting screws and remove the receiver-drier.
8 If a new receiver-drier is being installed, pour in the amount of refrigerant oil indicated in the specifications at the front of this Chapter.
9 Remove the old refrigerant line O-rings and replace with new ones. This should be done regardless of whether a new receiver-drier is being installed or not.
10 If a new receiver-drier is being installed, unscrew the safety pres-

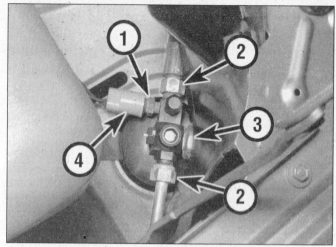

16.4 All models have the receiver-drier located behind the right headlight, although the configuration of lines and switches may vary

1 High-pressure switch   3 Low-pressure switch
2 Refrigerant lines   4 Electrical connector

sure switches and install them on the new unit before installation **(see illustration 16.4)**. **Note:** *Not all models have both the high and low pressure switches. Remove the switches appropriate to the receiver/drier in your vehicle.*
11 Lubricate O-rings with refrigerant oil before assembly.

Installation

12 Installation is the reverse of removal, but be sure to lubricate the O-rings on the fittings with refrigerant oil before connecting the fittings.
13 Have the system evacuated, recharged and leak tested by the shop that discharged it. **Note:** *The manufacturer recommends that all screwed-in fittings should be assembled with the use of a thread sealant. Specific sealant for the high or low pressure side of the system is available at an air conditioning repair shop or automotive parts store.*

17 Evaporator core - removal and installation

Warning: *The air conditioning system is under high pressure. DO NOT loosen any hose or line fittings or remove any components until after the system has been discharged by a dealer service department or licensed air conditioning repair shop. Always wear eye protection when disconnecting air conditioning system fitting*

1 Have the air conditioning system discharged (see Warning above).
2 Remove the trim panel on the sides of the center console (see Chapter 11)
3 Disconnect the electrical lead and remove the evaporator sensor.
4 Remove the evaporator cover exposing the refrigerant lines.
5 Disconnect the refrigerant lines from the evaporator core.
6 Remove the evaporator core from the case.
7 Installation is the reverse procedure of the removal.
8 Have the service department that discharged the system now evacuate and recharge the air conditioning system.

Chapter 4 Fuel and exhaust systems

Contents

Accelerator cable - check, adjustment and replacement	9
Air cleaner assembly - removal and installation	8
Air filter replacement	See Chapter 1
Airflow meter - check, removal and installation	13
Cold start injector and thermo-time switch - check and replacement	16
Exhaust system check	See Chapter 1
Exhaust system servicing - general information	19
Fuel filter replacement	See Chapter 1
Fuel injection system - check	12
Fuel injection system - general information	10
Fuel injection system - troubleshooting	11
Fuel injectors - check and replacement	17
Fuel lines and fittings - repair and replacement	5
Fuel pressure regulator - check and replacement	15
Fuel pressure relief procedure	2
Fuel pump, transfer pump and fuel level sending unit - check, removal and installation	4
Fuel pump/fuel pressure - check	3
Fuel system check	See Chapter 1
Fuel tank - removal and installation	6
Fuel tank cleaning and repair - general information	7
General information	1
Idle air stabilizer valve - check, adjustment and replacement	18
Throttle body - check, removal and installation	14

Specifications

Fuel pressure checks

Fuel system pressure (engine off, fuel system energized - **see illustrations 3.1a through 3.1d**)

325, 325e, 325es (Motronic)	36 to 37 psi
325i, 325is, 325iC (Motronic 1.1)	43 to 45 psi
318i (1984 and 1985 L-Jetronic)	43 to 45 psi
528e, 533i (up to 3/1987 Motronic)	36 to 37 psi
535i, 535is (Motronic 1.1)	43 to 44 psi
525i (Motronic 1.1)	39 to 40 psi

Pressure regulator check (at idle)
Vacuum hose attached

325, 325e, 325es (Motronic)	36 to 37 psi
325i, 325is, 325iC (Motronic 1.1)	43 to 45 psi
318i (1984 and 1985 L-Jetronic)	43 to 45 psi
528e, 533i (up to 3/1987 Motronic)	33 to 34 psi
535i, 535is (Motronic 1.1)	40 to 41 psi
525i (Motronic 1.1)	36 to 37 psi

Chapter 4 Fuel and exhaust systems

Pressure regulator check (at idle) (continued)
  Vacuum hose detached

325, 325e, 325es (Motronic)	39 to 40 psi
325i, 325is, 325iC (Motronic 1.1)	46 to 47 psi
318i (1984 and 1985 L-Jetronic)	46 to 47 psi
528e (up to 3/1987 Motronic)	39 to 40 psi
535i, 535is (Motronic 1.1)	46 to 47 psi
525i (Motronic 1.1)	42 to 43 psi
Fuel system hold pressure	30 psi
Fuel pump pressure (maximum)	92 to 100 psi
Fuel pump hold pressure	80 psi
Transfer pump pressure	4 to 5 psi

Injector resistance

318i (1984 and 1985 only)	14.5 to 17.5 ohms
325e and 325es (up to 9/1984)	14.5 to 17.5 ohms
325, 325e and 325es (10/1984 on)	2 to 3 ohms
325i, 325is and 325iC	14.5 to 17.5 ohms
528e (1982 through 3/1987)	2 to 3 ohms
528e (4/1987 on)	14.5 to 17.5 ohms
533i	2 to 3 ohms
535i (1983 through 1/1986)	14.5 to 17.5 ohms
535i (2/1986 on)*	14.5 to 17.5 ohms
525i (Motronic 1.1)	14.5 to 17.5 ohms

*Note: *Although the resistance is the same as the early 535i model injectors, the part number is different for the later style injectors*

Accelerator cable freeplay .... 3/64 to 3/32 inch

Torque specifications

Throttle body nuts/bolts	14 to 19 ft-lbs
Fuel rail mounting bolts	80 to 97 in-lbs

1 General information

All engines are equipped with electronic fuel injection (EFI). The EFI system consists of three basic sub-systems - the air intake system, the fuel system and the electronic control system.

These models are equipped with either the L-Jetronic or the Motronic fuel injection system. 1984 and 1985 318i models are equipped with the L-Jetronic system. All other 3 and 5-Series models are equipped with the Motronic system. Partway through 1987, all models were equipped with the updated version, Motronic 1.1 system. The Motronic 1.1 system is easily distinguished from the Motronic system by the absence of a cold start injector.

The fuel pump relay on Motronic systems is activated from a ground signal from the Motronic control unit (ECU). The fuel pump operates for a few seconds when the ignition is first switched ON, and it continues to operate only when the engine is actually running.

Air intake system

The air intake system consists of the air filter housing, the airflow meter, the throttle body and the intake manifold. All components except the intake manifold are covered in this Chapter; for information on removing and installing the intake manifold, refer to Chapter 2A.

The throttle valve inside the throttle body is actuated by the accelerator cable. When you depress the accelerator pedal, the throttle plate opens and airflow through the intake system increases. A flap inside the airflow meter opens wider as the airflow increases. A throttle position switch attached to the pivot shaft of the flap detects the angle of the flap (how much it's open) and converts this to a voltage signal which it sends to the computer.

Fuel system

An electric fuel pump supplies fuel under constant pressure to the fuel rail, which distributes fuel to the injectors. The electric fuel pump is located inside the fuel tank on newer models (Motronic 1.1 systems manufactured after 3/1987) or beside the fuel tank on early models (L-Jetronic and Motronic systems between 1982 through 3/1987). The early models are also equipped with a transfer pump located in the fuel tank. The transfer pump acts as an aid to the larger main pump for delivering the necessary pressure. A fuel pressure regulator controls the pressure in the fuel system. The fuel system is also equipped with a fuel pulsation damper located near the fuel filter. The damper reduces the pressure pulsations caused by fuel pump operation and the opening and closing of the injectors. The amount of fuel injected into the intake ports is precisely controlled by an Electronic Control Unit (ECU or computer).

Some later 5-Series models are equipped with a fuel cooler in the return line.

Electronic control system

Besides altering the injector opening duration as described above, the electronic control unit performs a number of other tasks related to fuel and emissions control. It accomplishes these tasks by using data relayed to it by a wide array of information sensors located throughout the engine compartment, comparing this information to its stored map and altering engine operation by controlling a number of different actuators. Since special equipment is required, most troubleshooting and repairing of the electronic control system is beyond the scope of the home mechanic. Additional information and testing procedures on the emissions system components (oxygen sensor, coolant temperature sensor, EVAP system, etc.) is contained in Chapter 6.

Warranty information

These vehicles are covered by a Federally-mandated extended warranty (5 years or 50,000 miles at the time this manual was written) which covers nearly every fuel system component in this Chapter. Before working on the fuel system, check with a dealer service department for warranty terms on your particular vehicle.

2 Fuel pressure relief procedure

Refer to illustrations 2.1a, 2.1b, 2.1c, 2.1d and 2.4
Warning: *Gasoline is extremely flammable, so take extra precautions when you work on any part of the fuel system. Don't smoke or allow open flames or bare light bulbs near the work area. Also, don't work in a garage where a natural gas-type appliance with a pilot light is pre-*

Chapter 4 Fuel and exhaust systems

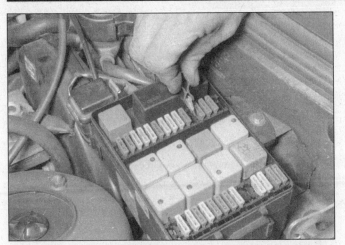

2.1a Remove fuse #11 on 325is models to disable the fuel pump

2.1b Remove fuse #11 on 318i models to disable the fuel pump

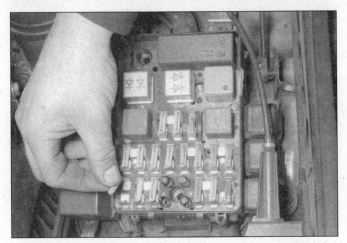

2.1c Remove the corner fuse on 528e models to disable the fuel pump

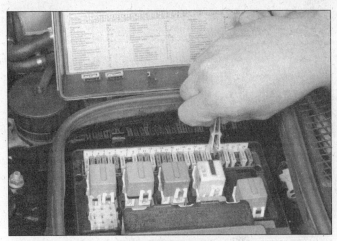

2.1d Remove fuse #23 on 525i and 535i models to disable the fuel pump

sent. Finally, when you perform any kind of work on the fuel system, wear safety glasses and have a Class B type fire extinguisher on hand. If you spill any fuel on your skin, clean it off immediately with soap and water.

1 Remove the fuel pump fuse from the main fuse panel **(see illustrations)**. **Note:** *Consult your owner's manual for the exact location of the fuel pump fuse if the information is not stamped onto the fuse cover.*

2 Start the engine and wait for the engine to stall, then turn off the ignition key.

3 Remove the fuel filler cap to relieve the fuel tank pressure.

4 The fuel system is now depressurized. **Note:** *Place a rag around the fuel line before removing the hose clamp to prevent any residual fuel from spilling onto the engine* **(see illustration)**.

5 Disconnect the cable from the negative terminal of the battery before working on any part of the system. **Caution:** *If the radio in your vehicle is equipped with an anti-theft system, make sure you have the correct activation code before disconnecting the battery, Refer to the information on page 0-7 at the front of this manual before detaching the cable.* **Note:** *If, after connecting the battery, the wrong language appears on the instrument panel display, refer to page 0-7 for the language resetting procedure.*

3 Fuel pump/fuel pressure - check

Warning: *Gasoline is extremely flammable, so take extra precautions when you work on any part of the fuel system. Don't smoke or allow open flames or bare light bulbs near the work area. Also, don't work in*

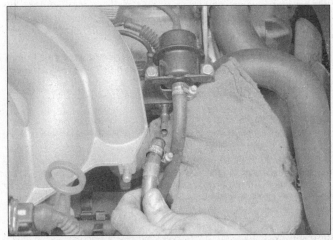

2.4 Be sure to place a shop rag under and around any fuel line when disconnecting

a garage where a natural gas-type appliance with a pilot light is present. Finally, when you perform any kind of work on the fuel system, wear safety glasses and have a Class B type fire extinguisher on hand. If you spill any fuel on your skin, clean it off immediately with soap and water.

Note 1: *The electric fuel pump is located inside the fuel tank on later models (Motronic 1.1 systems manufactured after 3/1987) or beside the fuel tank on early models (L-Jetronic and Motronic systems be-*

Chapter 4 Fuel and exhaust systems

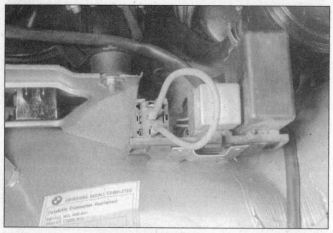

3.1a On 318i models (L-Jetronic system), use a jumper wire and bridge the terminals on the connector that correspond to the fuel pump relay pins #30 and #87b

3.1b On 325is models (Motronic 1.1 system), use a jumper wire and bridge the terminals on the connector that correspond to the fuel pump relay pins #30 and #87

3.1c On all 1989 and later 525 and 535 models, remove the four bolts and the protective cover to gain access to the fuel pump relay and ECU . . .

3.1d . . . then, use a jumper wire and bridge the terminals on the connector that correspond to fuel pump relay pins #30 and #87

tween 1982 through 3/1987). Early models are also equipped with a transfer pump located in the fuel tank. The transfer pump feeds the main pump but can't generate the high pressure required by the system.
Note 2: *The fuel pump relay on Motronic systems is activated from a ground signal from the Motronic control unit (ECU). The fuel pump operates for a few seconds when the ignition is first switched ON, and it continues to operate only when the engine is actually running.*
Note 3: *The following checks assume the fuel filter is in good condition. If you doubt the condition of your fuel filter, replace it (see Chapter 1).*
Note 4: *In order to get accurate test results, it is recommended that the fuel pressure be checked from both the main fuel pump and transfer pump on L-Jetronic and Motronic systems.*

Fuel pump/transfer pump operational check

Refer to illustrations 3.1a, 3.1b, 3.1c and 3.1d
1   Bridge the connector terminals that correspond to the fuel pump relay pins #30 and #87b on 1984 and 1985 318i models (four-cylinder engines) **(see illustration)** with a suitable jumper wire. Bridge the connector terminals that correspond to the fuel pump relay pins #30 and #87 on all six-cylinder engines **(see illustrations)** with a suitable jumper wire. **Note**: *A factory tool (BMW # 61 3 050) is available for use as a jumper wire. This tool has a built-in toggle switch and a 15 amp fuse to prevent any wiring damage caused by repeated connecting and disconnecting.*
2   Have an assistant turn the ignition key to On while you listen at the fuel tank (the fuel pump is inside the fuel tank on later Motronic 1.1 systems). You should hear a whirring sound for a couple of seconds. **Note**: *This test applies to the transfer pump also. If there is no whirring sound, there is a problem in the fuel pump circuit. Check the fuel pump main fuse and relay first (see Chapter 12). If the main relay is OK, test the fuel pump relay (see Step 32).*

Fuel system pressure check

Refer to illustrations 3.5 and 3.6
3   Relieve the system fuel pressure (see Section 2).
4   Detach the cable from the negative battery terminal. **Caution**: *If the radio in your vehicle is equipped with an anti-theft system, make sure you have the correct activation code before disconnecting the battery. Refer to the information on page 0-7 at the front of this manual before detaching the cable.* **Note**: *If, after connecting the battery, the wrong language appears on the instrument panel display, refer to page 0-7 for the language resetting procedure.*
5   Detach the fuel feed line from the fuel rail on L-Jetronic **(see illustration)** and Motronic systems (1982 through 3/1987) or from the fuel filter to the main fuel line on Motronic 1.1 systems (3/1987 through 1992).
6   Using a tee (three-way) fitting, a short section of high-pressure fuel hose and clamps, attach a fuel pressure gauge without disturbing normal fuel flow **(see illustration)**. **Warning**: *Do not use a plastic tee fitting for this test. It won't be able to withstand the fuel system pressure.*
7   Attach the cable to the negative battery terminal.

Chapter 4 Fuel and exhaust systems

3.5 Disconnect the fuel feed line (arrow) from the fuel rail (L-Jetronic system shown)

3.6 ... and attach the gauge between the fuel feed line and the fuel rail using a tee fitting

8   Bridge the terminals of the fuel pump relay using a jumper wire **(see illustrations 3.1a through 3.1d)**.
9   Turn the ignition switch to ON.
10 Note the fuel pressure and compare it with the pressure listed in this Chapter's Specifications.
11 If the system fuel pressure is less than specified:
  a) Inspect the system for a fuel leak. Repair any leaks and recheck the fuel pressure.
  b) If there are no leaks, install a new fuel filter and recheck the fuel pressure.
  c) If the pressure is still low, check the fuel pump pressure (see below) and the fuel pressure regulator (see Section 17).
12 If the pressure is higher than specified, check the fuel return line for an obstruction. If the line is not obstructed, replace the fuel pressure regulator.
13 Turn the ignition switch to Off, wait five minutes and look at the gauge. Compare the reading with the system hold pressure listed in this Chapter's Specifications. If the hold pressure is less than specified:
  a) Inspect the fuel system for a fuel leak. Repair any leaks and recheck the fuel pressure.
  b) Check the fuel pump pressure (see below).
  c) Check the fuel pressure regulator (see Section 17).
  d) Check the injectors (see Section 19).

Fuel pump pressure check

Warning: *For this test it is necessary to use a fuel pressure gauge with a bleeder valve in order to relieve the high fuel pressure. After the test is completed, the normal procedure for pressure relief will not work because the gauge is connected directly to the fuel pump.*

14 Relieve the fuel pressure (see Section 2).
15 Detach the cable from the negative battery terminal. **Caution**: *If the radio in your vehicle is equipped with an anti-theft system, make sure you have the correct activation code before disconnecting the battery. Refer to the information on page 0-7 at the front of this manual before detaching the cable.* **Note**: *If, after connecting the battery, the wrong language appears on the instrument panel display, refer to page 0-7 for the language resetting procedure.*
16 Detach the fuel feed hose from the fuel rail and attach a fuel pressure gauge directly to the hose. **Note**: *If the tee fitting is still connected to the gauge, be sure to plug the open end.*
17 Attach the cable to the negative battery terminal.
18 Using a jumper wire, bridge the terminals of the fuel pump relay **(see illustrations 3.1a through 3.1d)**.
19 Turn the ignition switch to On to operate the fuel pump.
20 Note the pressure reading on the gauge and compare the reading to the fuel pressure listed in this Chapter's Specifications.
21 If the indicated pressure is less than specified, inspect the fuel line for leaks between the pump and gauge. If no leaks are found, replace the fuel pump.
22 Turn the ignition key to Off and wait five minutes. Note the reading on the gauge and compare it to the fuel pump hold pressure listed in this Chapter's Specifications. If the hold pressure is less than specified, check the fuel lines between the pump and gauge for leaks. If no leaks are found, replace the fuel pump.
23 Remove the jumper wire. Relieve the fuel pressure by opening the bleeder valve on the gauge and directing the fuel into an approved fuel container. Remove the gauge and reconnect the fuel line.

Transfer pump pressure check

24 Relieve the system fuel pressure (see Section 2).
25 Detach the cable from the negative battery terminal. **Caution**: *If the radio in your vehicle is equipped with an anti-theft system, make sure you have the correct activation code before disconnecting the battery. Refer to the information on page 0-7 at the front of this manual before detaching the cable.* **Note**: *If, after connecting the battery, the wrong language appears on the instrument panel display, refer to page 0-7 for the language resetting procedure.*
26 Remove the transfer pump access plate (on some models it's located under the rear seat cushion - on others it's located under the carpet in the trunk). Detach the output hose on the transfer pump and attach a fuel pressure gauge to the outlet pipe.
27 Attach the cable to the negative battery terminal.
28 Using a jumper wire, bridge the terminals of the fuel pump relay **(see illustrations 3.1a through 3.1d)**.
29 Turn the ignition switch to On to operate the fuel pump.
30 Note the pressure reading on the gauge and compare the reading to the value listed in this Chapter's Specifications.
31 If the indicated pressure is less than specified, replace the transfer pump with a new one.

Fuel pump relay check

32 Turn the ignition key to the ON position.
33 With the relay intact, probe the indicated terminals from the backside of the relay electrical connector with a voltmeter. Check for battery voltage at terminal 30 (six-cylinder engines) or terminal 15 (1984 and 1985 318i models only - four-cylinder engines). **Note**: *On six-cylinder models with trunk-mounted batteries, check for a faulty fusible link. The 50 amp link is about 6 inches from the battery in a black wire.*
34 Turn the ignition key OFF and disconnect the relay from the electrical connector. Probe the connector terminals that correspond to fuel pump relay pins #85 (-) and 86(+) on all six-cylinder models or terminal 50 and ground (1984 and 1985 318i models) with a voltmeter. Have an assistant crank the engine over and observe the voltage read-

4-6   Chapter 4   Fuel and exhaust systems

4.4 Lift up the rubber boots (arrows) and detach the electrical connectors from the fuel pump

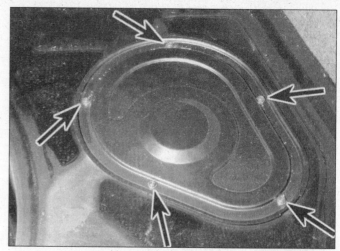

4.10 Remove the fuel pump cover screws (arrows) and lift the cover off the floor of the vehicle (on some models the fuel pump cover is located under the rear seat - on other models it's located in the trunk)

ing. There should be battery voltage.
35   If there is no voltage, check the fuse(s) and the wiring circuit for the fuel pump relay. If the voltage readings are correct and the fuel pump only runs with the jumper wire in place (see Step 1), then replace the relay with a new part.
36   If the fuel pump still does not run, check for the proper voltage at the fuel pump terminals (see Section 4). If necessary, replace the fuel pump.

4   Fuel pump, transfer pump and fuel level sending unit - check, removal and installation

Warning: *Gasoline is extremely flammable, so take extra precautions when you work on any part of the fuel system. Don't smoke or allow open flames or bare light bulbs near the work area. Also, don't work in a garage where a natural gas-type appliance with a pilot light is present. Finally, when you perform any kind of work on the fuel system, wear safety glasses and have a Class B type fire extinguisher on hand. If you spill any fuel on your skin, clean it off immediately with soap and water.*
Note 1: *The electric fuel pump is located inside the fuel tank on later models (Motronic 1.1 systems manufactured after 3/1987) or adjacent to the fuel tank on early models (L-Jetronic and Motronic systems between 1982 through 3/1987). The early models are also equipped with a transfer pump located in the fuel tank. The transfer pump feeds the larger main pump, which delivers the high pressure required for proper fuel system operation.*
Note 2: *The fuel level sending unit is located in the fuel tank with the transfer pump on early models (1982 through 3/1987) or with the main fuel pump on late models (3/1987 through 1992).*
1   Relieve the system fuel pressure (see Section 2) and remove the fuel tank filler cap to relieve pressure in the tank.
2   Disconnect the cable from the negative battery terminal. **Caution**: *If the radio in your vehicle is equipped with an anti-theft system, make sure you have the correct activation code before disconnecting the battery, Refer to the information on page 0-7 at the front of this manual before detaching the cable. Note: If, after connecting the battery, the wrong language appears on the instrument panel display, refer to page 0-7 for the language resetting procedure.*

Fuel pump (externally mounted)
Refer to illustration 4.4
3   Raise the vehicle and support it securely on jackstands.
4   Remove the two rubber boots that protect the fuel pump connectors and disconnect the wires from the pump **(see illustration)**.

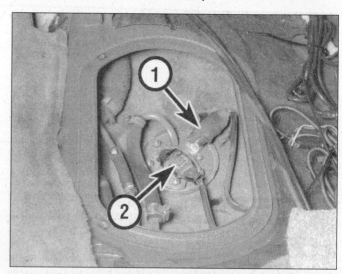

4.12a Transfer pump and fuel level sending unit on a Motronic system

1   Transfer pump electrical connector
2   Fuel level sending unit electrical connector

5   Use line clamps and pinch shut the fuel hoses on each side of the fuel pump. If you don't have line clamps, wrap the hoses with rags and clamp them shut with locking pliers, tightened just enough to prevent fuel from flowing out.
6   Loosen the clamps and disconnect the hoses from the pump.
7   Remove the fuel pump mounting screws and clamps and separate the fuel pump from the underside of the body.
8   Installation is the reverse of removal.

Fuel pump (in-tank) or transfer pump
Refer to illustrations 4.10, 4.12a, 4.12b and 4.13a through 4.13e
9   On some models access to the fuel pump is gained by removing the rear seat cushion. On other models access is gained by opening the trunk and removing the carpet.
10   Remove the screws from the fuel pump access cover **(see illustration)**.
11   Remove the cover.
12   Locate the fuel pump and sending unit electrical connectors **(see illustrations)** and unplug them. Also, disconnect the fuel inlet and return lines.

Chapter 4 Fuel and exhaust systems

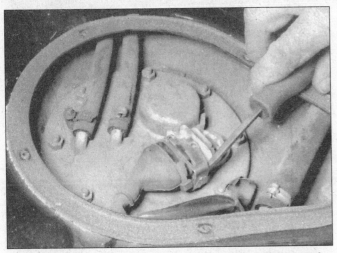

4.12b To unplug the connector, pry the bracket until the notch aligns with the slot on the retaining clip and release the connector from the assembly

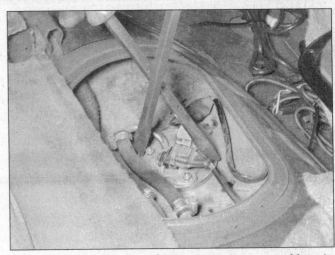

4.13a On models without mounting nuts, use two screwdrivers to rotate the assembly out of the notches

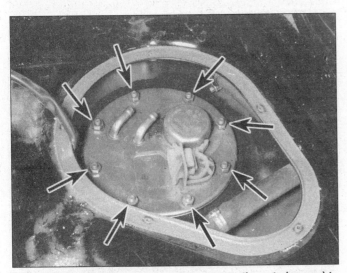

4.13b On models with mounting nuts, remove the nuts (arrows) to release the assembly from the fuel tank

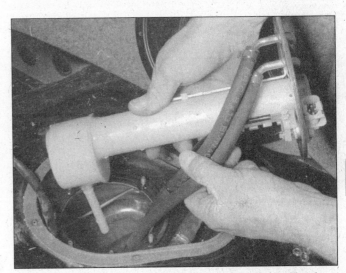

4.13c On Motronic 1.1 systems, first remove the fuel level sending unit . . .

13 On some models the assembly must be rotated counterclockwise to disengage the locking lugs from the fuel tank (see illustration). On other models, the assembly is secured to the tank with nuts (see illustration). Carefully lift the assembly from the fuel tank (see illustrations). It may be necessary to slightly twist the assembly to get the float to clear the opening.
14 On early models, remove the transfer pump mounting screws and clamps and separate the transfer pump from the assembly.
15 Installation is the reverse of removal. If the gasket between the fuel pump and fuel tank is dried, cracked or damaged, replace it.

Fuel level sending unit - check and replacement
Refer to illustrations 4.17a and 4.17b
16 Remove the main fuel pump or transfer pump (see previous steps) along with the fuel level sending unit.
17 Use an ohmmeter across the designated terminals and check for the correct resistance. On L-Jetronic and Motronic systems, follow the chart below. On Motronic 1.1 systems, connect the ohmmeter probes onto the fuel level sending unit terminals that correspond to pins #1 and #3 on the electrical connector (see illustrations). The resistance should decrease as the plunger rises.

4.13d . . . then pull the fuel pump straight up and out of the turret at the bottom of the fuel tank (keep all the fuel lines intact)

Chapter 4 Fuel and exhaust systems

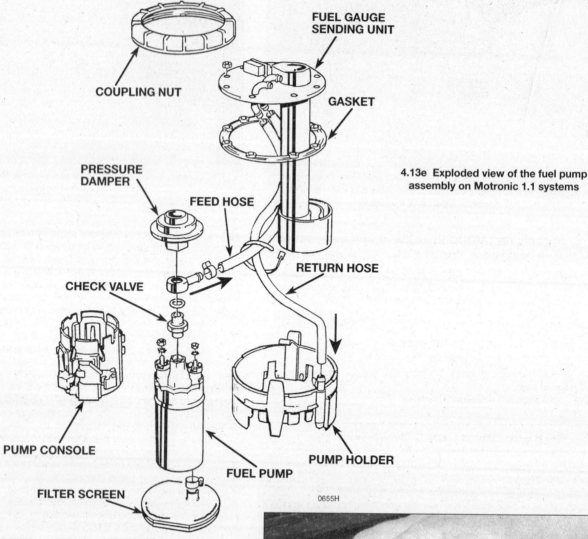

4.13e Exploded view of the fuel pump assembly on Motronic 1.1 systems

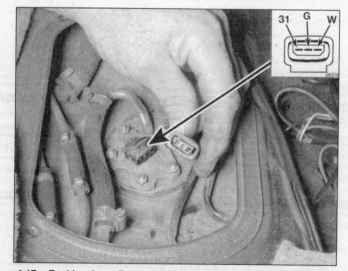

4.17a Fuel level sending unit terminal designations on L-Jetronic and Motronic systems

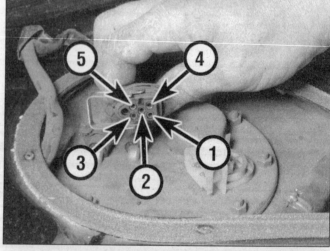

4.17b Fuel level sending unit terminal designations on Motronic 1.1 systems - when checking resistance on the sending unit, touch the ohmmeter probes to the terminals on the fuel pump/sending unit which correspond to terminals 1 and 3 of the connector

1. Fuel level sending unit ground
2. Warning lamp
3. Sending unit
4. Fuel pump ground
5. Fuel pump

Chapter 4 Fuel and exhaust systems

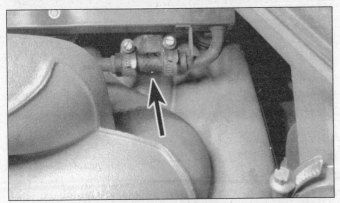

5.2 When checking the fuel lines, don't overlook these short sections of fuel hose attached to the main fuel rail - they're a common source of fuel leaks

6.4 Loosen the drain plug (arrow) and drain the fuel into an approved gasoline container

L-Jetronic and Motronic sending unit diagnostics

Connector designations	Float position	Resistance (ohms)
G and 31	slowly moving from the EMPTY position to the FULL position	resistance slowly decreases
	EMPTY	71.7 +- 2.3 ohms
	FULL	3.2 +- 0.7 ohms
W and 31	EMPTY (low fuel warning)	continuity

18 If the resistance readings are incorrect, replace the sending unit with a new part.
19 Installation is the reverse of removal.

5 Fuel lines and fittings - repair and replacement

Refer to illustration 5.2

Warning: *Gasoline is extremely flammable, so take extra precautions when you work on any part of the fuel system. Don't smoke or allow open flames or bare light bulbs near the work area, and don't work in a garage where a natural gas-type appliance (such as a water heater or clothes dryer) with a pilot light is present. If you spill any fuel on your skin, rinse it off immediately with soap and water. When you perform any kind of work on the fuel system, wear safety glasses and have a Class B type fire extinguisher on hand.*

1 Always relieve the fuel pressure and disconnect the negative battery cable before servicing fuel lines or fittings (see Section 2).
2 The fuel feed, return and vapor lines extend from the fuel tank to the engine compartment. The lines are secured to the underbody with clip and screw assemblies. These lines must be occasionally inspected for leaks, kinks and dents **(see illustration)**.
3 If evidence of dirt is found in the system or fuel filter during disassembly, the line should be disconnected and blown out. Check the fuel strainer on the in-tank fuel pump for damage and deterioration.
4 Because fuel lines used on fuel-injected vehicles are under high pressure, they require special consideration. If replacement of a rigid fuel line or emission line is called for, use welded steel tubing meeting BMW specification or its equivalent. Don't use copper or aluminum tubing to replace steel tubing. These materials cannot withstand normal vehicle vibration.
5 When replacing fuel hose, use only hose approved for high-pressure fuel injection applications.

6 Fuel tank - removal and installation

Refer to illustrations 6.4, 6.6, 6.10a and 6.10b

Warning: *Gasoline is extremely flammable, so take extra precautions when you work on any part of the fuel system. Don't smoke or allow open flames or bare light bulbs near the work area. Also, don't work in a garage where a natural gas-type appliance with a pilot light is present. Finally, when you perform any kind of work on the fuel system, wear safety glasses and have a Class B type fire extinguisher on hand. If you spill any fuel on your skin, clean it off immediately with soap and water.*
Note: *To avoid draining large amounts of fuel, make sure the fuel tank is nearly empty (if possible) before beginning this procedure.*

1 Remove the fuel tank filler cap to relieve fuel tank pressure.
2 Relieve the system fuel pressure (see Section 2).
3 Detach the cable from the negative battery terminal. **Caution**: *If the radio in your vehicle is equipped with an anti-theft system, make sure you have the correct activation code before disconnecting the battery, Refer to the information on page 0-7 at the front of this manual before detaching the cable.* **Note**: *If, after connecting the battery, the wrong language appears on the instrument panel display, refer to page 0-7 for the language resetting procedure.*
4 Remove the drain plug **(see illustration)** and drain the fuel into an approved gasoline container.
5 Unplug the fuel pump electrical connector and detach the fuel feed, return and vapor hoses (see Section 4).
6 Remove the fuel tank shield **(see illustration)**.
7 Detach the fuel filler neck and breather hoses.
8 Raise the vehicle and place it securely on jackstands.
9 Support the tank with a floor jack. Position a block of wood between the jack head and the fuel tank to protect the tank.
10 Remove the mounting bolts at the corners of the fuel tank and unbolt the retaining straps **(see illustrations)**. Pivot the straps down until they're hanging out of the way.
11 Lower the tank just enough so you can see the top and make sure you have detached everything. Finish lowering the tank and remove it from the vehicle.
12 Installation is the reverse of removal.

7 Fuel tank cleaning and repair - general information

1 All repairs to the fuel tank or filler neck should be carried out by a professional who has experience in this critical and potentially dangerous work. Even after cleaning and flushing of the fuel system, explosive fumes can remain and ignite during repair of the tank.
2 If the fuel tank is removed from the vehicle, it should not be placed in an area where sparks or open flames could ignite the fumes coming out of the tank. Be especially careful inside garages where a natural gas-type appliance is located, because the pilot light could cause an explosion.

8 Air cleaner assembly - removal and installation

Refer to illustration 8.6

1 Detach the cable from the negative battery terminal. **Caution**: *If*

4-10 Chapter 4 Fuel and exhaust systems

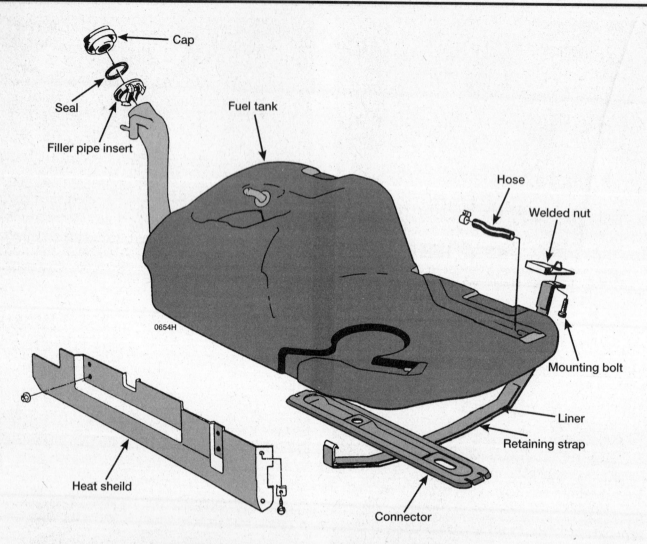

6.6 Exploded view of the fuel tank assembly on later 5-Series models

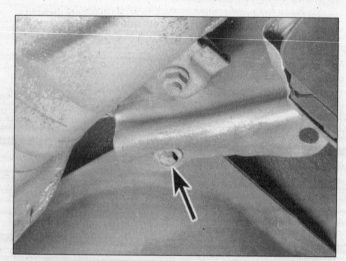

6.10a Remove the bolts from the side of the fuel tank

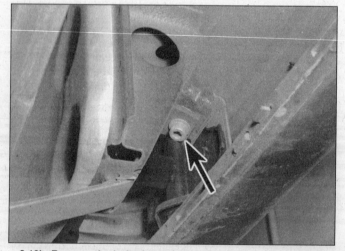

6.10b Remove the bolts (arrow) from the straps and lower the fuel tank

the radio in your vehicle is equipped with an anti-theft system, make sure you have the correct activation code before disconnecting the battery. Refer to the information on page 0-7 at the front of this manual before detaching the cable. **Note**: If, after connecting the battery, the wrong language appears on the instrument panel display, refer to page 0-7 for the language resetting procedure.

2 Detach the air intake duct from the front side of the air cleaner.
3 Detach the duct between the air cleaner and the throttle body.
4 Remove the air filter (see Chapter 1).
5 Unplug the electrical connector from the airflow meter (see Section 12).

Chapter 4 Fuel and exhaust systems

8.6 Remove the two nuts (arrows) from the air cleaner assembly (Motronic 1.1 system shown) and lift it off its mounts

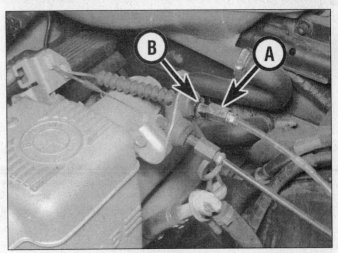

9.5 To adjust the accelerator cable freeplay, hold nut B stationary and turn nut A

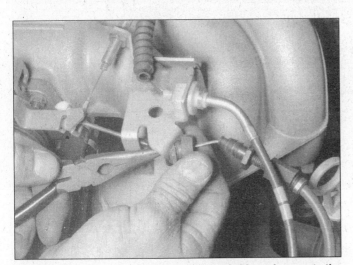

9.8 Push the rubber grommet from the backside and separate the cable from the bracket

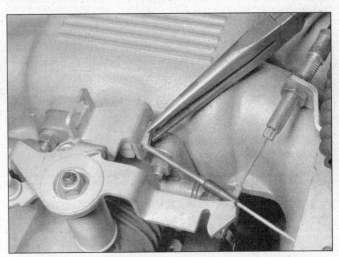

9.9 Pinch the plastic retainer and push it through the bracket recess on the throttle valve

6 Remove the air cleaner mounting bolts **(see illustration)** and lift the air cleaner assembly from the engine compartment.
7 Installation is the reverse of removal.

9 Accelerator cable - check, adjustment and replacement

Check

1 Separate the air intake duct from the throttle body.
2 Have an assistant depress the accelerator pedal to the floor while you watch the throttle valve. It should move to the fully open (horizontal) position.
3 Release the accelerator pedal and make sure the throttle valve returns smoothly to the fully closed position. The throttle valve should not contact the body at any time during its movement or else the unit must be replaced.

Adjustment

Refer to illustration 9.5

4 Warm the engine to normal operating temperature and turn it off. Depress the accelerator pedal to the floor twice, then check the cable freeplay at the throttle body. Compare it to the value listed in this Chapter's Specifications.

5 If the freeplay isn't within specifications, adjust it by turning nut A **(see illustration)**.
6 Have an assistant help you verify the throttle valve is fully open when the accelerator pedal is depressed to the floor.

Replacement

Refer to illustration 9.8, 9.9, 9.10a and 9.10b

Note: *You'll need a flashlight for the under-dash portion of the following procedure.*

7 Detach the cable from the negative battery terminal. **Caution:** *If the radio in your vehicle is equipped with an anti-theft system, make sure you have the correct activation code before disconnecting the battery. Refer to the information on page 0-7 at the front of this manual before detaching the cable.* **Note**: *If, after connecting the battery, the wrong language appears on the instrument panel display, refer to page 0-7 for the language resetting procedure.*
8 Loosen the cable adjuster locknuts and detach the cable from its support bracket located on the intake manifold **(see illustration)**.
9 Pinch the plastic retainer with a pair of needle-nose pliers and push it out of the bracket **(see illustration)**.
10 Pull the cable down through the slot and away from the bracket **(see illustrations)**.
11 Working from underneath the driver's side of the dash, reach up and detach the throttle cable from the top of the accelerator pedal.

Chapter 4 Fuel and exhaust systems

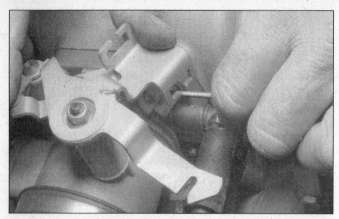

9.10a Rotate the throttle valve and remove the cable end from the slotted portion of the valve

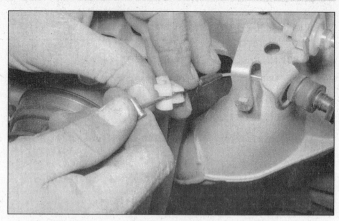

9.10b After the cable is off the throttle valve assembly, remove the plastic retainer from the cable

12 Pull the cable through the firewall, from the engine compartment side.
13 Installation is the reverse of removal. Be sure to adjust the cable as described in Steps 4 through 6.

10 Fuel injection system - general information

All models covered by this manual are equipped with an Electronic Fuel Injection (EFI) system. The EFI system is composed of three basic sub systems: fuel system, air intake system and electronic control system.

Fuel system

An electric fuel pump located inside the fuel tank (Motronic 1.1 systems) or beside the fuel tank (L-Jetronic or Motronic systems) supplies fuel under constant pressure to the fuel rail, which distributes fuel evenly to all injectors. From the fuel rail, fuel is injected into the intake ports, just above the intake valves, by the fuel injectors. The amount of fuel supplied by the injectors is precisely controlled by an Electronic Control Unit (ECU). An additional injector, known as the cold start injector (L-Jetronic and Motronic systems only), supplies extra fuel into the intake manifold for starting. A pressure regulator controls system pressure in relation to intake manifold vacuum. A fuel filter between the fuel pump and the fuel rail filters fuel to protect the components of the system.

Air intake system

The air intake system consists of an air filter housing, an airflow meter, a throttle body, the intake manifold and the associated ducting. The airflow meter is an information gathering device for the ECU. These models are equipped with the vane-type airflow meter. A potentiometer measures intake airflow and a temperature sensor measures intake air temperature. This information helps the ECU determine the amount (duration) of fuel to be injected by the injectors. The throttle plate inside the throttle body is controlled by the driver. As the throttle plate opens, the amount of air that can pass through the system increases, so the potentiometer opens further and the ECU signals the injectors to increase the amount of fuel delivered to the intake ports.

Electronic control system

The computer control system controls the EFI and other systems by means of an Electronic Control Unit (ECU). The ECU receives signals from a number of information sensors which monitor such variables as intake air volume, intake air temperature, coolant temperature, engine rpm, acceleration/deceleration and exhaust oxygen content. These signals help the ECU determine the injection duration necessary for the optimum air/fuel ratio. These sensors and their corresponding ECU-controlled output actuators are located throughout the engine compartment. For further information regarding the ECU and its relationship to the engine electrical systems and ignition system, refer to Chapters 5 and 6.

These models are equipped with either an L-Jetronic EFI system (1984 and 1985 318i models) or a Motronic EFI system (all six-cylinder models). Later models are equipped with the updated version, Motronic 1.1.

L-Jetronic fuel-injection system

The Bosch L-Jetronic fuel injection system is used only on 1984 and 1985 318i models. It is an electronically controlled fuel injection system that utilizes one solenoid operated fuel injector per cylinder. The system is governed by an Electronic Control Unit (ECU) which processes information sent by various sensors, and in turn precisely meters the fuel to the cylinders by adjusting the amount of time that the injectors are open.

An electric fuel pump delivers fuel under high pressure to the injectors through the fuel feed line and an in-line filter. A pressure regulator keeps fuel available at an optimum pressure, allowing pressure to rise or fall depending on engine speed and load. Fuel in excess of injector needs is returned to the fuel tank by a separate line.

A sensor in the air intake duct constantly measures the mass of the incoming air, and through the ECU adjusts the fuel mixture to provide an optimum air/fuel ratio.

Another device, called the oxygen sensor, is mounted in the exhaust system and continually reads the oxygen content of the exhaust gas. This information is also used by the ECU to adjust the duration of injection, making it possible to meter the fuel very accurately to comply with strict emission control standards.

Other components incorporated in the system are the throttle valve (which controls airflow to the engine), the coolant temperature sensor, the throttle position switch, idle stabilizer valve (which bypasses air around the throttle plate to control idle speed) and associated relays and fuses.

Motronic fuel injection system

The Motronic engine management system is used on vehicles equipped with the six-cylinder engine. The Motronic system combines the fuel control of the L-Jetronic fuel injection system with the control of ignition timing, idle speed and emissions into one control unit.

The fuel injection and idle speed control functions of Motronic are similar to those used on the L-Jetronic system described above. Where the Motronic system differs from L-Jetronic is its "adaptive circuitry" in the oxygen sensor system. Adaptive circuitry enables the oxygen sensor system to adjust the operating range of fuel metering in accordance with subtle changes in operating conditions caused by such things as normal engine wear, vacuum leaks, changes in altitude, etc. It also differs in the type and function of the sensors. Motronic systems incorporate speed and reference sensors located in the bellhousing (Motronic) or on the front pulley (Motronic 1.1) to relay information to the ECU for the purposes of controlling the ignition timing. As a result, the timing is not adjustable and is controlled by the computer. For more information on the Motronic system, see Chapter 6.

Chapter 4 Fuel and exhaust systems

4-13

11 Fuel injection system - troubleshooting

L-Jetronic fuel injection system

Symptom	Probable cause	Corrective action
Engine starts hard or fails to start when cold	Cold start valve or thermo-time switch faulty	Test cold start valve and thermo-time switch. Replace faulty parts (see Section 16)
	Fuel pump inoperative	Check fuel pump fuse and fuel pump relay (see Sections 3 and 4)
	Airflow meter flap (door) binds or stuck in open position	Inspect the airflow meter for damage (see Section 13)
	Fuel pressure incorrect	Test system pressure (see Section 3). Test fuel pressure regulator (Section 15)
	Large intake leaks	Inspect all vacuum lines, air ducts and oil filler and dipstick seals
	Fuel injectors clogged or not operating	Check fuel injectors (see Section 17) and wiring harness
	Coolant temperature sensor faulty or wiring problem	Test coolant temperature sensor (see Chapter 6, section 4)
	TPS incorrectly adjusted	Check the adjustment on the TPS (see Chapter 6, Section 4)
	Incorrect ignition timing	Check ignition timing (see Chapter 5). Check vacuum advance system
	Bad fuel	Check the fuel and drain the tank if necessary
	Faulty ECU	Have the ECU tested at a dealer service department or other repair shop
Engine starts hard or fails to start when warm	Cold start valve leaking or operating continuously	Test cold start valve and thermo-time switch (see Section 16)
	Fuel pressure incorrect	Test fuel pump(s). Replace if necessary (see Section 3)
	Insufficient residual fuel pressure	Test residual fuel pressure. Replace fuel pump or fuel accumulator as necessary (see Section 16)
	Fuel leak(s)	Inspect fuel lines and fuel injectors for leaks. Correct leaks as required (see Chapter 4)
	Coolant temperature sensor faulty or wiring problem	Test coolant temperature sensor (see Chapter 6, Section 4)
	Vapor lock (warm weather)	Check fuel pressure (see Section 3)
	EVAP system faulty	Check EVAP system (see Chapter 6, Section 6)
	Incorrect ignition timing	Check ignition timing (see Chapter 5). Check vacuum advance system
	Faulty ECU	Have the ECU tested at a dealer service department or other repair shop
	Idle speed control system faulty	Test the idle air stabilizer valve (see Section 18)
Engine misses and hesitates under load	Fuel injector clogged or faulty	Test fuel injectors. Check for clogged injector lines. Replace faulty injectors (see Section 17)
	Fuel pressure incorrect	Test fuel system pressure (see Section 3). Test fuel pressure regulator (see Section 15)
	Fuel leak(s)	Inspect fuel lines and fuel injectors for leaks (see Chapter 4)
	Engine maintenance	Tune-up engine (see Chapter 1). Check the distributor cap, rotor, ignition wires and spark plugs and replace any faulty components

L-Jetronic fuel injection system (continued)

Symptom	Probable cause	Corrective action
Engine misses and hesitates under load (continued)	Airflow meter flap (door) binds or stuck in open position	Inspect the airflow meter for damage (see Section 13)
	Large intake leaks	Inspect all vacuum lines, air ducts and oil filler and dipstick seals
Engine has erratic idle speed	Idle air stabilizer valve faulty	Check the idle air stabilizer valve (see Section 18)
	No power to the valve	Check the idle air stabilizer relay and wiring circuit (see Chapter 12)
	Vacuum advance system faulty	Check vacuum advance system and electronic vacuum advance relay
	Idle speed control unit faulty	Have the idle speed control unit checked by a dealership service department

Motronic and Motronic 1.1 fuel injection systems

Note: *On Motronic 1.1 systems, begin troubleshooting by checking for stored trouble codes (see Chapter 6).*

Symptom	Probable cause	Corrective action
Engine starts hard or fails to start when cold	Cold start valve or thermo-time switch faulty (Motronic system only)	Test cold start valve and thermo-time switch. Replace faulty parts (see Section 16)
	Fuel pump not running	Check fuel pump fuse and fuel pump relay (see Section 2 and 3)
	Airflow meter flap (door) binds or stuck in open position	Inspect the airflow meter for damage (see Section 13)
	Fuel pressure incorrect	Test system pressure (see Section 3)
	Large intake leaks	Inspect all vacuum lines, air ducts and oil filler and dipstick seals
	Fuel injectors clogged or not operating	Check fuel injectors (see Section 17) and wiring harness
	Coolant temperature sensor faulty or wiring problem	Test coolant temperature sensor (see Chapter 6, Section 4)
	TPS incorrectly adjusted	Check the adjustment of the TPS (see Chapter 6, Section 4)
	Bad fuel	Check the fuel and drain the tank if necessary
	Faulty ECU	Have the ECU tested at a dealer service department or other repair shop
	Reference signal missing	Faulty reference sensor, flywheel or reference pin missing (see Chapter 5)
Engine starts hard or fails to start when warm	Cold start valve leaking or operating continuously (Motronic system only)	Test cold start valve and thermo-time switch (see Sections 16)
	Fuel pressure incorrect	Test fuel pressure (see Section 3)
	Insufficient residual fuel pressure	Test fuel system hold pressure (see Section 3)
	Fuel leak(s)	Inspect fuel lines and fuel injectors for leaks. Correct leaks as required
	Coolant temperature sensor faulty or wiring problem	Test coolant temperature sensor (see Chapter 6, Section 4)
	Vapor lock (in warm weather)	Check fuel pressure (see Section 3)
	EVAP system faulty	Check EVAP system (see Chapter 6, Section 6)

Chapter 4 Fuel and exhaust systems

Symptom	Probable cause	Corrective action
Engine starts hard or fails to start when warm	Faulty ECU	Have the ECU tested at a dealer service department or other repair shop
	Idle speed control system faulty	Test the idle air stabilizer valve (see Section 18)
	Oxygen sensor system faulty	Check the oxygen sensor (see Chapter 6, Section 4)
Engine misses and hesitates under load	Fuel injector clogged	Test fuel injectors. Check for clogged injector lines. Replace faulty injectors (see Section 17)
	Fuel pressure incorrect	Test fuel system pressure (see Section 3). Test fuel pressure regulator (see Section 15)
	Fuel leak(s)	Inspect fuel lines and fuel injectors for leaks (see Chapter 4)
	Engine maintenance	Tune-up engine (see Chapter 1). Check the cap, rotor, ignition wires and spark plugs and replace any faulty components
	Airflow meter flap (door) binds or stuck in open position	Inspect the airflow meter for damage (see Section 13)
	Large intake leaks	Inspect all vacuum lines, air ducts and oil filler and dipstick seals
	TPS out of adjustment	Adjust the TPS (see Chapter 6)
Engine idles too fast	Accelerator pedal, cable or throttle valve binding	Inspect for worn or broken parts, kinked cable or other damage. Replace faulty parts
	Air leaking past throttle valve	Inspect throttle valve and adjust or replace as required
Engine has erratic idle speed	Idle air stabilizer valve faulty	Check the idle air stabilizer valve (see Section 18)
	No power to the valve	Check the idle air stabilizer relay and wiring circuit (see Chapter 12)
	Idle speed control unit faulty	Have the idle speed control unit checked by a dealership service department
Low power	Coolant temperature sensor faulty or wire to sensor broken	Test coolant temperature sensor and wiring. Repair wiring or replace sensor if faulty (see Chapter 6)
	Fuel pressure incorrect	Check fuel pressure on the main pump and transfer pump (if equipped) (see Section 2 and 3)
	Throttle plate not opening fully	Check accelerator cable adjustment to make sure throttle is opening fully. Adjust cable if necessary (see Section 9)
Poor fuel mileage	Cold start valve leaking (Motronic system only)	Test and, if necessary, replace cold start injector (see Section 16)
	Oxygen sensor	Test the oxygen sensor (see Chapter 6, Section 4))
	Sticking parking brake	Check the parking brake system (see Chapter 9)

12 Fuel injection system - check

Refer to illustration 12.7

Warning: *Gasoline is extremely flammable, so take extra precautions when you work on any part of the fuel system. Don't smoke or allow open flames or bare light bulbs near the work area, and don't work in a garage where a natural gas-type appliance (such as a water heater or clothes dryer) with a pilot light is present. If you spill any fuel on your skin, rinse it off immediately with soap and water. When you perform any kind of work on the fuel system, wear safety glasses and have a Class B type fire extinguisher on hand.*

1   Check the ground wire connections. Check all wiring harness connectors that are related to the system. Loose connectors and poor grounds can cause many problems that resemble more serious malfunctions.
2   Make sure the battery is fully charged, as the control unit and sensors depend on an accurate supply voltage in order to properly meter the fuel.
3   Check the air filter element - a dirty or partially blocked filter will severely impede performance and economy (see Chapter 1).
4   If a blown fuse is found, replace it and see if it blows again. If it does, search for a grounded wire in the harness related to the system.
5   Check the air intake duct from the airflow meter to the intake

Chapter 4 Fuel and exhaust systems

12.7 Use a stethoscope or screwdriver to determine if the injectors are working properly - they should make a steady clicking sound that rises and falls with engine speed changes

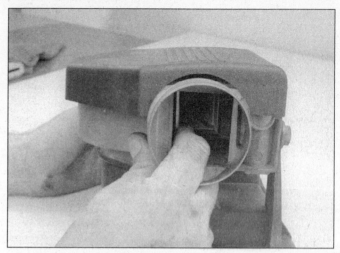

13.1 Check for binding of the flap in the airflow meter as it nears closing position or full throttle position. Any hesitation or binding will cause erratic idle conditions, rich fuel mixture or poor acceleration and throttle response (airflow meter removed for clarity)

13.3 Position the ohmmeter probes onto terminals #7 and #8 of the airflow meter and check for a smooth change in resistance as the vane door of the airflow meter is slowly opened and closed

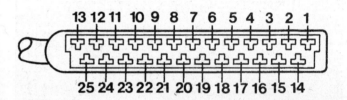

13.5 Locate the ECU under the glove box and unplug the electrical connector. Probe terminals #7 and #8 and perform the same resistance checks as in Step 3. The test results should be the same.

manifold for leaks, which will result in an excessively lean mixture. Also check the condition of the vacuum hoses connected to the intake manifold.

6 Remove the air intake duct from the throttle body and check for dirt, carbon and other residue build-up. If it's dirty, clean it with carburetor cleaner and a toothbrush.

7 With the engine running, place a screwdriver or a stethoscope against each injector, one at a time, and listen for a clicking sound, indicating operation **(see illustration)**.

8 Check the fuel system pressure (see Section 3).

9 Check for trouble codes stored in the ECU (see Chapter 6).

13 Airflow meter - check, removal and installation

Check

L-Jetronic systems

Refer to illustrations 13.1, 13.3 and 13.5

1 Remove the duct from the intake end of the airflow meter. Carefully open and close the sensor flap and check for binding. The flap can bend during a backfire and cause incorrect resistance readings. The flap will bind and stick in a partially open position, causing the engine to run rich and stall when it returns to idle.

2 Disconnect the electrical connector from the airflow meter.

3 Using an ohmmeter, check the resistance between terminals #7 and #8 **(see illustration)**. The resistance should increase steadily (without any flat-spots) as the sensor flap is slowly moved from the fully CLOSED position to the fully OPEN position.

4 Also, check the intake air temperature sensor (inside airflow meter). Using an ohmmeter, probe terminals #8 and #9 **(see illustration 13.3)** and check for the proper resistance. The resistance should be 2,200 to 2,700 ohms at 68-degrees F.

5 If the resistance readings are correct, check the wiring harness (see Chapter 12). Plug in the connector to the airflow meter. Disconnect the electrical connector from the ECU (located under the glove box) and probe terminals #7 and #8 **(see illustration)** with an ohmmeter. Carefully move the door of the airflow meter and observe the change in resistance as it move from closed to fully open. The test results should be the same as Step 3. If there are any differences in the test results, there is a possible shorted or open wire in the harness.

Motronic and Motronic 1.1 systems

Refer to illustrations 13.6, 13.7a and 13.7b

6 Remove the access cover for the ECU (computer) (see Chapter 6) and disconnect the harness connector **(see illustration)**.

7 Using an ohmmeter, probe the designated terminals of the ECU electrical connector **(see illustrations)** and check for the proper change in resistance. On Motronic systems, probe terminals #7 and #9. The resistance should increase steadily (without any flat spots) as the sensor flap is slowly moved from the fully CLOSED position to the fully OPEN position. On Motronic 1.1 systems, probe terminals #7 and #12. The resistance should increase steadily (without any flat-spots) as the sensor flap is slowly moved from the fully CLOSED position to the

Chapter 4 Fuel and exhaust systems

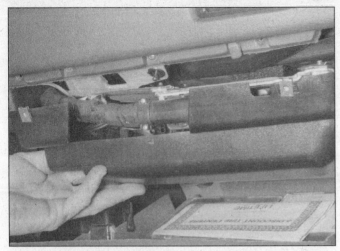

13.6 Remove the under-dash panel near the glove box to gain access to the ECU on Motronic systems

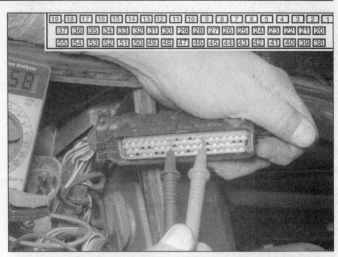

13.7a Position the ohmmeter probes on terminals #7 and #12 (Motronic 1.1 systems) of the ECU connector and check for a smooth increase and decrease in resistance as the door on the airflow meter is slowly opened and closed

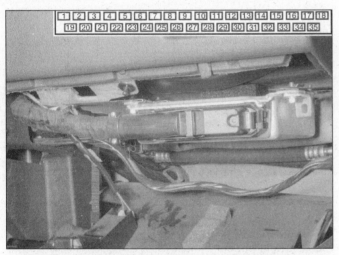

13.7b Unplug the connector, position the ohmmeter probes on terminals #7 and #9 (Motronic systems) and check for a smooth increase and decrease in resistance as the door on the airflow meter is slowly opened and closed

13.11a Push the tab and remove the air horn from inside the air cleaner assembly

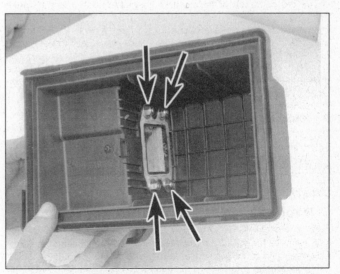

13.11b Remove the nuts (arrows) from the air cleaner housing and detach the airflow meter

fully OPEN position. **Note**: *Motronic systems are distinguished from the others by the 35-pin electrical connector for the ECU, while Motronic 1.1 systems use a 55-pin electrical connector.*

8 If the resistance readings are incorrect, check the wiring harness.

Removal and installation (all engines)

Refer to illustrations 13.11a and 13.11b

9 Disconnect the electrical connector from the airflow meter.
10 Remove the air cleaner assembly (see Section 8).
11 Remove the nuts **(see illustrations)** and lift the airflow meter from the engine compartment or from the air cleaner assembly.
12 Installation is the reverse of removal.

14 Throttle body - check, removal and installation

Check

1 Detach the air intake duct from the throttle body (see Section 8) and move the duct out of the way.
2 Have an assistant depress the throttle pedal while you watch the throttle valve. Check that the throttle valve moves smoothly when the

14.11 Remove the nuts (arrows) and lift the throttle body from the intake manifold (the lower two nuts are hidden from view)

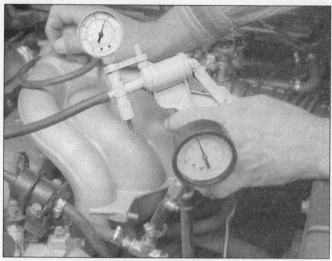

15.5 Carefully watch the fuel pressure gauge as vacuum is applied (fuel pressure should decrease as vacuum increases)

throttle is moved from closed (idle position) to fully open (wide open throttle).

3   If the throttle valve is not working properly, replace the throttle body unit.

Removal and installation

Refer to illustration 14.11

Warning: *Wait until the engine is completely cool before beginning this procedure.*

4   Detach the cable from the negative battery terminal. **Caution**: *If the radio in your vehicle is equipped with an anti-theft system, make sure you have the correct activation code before disconnecting the battery, Refer to the information on page 0-7 at the front of this manual before detaching the cable.* **Note**: *If, after connecting the battery, the wrong language appears on the instrument panel display, refer to page 0-7 for the language resetting procedure.*

5   Detach the air intake duct from the throttle body and set it aside.

6   Detach the accelerator cable from the throttle body (see Section 9).

7   Detach the cruise control cable, if equipped.

8   Clearly label all electrical connectors (TPS, cold start injector, idle air stabilizer, etc.), then unplug them.

9   Clearly label all vacuum hoses, then detach them.

10   Unscrew the radiator or expansion tank cap to relieve any residual pressure in the cooling system, then reinstall it. Clamp shut the coolant hoses, then loosen the hose clamps and detach the hoses. Be prepared for some coolant leakage.

11   Remove the throttle body mounting nuts (upper) and bolts (lower) and detach the throttle body from the air intake plenum **(see illustration)**.

12   Cover the air intake plenum opening with a clean cloth to prevent dust or dirt from entering while the throttle body is removed.

13   Installation is the reverse of removal. Be sure to tighten the throttle body mounting nuts to the torque listed in this Chapter's Specifications and adjust the throttle cable (see Section 9) when you're done.

15 Fuel pressure regulator - check and replacement

Warning: *Gasoline is extremely flammable, so take extra precautions when you work on any part of the fuel system. Don't smoke or allow open flames or bare light bulbs near the work area, and don't work in a garage where a natural gas-type appliance (such as a water heater or clothes dryer) with a pilot light is present. If you spill any fuel on your skin, rinse it off immediately with soap and water. When you perform any kind of work on the fuel system, wear safety glasses and have a Class B type fire extinguisher on hand.*

Check

Refer to illustration 15.5

1   Relieve the fuel system pressure (see Section 2).

2   Detach the cable from the negative battery terminal. **Caution**: *If the radio in your vehicle is equipped with an anti-theft system, make sure you have the correct activation code before disconnecting the battery, Refer to the information on page 0-7 at the front of this manual before detaching the cable.* **Note**: *If, after connecting the battery, the wrong language appears on the instrument panel display, refer to page 0-7 for the language resetting procedure.*

3   Disconnect the fuel line and install a fuel pressure gauge (see Section 3). Reattach the cable to the battery.

4   Pressurize the fuel system and check for leakage around the gauge connections.

5   Connect a vacuum pump to the fuel pressure regulator **(see illustration)**.

6   Read the fuel pressure gauge with vacuum applied to the pressure regulator and also with no vacuum applied. The fuel pressure should DECREASE as vacuum INCREASES. Compare your readings with the values listed in this Chapter's Specifications.

7   Reconnect the vacuum hose to the regulator and check the fuel pressure at idle, comparing your reading with the value listed in this Chapter's Specifications. Disconnect the vacuum hose and watch the gauge - the pressure should jump up to the maximum specified pressure as soon as the hose is disconnected.

8   If the fuel pressure is low, pinch the fuel return line shut and watch the gauge. If the pressure doesn't rise, the fuel pump is defective or there is a restriction in the fuel feed line. If the pressure rises sharply, replace the pressure regulator.

9   If the indicated fuel pressure is too high, disconnect the fuel return line and blow through it to check for a blockage. If there is no blockage, replace the fuel pressure regulator.

10   If the pressure doesn't fluctuate as described in Step 7, connect a vacuum gauge to the pressure regulator vacuum hose and check for vacuum.

11   If there is vacuum present, replace the fuel pressure regulator.

12   If there isn't any reading on the gauge, check the hose and its port for a leak or a restriction.

Replacement

Refer to illustration 15.15

13   Relieve the system fuel pressure (see Section 2).

14   Detach the cable from the negative battery terminal. **Caution**: *If the radio in your vehicle is equipped with an anti-theft system, make sure you have the correct activation code before disconnecting the battery, Refer to the information on page 0-7 at the front of this manual*

Chapter 4 Fuel and exhaust systems

15.15 Remove the two bolts (arrows) and remove the fuel pressure regulator from the fuel rail (535i shown, others similar)

16.1 Location of the cold start injector electrical connector on a 318i model. Most cold start injectors are mounted in the intake manifold

16.2 Watch for a steady, conical-shaped spray of fuel when the ignition key is switched ON

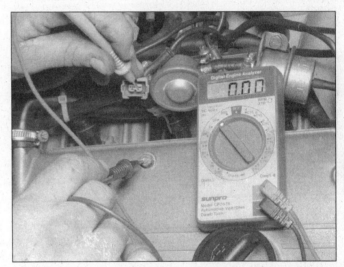

16.3 Check for a voltage signal (about 12-volts) at the cold start injector connector when the ignition key is switched ON

before detaching the cable. **Note**: *If, after connecting the battery, the wrong language appears on the instrument panel display, refer to page 0-7 for the language resetting procedure.*

15 Detach the vacuum hose and fuel return hose from the pressure regulator, then unscrew the mounting bolts **(see illustration)**.
16 Remove the pressure regulator.
17 Installation is the reverse of removal. Be sure to use a new O-ring. Coat the O-ring with a light film of engine oil prior to installation.
18 Check for fuel leaks after installing the pressure regulator.

16 Cold start injector and thermo-time switch - check and replacement

Warning: *Gasoline is extremely flammable, so take extra precautions when you work on any part of the fuel system. Don't smoke or allow open flames or bare light bulbs near the work area, and don't work in a garage where a natural gas-type appliance (such as a water heater or clothes dryer) with a pilot light is present. If you spill any fuel on your skin, rinse it off immediately with soap and water. When you perform any kind of work on the fuel system, wear safety glasses and have a Class B type fire extinguisher on hand.*

Check

Cold start injector

Refer to illustrations 16.1, 16.2 and 16.3

1 Make sure engine coolant is below 86-degrees F. Preferably, the engine should sit for several hours before performing this check. Disconnect the electrical connector from the cold start valve **(see illustration)** and move it aside, away from the work area - there will be fuel vapor present. Remove the two screws holding the valve to the air intake plenum and take the valve out. The fuel line must be left connected to the valve. Wipe the nozzle of the valve. Disable the ignition system (see Chapter 2 Part B, Section 3). Turn the ignition On and operate the fuel pump for one minute **(see illustrations 3.1a through 3.1d)**. There must be no fuel dripping from the nozzle. If there is, the valve is faulty and must be replaced. Switch off the ignition.
2 Now direct the nozzle of the valve into a can or jar. Reconnect the electrical connector to the valve. Have an assistant turn the ignition On and operate the starter. The valve should squirt a conical shaped spray into the jar **(see illustration)**. If the spray pattern is good the valve is working properly. If the spray pattern is irregular, the valve is damaged and should be replaced.
3 If the cold start injector does not spray any fuel, check for a voltage signal at the electrical connector for the cold start valve **(see illustration)**. If there is no voltage, check the thermo-time switch.

16.4 Check for a voltage signal on the black/yellow wire of the thermo-time switch when the ignition is ON

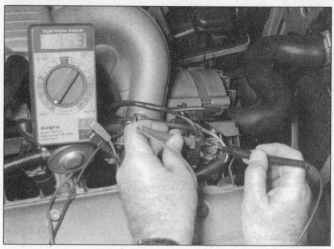

16.6 Check the resistance of the thermo-time switch with the engine coolant temperature below 86 degrees F. There should be continuity.

Thermo-time switch

Refer to illustrations 16.4 and 16.6

4 The thermo-time switch detects the temperature of the engine and controls the action of the cold start injector. It is usually located up front, near the coolant temperature sensor. Make sure engine coolant is below 86-degrees F. Preferably the engine should sit for several hours. Disable the ignition system (see Chapter 2A). Pull back the rubber boot from the thermo-time switch **(see illustration)** and probe the terminal (black/yellow wire) of the connector with a voltmeter.

5 Have an assistant turn the ignition to START and observe that the voltmeter registers a voltage signal the moment the key is turned ON. This signal should last approximately 6 to 10 seconds depending on the temperature of the engine.

6 If the voltage is correct, check the thermo-time switch. Unplug the electrical connector and, using an ohmmeter, check for continuity between the terminals of the thermo-time switch **(see illustration)**. Continuity should exist.

7 Next, warm up the engine above 105-degrees F and confirm that there is no continuity between the terminals. If there is, the switch is faulty and must be replaced. **Note:** *On 5-Series models, there are several versions of the thermo-time switch. Each one is stamped with an opening temperature and maximum duration. On 1982 528e and 1983 533i models, excessively long cranking times may develop at warm temperatures. If this problem occurs, replace the thermo-time switch with a warmer version.*

Replacement

Cold start injector

8 Relieve the fuel pressure (see Section 2).
9 Disconnect the electrical connector from the cold start injector.
10 On models so equipped, use a box-end or socket wrench and remove the fuel line fitting connected to the cold start valve. On other models, simply loosen the hose clamp and detach the hose from the valve.
11 Remove the Allen bolts that retain the cold start valve to the air intake plenum and remove the valve.
12 Installation is the reverse of removal, but be sure to clean the mating surfaces and use a new gasket.

Thermo-time switch

Warning: *Wait until the engine is completely cool before beginning this procedure. Also, remove the expansion tank or radiator cap to relieve any residual pressure in the cooling system.*

13 Prepare the new thermo-time switch for installation by wrapping the threads with Teflon tape or applying a light coat of thread sealant to them.
14 Disconnect the electrical connector from the thermo-time switch.

15 Use a deep socket and unscrew the switch. Once the switch is removed coolant will start to leak out, so install the new switch as quickly as possible. Tighten the switch securely and plug in the electrical connector.

17 Fuel injectors - check and replacement

Warning: *Gasoline is extremely flammable, so take extra precautions when you work on any part of the fuel system. Don't smoke or allow open flames or bare light bulbs near the work area, and don't work in a garage where a natural gas-type appliance (such as a water heater or clothes dryer) with a pilot light is present. If you spill any fuel on your skin, rinse it off immediately with soap and water. When you perform any kind of work on the fuel system, wear safety glasses and have a Class B type fire extinguisher on hand.*

Check

In-vehicle check

Refer to illustration 17.5

1 Using a mechanic's stethoscope (available at most auto parts stores), check for a clicking sound at each of the injectors while the engine is idling **(see illustration 12.7)**. The injectors should make a steady clicking sound if they are operating properly.
2 Increase the engine speed above 3500 RPM. The clicking sound should rise with engine speed.
3 If you don't have a stethoscope, a screwdriver can be used. Place the tip of the screwdriver against the injector and press your ear against the handle.
4 If an injector isn't functioning (not clicking), purchase a special injector test light (sometimes called a "noid" light) and install it into the injector electrical connector. Start the engine and make sure the light flashes. If it does, the injector is receiving the proper voltage, so the injector itself must be faulty.
5 Unplug each injector connector and check the resistance of the injector **(see illustration)**. Check your readings with the values listed in this Chapter's Specifications. Replace any that do not have the correct amount of resistance.

Volume test

6 Because a special injection checker is required to test injector volume, this procedure is beyond the scope of the home mechanic. Have the injector volume test performed by a dealer service department or other repair shop.

Replacement

Refer to illustration 17.8

7 Unplug the main electrical connector for the fuel injector wiring

Chapter 4 Fuel and exhaust systems

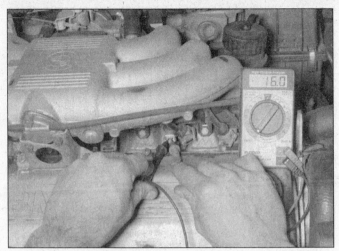

17.5 Check the resistance of each of the fuel injectors and compare it to the Specifications

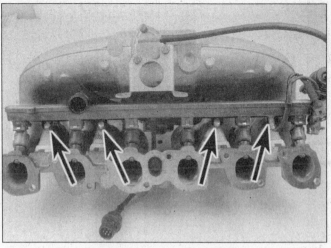

17.8 Remove the bolts (arrows) and separate the fuel rail and injectors from the intake manifold

harness. Remove the intake manifold (see Chapter 2A).
8   Detach the fuel hoses from the fuel rail and remove the fuel rail mounting bolts **(see illustration)**.
9   Lift the fuel rail/injector assembly from the intake manifold.
10   Unplug the electrical connectors from the fuel injectors. Detach the injectors from the fuel rail.
11   Installation is the reverse of removal. Be sure to replace all O-rings with new ones. Coat them with a light film of engine oil to prevent damage to the O-rings during installation. Be sure to pressurize the fuel system and check for leaks before starting the engine.

18   Idle air stabilizer valve - check, adjustment and replacement

Refer to illustrations 18.9, 18.19, 18.21 and 18.26
1   The idle air stabilizer system works to maintain engine idle speed within a 200 rpm range regardless of varying engine loads at idle. An electrically operated valve allows a small amount of air to flow past the throttle plate to raise the idle speed whenever the idle speed drops below approximately 750 rpm. If the idle speed rises above approximately 950 rpm, the idle air stabilizer valve closes and stops extra air from bypassing the throttle plate, reducing the idle speed.
2   L-Jetronic (1984 and 1985 318i only) systems are equipped with a separate idle speed control unit (computer) located under the glove box. The idle air stabilizer valve is equipped with an adjusting screw. Early models are equipped with plastic valves but they still can be adjusted by removing the hose and inserting a very thin screwdriver inside.
3   Motronic systems are also equipped with a separate idle speed control unit (computer) located under the glove box. The idle air stabilizer valve is equipped with an adjusting screw.
4   Motronic 1.1 systems control the idle air stabilizer valve with the ECU. These systems do not have any provisions for adjustment.

Preliminary check
5   Before performing any checks on the idle air stabilizer valve, make sure these criteria are met:
   a) The engine has reached operating temperature (140-degrees F)
   b) Turn off all electrical accessories air conditioning, heater controls, headlights, auxiliary cooling fan, etc.)
   c) TPS sensor must be operating correctly (see Chapter 6)
   d) There must not be any exhaust leaks
   e) There must not be any vacuum leaks
   f) The oxygen sensor must be operating properly (see Chapter 6)
6   Connect a tachometer in accordance with the tool manufacturer's instructions. **Caution:** *The ignition must be off before making any electrical connections.*
7   The idle air stabilizer valve operates continuously when the igni-

18.9 Check the resistance of the idle air stabilizer valve - it should be 9 to 10 ohms (L-Jetronic systems)

tion is on. Start the engine and make sure the valve is vibrating and humming slightly.

L-Jetronic system
Check
8   With the engine running, disconnect the electrical connector from the valve. The idle speed should increase to about 2,000 rpm.
9   If the idle speed does not increase, turn the engine off. Using an ohmmeter, check the resistance across the terminals of the valve **(see illustration)**. There should be 9 to 10 ohms with the ambient air temperature at about 73-degrees F.
10   Using a pair of jumper wires, apply battery voltage to the valve and confirm that the valve closes tightly. When the voltage is removed, the valve should reopen.
11   If the idle air stabilizer valve fails any of the tests, replace it with a new one.
12   If the idle air stabilizer valve passes the tests, check the control current.
13   Unplug the electrical connector from the valve. Using a jumper wire, connect one terminal of the electrical connector to one of the terminals on the valve, Connect an ammeter (0 to 1,000 mA range) between the other terminal on the electrical connector and the remaining terminal on the valve. Start the engine and allow it to idle. With the engine running, the current reading should be between 400 and 500 mA.

18.19 Location of the adjustment screw on the idle air stabilizer valve (L-Jetronic system)

18.21 Check the idle air stabilizer valve resistance on the two outer terminals on Motronic 1.1 systems - it should be about 40 ohms

Adjust the valve if the current reading is not as specified (see Step 15). **Note**: *The idle air stabilizer current will fluctuate between 400 and 1,100 mA if the engine is too cold, if the coolant temperature sensor is faulty, if the idle speed needs to be adjusted, if there is an engine vacuum leak or if electrical accessories are on.*

14 If there is no current reading, have the idle speed control unit diagnosed by a dealer service department or other repair shop. **Note**: *The idle air stabilizer control unit (located under the glove box) can develop an electrical connector problem that intermittently turns the valve ON and OFF. Check the connector very carefully before replacing any parts. Sometimes a new control unit will fix the problem only temporarily.*

Adjustment

15 With the engine off, connect a tachometer in accordance with the tool manufacturer's instructions.
16 Make sure the ignition timing is correct (see Chapter 1).
17 Connect an ammeter to the valve (see Step 13).
18 With the engine running, the current reading should be 450 to 470 mA at 850 to 900 rpm (manual transmission, or 460 to 480 mA at 850 to 900 rpm (automatic transmission).
19 If the control current is not correct, turn the adjusting screw until it is within the specified range **(see illustration)**. **Note**: *On metal type valves, the adjusting screw is external. On plastic type valves, the adjustment screw is internal and can be reached by removing the hose at the end of the valve.*

Motronic and Motronic 1.1 systems

Check

Note: *There are two types of idle air stabilizer valves on these systems; early models (before 3/1987) are usually equipped with a two-wire valve (Motronic) while later models (4/1987 and on) are equipped with a three-wire valve (Motronic 1.1).*

20 With the engine running, disconnect the electrical connector from the valve. The idle speed should increase to about 2,000 rpm.
21 If the idle speed does not increase:
   a) Two-wire valve - Using a pair of jumper wires, apply battery voltage to the valve and confirm that the valve closes tightly. When the voltage is removed, the valve should re-open. Also, check the resistance of the valve **(see illustration 18.9)**. The resistance should be about 9 or 10 ohms.
   b) Three-wire valve - Turn the engine off and unplug the electrical connector from the valve. Using an ohmmeter, check the resistance on the two outer terminals of the valve. **(see illustration)**. There should be about 40 ohms of resistance. Check the resistance on the center and outside terminals of the valve. They both should be about 20 ohms of resistance.

22 If the idle air stabilizer valve fails any of the tests, replace it with a new one.
23 If the idle air stabilizer valve tests are all correct, check the control current (two-wire valve) or the voltage (three-wire valve).
24 On two-wire valves, connect an ammeter (0 to 1,000 mA range) as described in Step 13. Start the engine and allow it to idle. With the engine running, the current reading should be between 400 and 500 mA. Adjust the valve if the current reading is not as specified. **Note**: *The idle air stabilizer current will fluctuate between 400 and 1,100 mA if the engine is too cold, if the coolant temperature sensor is faulty, if there is an engine vacuum leak or if electrical accessories are on.*
25 If there is no current reading, have the idle speed control unit (under glove box) diagnosed by a dealer service department or other repair shop.
26 On three-wire valves, check for voltage at the electrical connector. With the ignition key ON, there should be battery voltage present at the center terminal **(see illustration)**. There should be about 10 volts between the center terminal and each of the outer terminals.
27 If there is no voltage reading, have the idle speed control unit (Motronic) or the ECU (Motronic 1.1) diagnosed by a dealer service department or other repair shop.

Adjustment (Motronic only)

28 With the engine off, connect a tachometer to the ignition system (see Step 3).
29 Make sure the ignition timing is correct (see Chapter 1).
30 Connect an ammeter to the valve (as described in Step 13).
31 With the engine running, the current draw should be 450 to 470 mA at 700 to 750 rpm.
32 If the control current is not correct, turn the adjusting screw until it is within the specified range. **Note**: *Turn the idle air bypass screw clockwise to INCREASE the current or counterclockwise to DECREASE the current.*

Replacement

33 Remove the electrical connector and the bracket from the idle air stabilizer valve and remove the valve.
34 Installation is the reverse of removal.

19 Exhaust system servicing - general information

Warning: *Inspect or repair exhaust system components only after enough time has elapsed after driving the vehicle to allow the system components to cool completely. Also, when working under the vehicle make sure it is securely supported by jackstands.*

Chapter 4 Fuel and exhaust systems

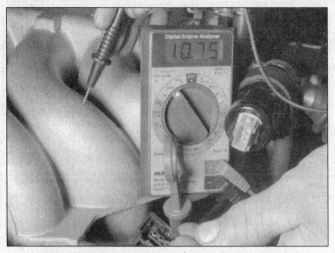

18.26 Check for battery voltage on the center terminal

19.1 A typical exhaust system rubber hanger

Muffler and pipes

Refer to illustration 19.1

1   The exhaust system consists of the exhaust manifold, catalytic converter, mufflers and all connecting pipes, brackets, hangers **(see illustration)** and clamps. The exhaust system is attached to the body with mounting brackets and rubber hangers. If any of the parts are improperly installed, excessive noise and vibration may be transmitted to the body.

2   Inspect the exhaust system regularly to keep it safe and quiet. Look for any damaged or bent parts, open seams, holes, loose connections, excessive corrosion or other defects which could allow exhaust fumes to enter the vehicle. Deteriorated exhaust system components should not be repaired; they should be replaced with new parts.

3   If the exhaust system components are extremely corroded or rusted together, welding equipment will probably be required to remove them. The convenient way to accomplish this is to have a muffler repair shop remove the corroded sections with a cutting torch. If, however, you want to save money by doing it yourself (and you don't have a welding outfit with a cutting torch), cut off the old components with a hacksaw. If you have compressed air, a special pneumatic cutting chisel can also be used. If you decide to tackle the job at home, be sure to wear safety goggles to protect your eyes from metal chips and use work gloves to protect your hands.

4   Here are some simple guidelines to follow when repairing the exhaust system:

   a) Work from the back to the front of the vehicle when removing exhaust system components.
   b) Apply penetrating oil to the exhaust system fasteners to make them easier to remove.
   c) Use new gaskets, hangers and clamps when installing exhaust system components.
   d) Apply anti-seize compound to the threads of all exhaust system fasteners during reassembly.
   e) Be sure to allow sufficient clearance between newly installed parts and all points on the under body to avoid overheating the floor pan and possibly damaging the interior carpet and insulation. Pay particularly close attention to the catalytic converters and heat shields.

Catalytic converters

5   Although the catalytic converters are emissions-related components, they are discussed here because, physically, they're integral parts of the exhaust system. Always check the converters whenever you raise the vehicle to inspect or service the exhaust system.

6   Raise the vehicle and place it securely on jackstands.

7   Visually inspect all catalytic converters on the vehicle for cracks or damage.

8   Check all converters for tightness.

9   Check the insulation covers welded onto the catalytic converters for damage or a loose fit.

10   Start the engine and run it at idle speed. Check all converter connections for exhaust gas leakage.

Notes

Chapter 5 Engine electrical systems

Contents

Air gap – check and adjustment (318i models)	11
Alternator brushes – check and replacement	17
Alternator – removal and installation	15
Battery cables – check and replacement	4
Battery check and maintenance	See Chapter 1
Battery – emergency jump starting	2
Battery – removal and installation	3
Charging system – check	14
Charging system – general information and precautions	13
Distributor – removal and installation	8
Drivebelt check, adjustment and replacement	See Chapter 1
General information	1
Ignition speed sensors – check and replacement (Motronic ignition systems)	12
Ignition system – check	6
Ignition coil – check and replacement	9
Ignition system – general information and precautions	5
Ignition timing – adjustment (318i models)	7
Impulse generator and ignition control unit – check and replacement (318i models only)	10
Spark plug replacement	See Chapter 1
Spark plug wire check and replacement	See Chapter 1
Starter motor – in-vehicle check	19
Starter motor – removal and installation	20
Starter solenoid – removal and installation	21
Starting system – general information and precautions	18
Voltage regulator – replacement	16

Specifications

General

Coil primary resistance
   L-Jetronic systems... 0.82 ohms
   Motronic and Motronic 1.1 systems .... 0.50 ohms
Coil secondary resistance
   L-Jetronic systems... 8,250 ohms
   Motronic and Motronic 1.1 systems .... 5,000 to 6,000 ohms

Distributor (318i models only)

Air gap .... 0.012 to 0.028 inch
Ignition timing (vacuum line disconnected at distributor)
   no. 0 237 002 080 distributor .... 15 degrees BTDC at 2,000 rpm
   no. 0 237 002 096 distributor .... 26 degrees BTDC at 4,000 rpm
Idle speed
   No. 0 237 002 080 distributor .... 750 (±) 50 rpm
   No. 0 237 002 096 distributor
      Manual transmission... 850 (±) 50 rpm
      Automatic transmission .... 750 (±) 50 rpm
Pick-up coil/igniter resistance .... 900 to 1,200 ohms

Chapter 5 Engine electrical systems

3.3a Always detach the cable from the negative battery terminal first, then detach the positive cable – to remove the hold-down assembly, simply remove both nuts (arrows) (325is model shown) (battery is located in the trunk)

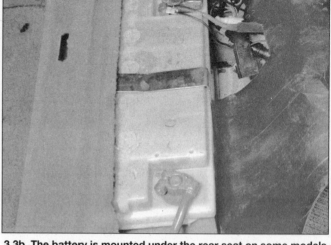

3.3b The battery is mounted under the rear seat on some models (535i shown)

1 General information

The engine electrical systems include all ignition, charging and starting components. Because of their engine-related functions, these components are discussed separately from chassis electrical devices such as the lights, the instruments, etc. (which are included in Chapter 12).

Always observe the following precautions when working on the electrical systems:

a) Be extremely careful when servicing engine electrical components. They are easily damaged if checked, connected or handled improperly.
b) Never leave the ignition switch on for long periods of time with the engine off.
c) Don't disconnect the battery cables while the engine is running.
d) Observe the rules when jump-starting your vehicle. Read the precautions at the front of this manual.
e) Always disconnect the negative cable first and hook it up last or the battery may be shorted by the tool being used to loosen the cable clamps.
f) Don't charge the battery with the cables connected to the terminals.

It's also a good idea to review the safety-related information regarding the engine electrical systems located in the *Safety first* section near the front of this manual before beginning any operation included in this Chapter.

Caution: *If the radio in your vehicle is equipped with an anti-theft system, make sure you have the correct activation code before disconnecting the battery, Refer to the information on page 0-7 at the front of this manual before detaching the cable.* **Note**: *If, after connecting the battery, the wrong language appears on the instrument panel display, refer to page 0-7 for the language resetting procedure.*

2 Battery – emergency jump starting

Refer to the *Booster battery (jump) starting* procedure at the front of this manual.

3 Battery – removal and installation

Refer to illustrations 3.3a and 3.3b

Note: *Some models are equipped with the battery in the trunk or under the rear seat. Consult the owners manual for the location of the battery.*

1 Disconnect the cable from the negative terminal of the battery. **Caution:** *If the radio in your vehicle is equipped with an anti-theft system, make sure you have the correct activation code before disconnecting the battery, Refer to the information on page 0-7 at the front of this manual before detaching the cable.* **Note**: *If, after connecting the battery, the wrong language appears on the instrument panel display, refer to page 0-7 for the language resetting procedure.*

2 Detach the cable from the positive terminal.
3 Remove the battery hold-down bracket **(see illustrations)** and lift out the battery. Be careful – it's heavy.
4 While the battery is out, inspect the carrier (tray) for corrosion (see Chapter 1).
5 If you are replacing the battery, make sure that you get one that's identical, with the same dimensions, amperage rating, cold cranking rating, etc.
6 Installation is the reverse of removal.

4 Battery cables – check and replacement

1 Periodically inspect the entire length of each battery cable for damage, cracked or burned insulation and corrosion. Poor battery cable connections can cause starting problems and decreased engine performance. **Caution:** *If the radio in your vehicle is equipped with an anti-theft system, make sure you have the correct activation code before disconnecting the battery, Refer to the information on page 0-7 at the front of this manual before detaching the cable.* **Note**: *If, after connecting the battery, the wrong language appears on the instrument panel display, refer to page 0-7 for the language resetting procedure.*

2 Check the cable-to-terminal connections at the ends of the cables for cracks, loose wire strands and corrosion. The presence of white, fluffy deposits under the insulation at the cable terminal connection is a sign that the cable is corroded and should be replaced. Check the terminals for distortion, missing mounting bolts and corrosion.

3 When removing the cables, always disconnect the negative cable first and hook it up last or the battery may be shorted by the tool used to loosen the cable clamps. Even if only the positive cable is being replaced, be sure to disconnect the negative cable from the battery first (see Chapter 1 for further information regarding battery cable removal).

4 Disconnect the old cables from the battery, then trace each of them to their opposite ends and detach them from the starter solenoid and ground terminals. Note the routing of each cable to ensure correct installation.

5 If you are replacing either or both of the old cables, take them with you when buying new cables. It is vitally important that you re-

Chapter 5 Engine electrical systems

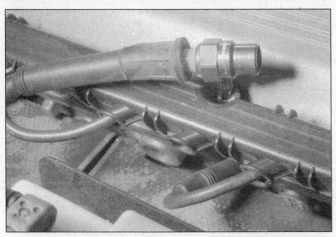

6.1 To use a calibrated ignition tester, simply disconnect a spark plug wire, clip the tester to a convenient ground (like a valve cover bolt or nut) and operate the starter – if there is enough power to fire the plug, sparks will be visible between the electrode tip and the tester body

place the cables with identical parts. Cables have characteristics that make them easy to identify: positive cables are usually red, larger in cross-section and have a larger diameter battery post clamp; ground cables are usually black, smaller in cross-section and have a slightly smaller diameter clamp for the negative post.

6 Clean the threads of the solenoid or ground connection with a wire brush to remove rust and corrosion. Apply a light coat of battery terminal corrosion inhibitor, or petroleum jelly, to the threads to prevent future corrosion.

7 Attach the cable to the solenoid or ground connection and tighten the mounting nut/bolt securely.

8 Before connecting a new cable to the battery, make sure that it reaches the battery post without having to be stretched.

9 Connect the positive cable first, followed by the negative cable.

5 Ignition system – general information and precautions

The ignition system includes the ignition switch, the battery, the distributor, the primary (low voltage) and secondary (high voltage) wiring circuits, the spark plugs and the spark plug wires. Early model 318i (1984 and 1985 models only) are equipped with a Transistorized Coil Ignition System (TCI-i) while all other models (six-cylinder engines) are equipped with the Motronic ignition system (DME).

When working on the ignition system, take the following precautions:
a) If the engine won't start, don't keep the ignition switch on for more than 10 seconds.
b) Never allow an ignition coil terminal to contact ground. Grounding the ignition coil can damage the igniter and/or the coil itself.
c) Don't disconnect the battery when the engine is running.
d) Make sure the igniter is properly grounded.

Transistorized Coil Ignition System (TCI-i)

This system is equipped with four major components; the impulse generator, the ignition control unit, the coil and the spark plugs. The impulse generator provides a timing signal for the ignition system. Equivalent to cam-actuated breaker points in a standard distributor, the impulse generator creates an A/C voltage signal every time the trigger wheel tabs pass the impulse generator tabs. When the ignition control unit (capacitive discharge unit) receives the voltage signal, it triggers a spark discharge from the coil by interrupting the primary coil circuit. The ignition dwell (coil charging time) is adjusted by the ignition control unit for the most intense spark. **Note:** *The air gap (distance between the impulse generator and trigger wheel tabs) can be adjusted (see Section 11).*

There are two types of control units on these models; Bosch (1984 and early 1985) or the Siemens/Telefunken control units. The two types can be distinguished by their electrical connectors. The Bosch type uses a single, flat connector at the bottom of the unit while the Siemens/Telefunken control unit uses two round electrical connectors at the front of the unit.

Ignition timing is mechanically adjusted (see Section 7) to change the engine speed and load conditions. A centrifugal advance unit that consists of spring-loaded rotating weights advances ignition timing as engine rpm's increase. The vacuum advance adjusts ignition timing to compensate for changes in engine load.

Motronic ignition systems

These systems, also known as Digital Motor Electronics (DME), incorporate all ignition and fuel injection functions into one central control unit or ECU (computer). The ignition timing is based on inputs the ECU receives for engine load, engine speed, ignition quality, coolant temperature and air intake temperature. The only function the distributor performs is the distribution of the high voltage signal to the individual spark plugs. The distributor is attached directly to the cylinder head. There is no mechanical spark advance system used on these systems.

Ignition timing is electronically controlled and is not adjustable on Motronic systems. During starting, a crankshaft position sensor (reference sensor) relays the crankshaft position to the ECU and an initial baseline ignition point is determined. Once the engine is running, the ignition timing is continually changing based on the various input signals to the ECU. Engine speed is signalled by a speed sensor. Early Motronic systems are equipped with the reference sensor and the speed sensor mounted on the bellhousing over the flywheel on the driver's side while later Motronic 1.1 systems are equipped with a single sensor (pulse sensor) mounted up-front over the crankshaft pulley. This sensor functions as a speed sensor as well as a reference sensor (angle). Refer to Section 12 for checking and replacing the ignition sensors. **Note:** *Some models are equipped with a TDC sensor mounted on the front of the engine. This sensor is strictly for the BMW service test unit and it is not part of the Motronic ignition system.*

Precautions

Certain precautions must be observed when working on a transistorized ignition system.
a) Do not disconnect the battery cables when the engine is running
b) Make sure the ignition control unit (1984 and 1985 318i models only) is always well grounded (see Section 10).
c) Keep water away from the distributor
d) If a tachometer is to be connected to the engine, always connect the tachometer positive (+) lead to the ignition coil negative terminal (-) and never to the distributor.
e) Do not allow the coil terminals to be grounded, as the impulse generator or coil could be damaged.
f) Do not leave the ignition switch on for more than ten minutes if the engine isn't running or will not start.

6 Ignition system – check

Refer to illustration 6.1

Warning: *Because of the high voltage generated by the ignition system, extreme care should be taken whenever an operation is performed involving ignition components. This not only includes the igniter (electronic ignition), coil, distributor and spark plug wires, but related components such as spark plug connectors, tachometer and other test equipment.*

1 If the engine turns over but will not start, disconnect the spark plug wire from any spark plug and attach it to a calibrated spark tester (available at most auto parts stores). **Note:** *There are two different types of spark testers. Be sure to specify electronic (breakerless) ignition. Connect the clip on the tester to a ground such as a metal bracket* **(see illustration)**.

2 If you are unable to obtain a calibrated spark tester, remove the

5-4 Chapter 5 Engine electrical systems

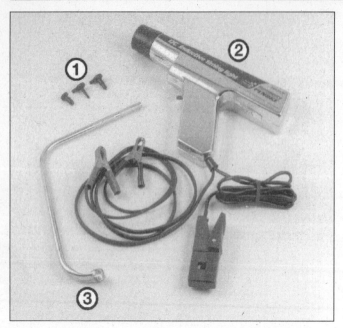

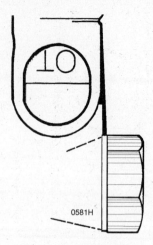

7.9a Timing mark on 1984 and 1985 318i models with distributors marked 0 237 002 080

7.1 Tools needed to check and adjust the ignition timing

1 **Vacuum plugs** - Vacuum hoses will, in most cases, have to be disconnected and plugged. Molded plugs in various shapes and sizes are available for this
2 **Inductive pick-up timing light** - Flashes a bright, concentrated beam of light when the number one spark plug fires. Connect the leads according to the instructions supplied with the light
3 **Distributor wrench** - On some models, the hold-down bolt for the distributor is difficult to reach and turn with conventional wrenches or sockets. A special wrench like this must be used

spark plug wire from one of the spark plugs. Using an insulated tool, hold the wire about 1/4-inch from a good ground.
3 Crank the engine and observe the tip of the tester or spark plug wire to see if a spark occurs. If bright blue, well-defined sparks occur, sufficient voltage is reaching the plugs to fire the engine. However, the plugs themselves may be fouled, so remove and check them as described in Chapter 1.
4 If there's no spark, check another wire in the same manner. A few sparks followed by no spark is the same condition as no spark at all.
5 If no spark occurs, remove the distributor cap and check the cap and rotor as described in Chapter 1. If moisture is present, use WD-40 (or something similar) to dry out the cap and rotor, then reinstall the cap and repeat the spark test.
6 If there's still no spark, disconnect the coil wire from the distributor, hold it about 1/4-inch from a good ground and crank the engine again.
7 If no spark occurs, check the primary wire connections at the coil to make sure they're clean and tight. Make any necessary repairs, then repeat the check again.
8 If sparks do occur, the distributor cap, rotor, plug wire(s) or spark plug(s) may be defective. If there's still no spark, the coil-to-cap wire may be bad. If a substitute wire doesn't make any difference, check the ignition coil (see Section 9). **Note:** *Refer to Section 10 and 11 for more test procedures on the distributors installed on 1984 and 1985 318i models.*

7 Ignition timing – adjustment (318i models)

Refer to illustrations 7.1, 7.9a and 7.9b
Warning: *The electric cooling fan on these models can activate at any time, even when the ignition is in the Off position. Disconnect the fan*

motor when working in the vicinity of the fan.
Note 1: *It is important to identify the distributor code number stamped on the side of the distributor body in order to establish the correct timing specifications and the timing mark symbols. 1984 (early models) are numbered 0 237 002 080 and 1984 and 1985 (late models) are numbered 0 237 002 096. Refer to the specifications listed in this Chapter for the correct settings.*
Note 2: *If the information in this Section differs from the Vehicle Emission Control Information label in the engine compartment of your vehicle, the label should be considered correct.*

1 Some special tools are required for this procedure **(see illustration)**. The engine must be at normal operating temperature and the air conditioner must be Off. Make sure the idle speed is correct.
2 Apply the parking brake and block the wheels to prevent movement of the vehicle. The transmission must be in Park (automatic) or Neutral (manual).
3 The timing marks on these models are located on the engine flywheel and are viewed through the timing check hole in the bellhousing.
4 Disconnect the vacuum hose from the distributor vacuum advance unit. It is not necessary to plug the line.
5 Connect a tachometer and timing light according to the tool manufacturer's instructions (an inductive pick-up timing light is preferred). Generally, the power leads are attached to the battery terminals and the pick-up lead is attached to the number one spark plug wire. The number one spark plug is the one closest to the drivebelt end of the engine. **Caution:** *If an inductive pick-up timing light isn't available, don't puncture the spark plug wire to attach the timing light pick-up lead. Instead, use an adapter between the spark plug and plug wire. If the insulation on the plug wire is damaged, the secondary voltage will jump to ground at the damaged point and the engine will misfire.*
6 With the ignition OFF, loosen the distributor clamp nut just enough to allow the distributor to pivot without any slipping.
7 Make sure the timing light wires are routed away from the drivebelts and fan, then start the engine.
8 Raise the engine rpm to the specified limit and then point the flashing timing light at the timing marks – be very careful of moving engine components.
9 The mark on the flywheel will appear stationary. If it's aligned with the specified point on the bellhousing, the ignition timing is correct **(see illustrations)**.
10 If the marks aren't aligned, adjustment is required. Turn the distributor very slowly until the marks are aligned.
11 Tighten the nut on the distributor clamp and recheck the timing.
12 Turn off the engine and remove the timing light (and adapter, if used). Reconnect and install any components which were disconnected or removed.

Chapter 5 Engine electrical systems

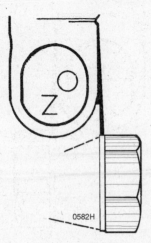

7.9b Timing mark on 1984 and 1985 318i models with distributors marked 0 237 002 096

8.7 Mark the relationship of the rotor to the distributor housing (arrow)

Installation

10 Before installing the distributor, make certain the number one piston is still at TDC on the compression stroke.
11 Insert the distributor into the engine with the adjusting clamp centered over the hold-down hole. Make sure the gear does not turn as the distributor is inserted.
12 Install the hold-down bolt. The marks previously made on the distributor housing and on the rotor and engine should line up before the bolt is tightened.
13 Install the distributor cap.
14 Connect the wiring for the distributor.
15 Install the spark plug wires.
16 Install the vacuum hoses as previously marked.
17 Adjust the ignition timing (see Section 7).

Motronic systems

Refer to illustrations 8.18 and 8.19

18 Remove the cover from the distributor **(see illustration)** and remove the distributor cap (see Chapter 1).
19 Use a small Allen wrench and remove the three screws from the rotor **(see illustration)**.
20 Remove the rotor. Installation is the reverse of removal.

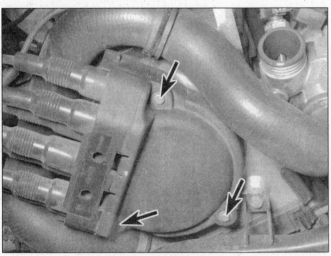

8.18 Remove the three bolts (arrows) from the distributor cap

8 Distributor – removal and installation

L-Jetronic (1984 and 1985 318i models only)

Removal

Refer to illustration 8.7

1 After carefully marking them, remove the coil wire and spark plug wires from the distributor cap (see Chapter 1).
2 Remove the number one spark plug (the one nearest you when you are standing in front of the engine).
3 Manually rotate the engine to Top Dead Center (TDC) on the compression stroke for number one piston (see Chapter 2A)
4 Carefully mark the vacuum hoses if more than one is present on your distributor.
5 Disconnect the vacuum hose(s).
6 Disconnect the primary wires from the distributor.
7 Mark the relationship of the rotor tip to the distributor housing **(see illustration)**. Also mark the relationship of the distributor housing to the engine.
8 Remove the hold-down bolt and clamp.
9 Remove the distributor. **Note:** *Do not rotate the engine with the distributor out.*

9 Ignition coil – check and replacement

Refer to illustrations 9.4a, 9.4b, 9.4c, 9.5a and 9.5b
Caution: *If the coil terminals touch a ground source, the coil and/or impulse generator could be damaged.*
Note: *On models equipped with Motronic 1.1, a faulty ECU can cause the ignition coil to become damaged. Be sure to test the ignition coil in the event of ECU failure and a no-start condition.*

1 Mark the wires and terminals with pieces of numbered tape, then remove the primary wires and the high-tension wire from the coil.
2 Remove the coil assembly from its mount, clean the outer case and check it for cracks and other damage.
3 Inspect the coil primary terminals and the coil tower terminal for corrosion. Clean them with a wire brush if any corrosion is found.
4 Check the coil primary resistance by attaching the leads of an ohmmeter to the primary terminals **(see illustrations)**. Compare the measured resistance to the Specifications listed in this Chapter.
5 Check the coil secondary resistance by hooking one of the ohmmeter leads to one of the primary terminals and the other ohmmeter lead to the coil high-tension terminal **(see illustrations)**. On L-Jetronic systems, connect the ohmmeter to the no.1 (-) coil terminal and the center tower. On Motronic and Motronic 1.1 systems, connect the ohmmeter to the no.15 (+) coil terminal and the center tower. Compare

8.19 Remove the rotor screws with an Allen wrench and pull the rotor off the shaft

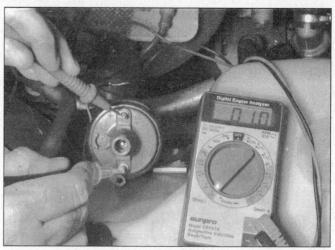

9.4a Using an ohmmeter, measure the resistance between the primary terminals of the ignition coil (L-Jetronic system shown)

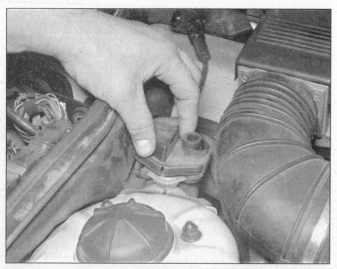

9.4b Some Motronic 1.1 systems use a different style coil. First, remove the coil cover and . . .

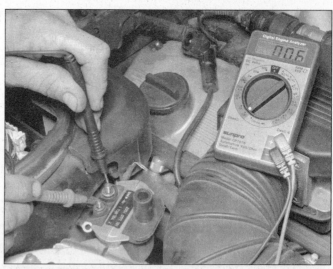

9.4c . . . using an ohmmeter, measure the resistance between the primary terminals of the coil (535i model shown)

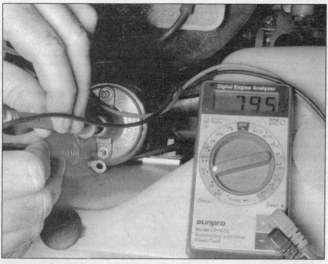

9.5a Using an ohmmeter, measure the secondary resistance of the coil (L-Jetronic system)

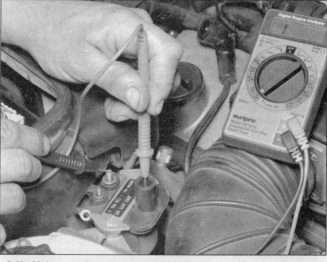

9.5b Using an ohmmeter, measure the secondary resistance of the coil (Motronic 1.1 system)

Chapter 5 Engine electrical systems

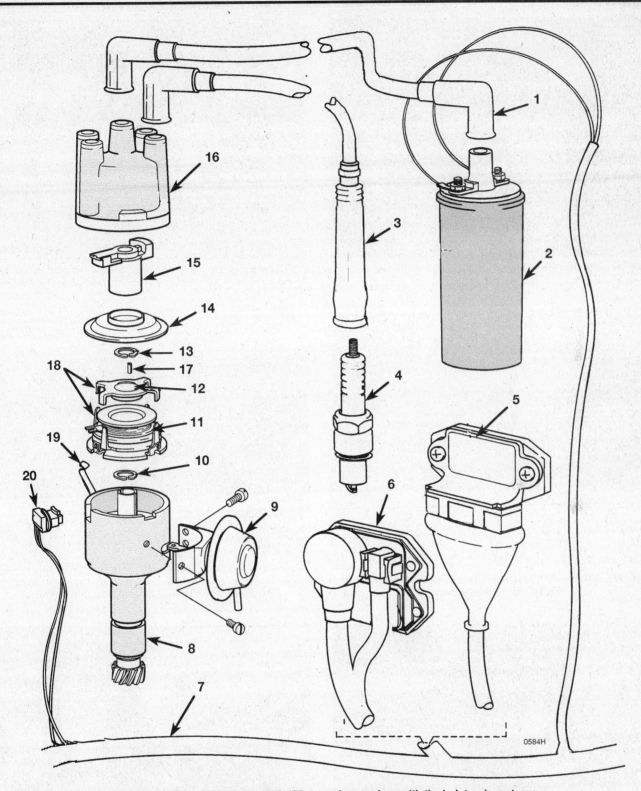

10.1 Schematic of the electronic ignition used on engines with the L-Jetronic system

1 Coil wire
2 Ignition coil
3 Spark plug wire
4 Spark plug
5 Ignition control unit (Bosch)
6 Ignition control unit (Siemens/Telefunken)
7 Wiring harness
8 Distributor housing with counterweights
9 Vacuum diaphragm
10 Clip
11 Impulse generator
12 Trigger wheel
13 Circlip
14 Dust shield
15 Ignition rotor
16 Distributor cap
17 Roll pin
18 Tabs
19 Clip
20 Impulse generator connector

5-8 Chapter 5 Engine electrical systems

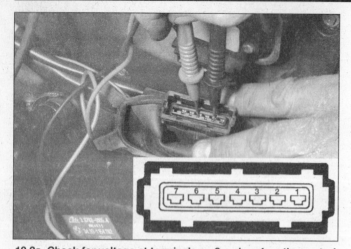

10.2a Check for voltage at terminals no.2 and no.4 on the control unit electrical connector (Bosch system shown)

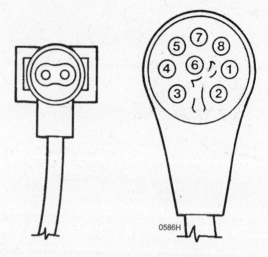

10.2b Check for voltage at terminals no.6 and no.3 on the control unit electrical connector (Siemens/Telefunken system shown)

the measured resistance to the Specifications listed in this Chapter.
6   If the measured resistances are not as specified, the coil is defective and should be replaced with a new one.
7   It is essential for proper ignition system operation that all coil terminals and wire leads to be kept clean and dry.
8   Install the coil in its mount and hook up the wires. Installation is the reverse of removal.

10 Impulse generator and ignition control unit – check and replacement (318i models only)

Refer to illustrations 10.1, 10.2a, 10.2b, 10.8, 10.18 and 10.19
1   The impulse generator and ignition control unit need to be tested in the event there is no spark at the spark plugs. Make sure the plug wires, ignition coil and spark plugs are working properly (see Section 6 and Section 9). There are two types of control units on these models; Bosch (1984 and early 1985) or the Siemens/Telefunken control units. The two types **(see illustration)** can be distinguished by their electrical connectors. The Bosch type uses a single, flat connector at the bottom of the unit while the Siemens/Telefunken control unit uses two, round electrical connectors at the front of the unit.

Check

Voltage supply and ground to ignition control unit
2   With the ignition key OFF, remove the harness connectors from the ignition control unit **(see illustrations)**. Connect a voltmeter between terminals no.2 and no.4 on Bosch systems or between terminals no.6 and no.3 on Siemens/Telefunken systems.
3   Turn the ignition key ON. There should be battery voltage on the designated terminals. If there is no voltage, check the wiring harness for an open circuit (see Chapter 12).
4   Using an ohmmeter, check for continuity between terminal no.2 (Bosch) or no.6 (Siemens/Telefunken) and body ground. Continuity should exist.
5   Using an ohmmeter, check for continuity between terminal no.4 (Bosch) or no.3 (Siemens/Telefunken) and terminal no.15 of the ignition coil. Continuity should exist.
6   If the readings are incorrect, repair the wiring harness.

Impulse generator signal
7   If the ignition control unit is receiving battery voltage, check the AC signal voltage coming from the impulse generator to the control unit.
8   Using a digital voltmeter;
  a) On Bosch systems, connect the positive probe to terminal no.5 of the control unit and the negative probe to terminal no.6 (see illustration).

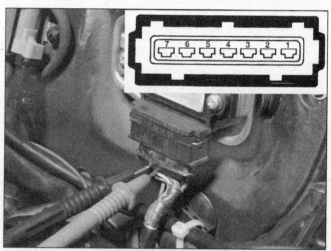

10.8 Back-probe the ignition control unit connector and check for signal voltage on terminals no.5 and no.6. It is very helpful to use angled probes

  b) On Siemens/Telefunken systems connect the positive probe to terminal (+) of the control unit (smaller connector) and the negative probe to terminal (-) **(see illustration 10.2b)**.
9   Have an assistant crank the engine over and observe that there is 1 to 2 volts AC present. If there is no voltage, check the wiring harness between the impulse generator (in the distributor) and the control unit. If the voltage is correct, check the impulse generator resistance.
Warning: *Do not crank the engine over for an excessive amount of time. If necessary, disconnect the cold start injector (see Chapter 4) electrical connector to stop the flow of fuel into the engine.*
10 To check the resistance in the impulse generator;
  a) On Bosch units, measure the resistance between terminals no.5 and no.6 **(see illustration 10.2a)**. The readings should be 1,000 to 1,200 ohms.
  b) On Siemens/Telefunken units, measure the resistance between the terminals of the small connector **(see illustration 10.2b)**. The readings should be 1,000 to 1,200 ohms.
11 If the resistance readings are incorrect, replace the impulse generator. If the resistance readings for the impulse generator are correct and the control unit voltages (supply voltage [Steps 1 through 6] and signal voltage [Steps 7 through 9]) are incorrect, replace the control unit.

Chapter 5 Engine electrical systems

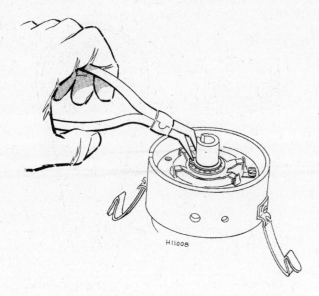

10.18 Use snap-ring pliers and remove the snap-ring from the distributor shaft

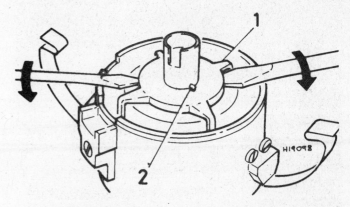

10.19 Carefully pry the trigger wheel off the distributor shaft

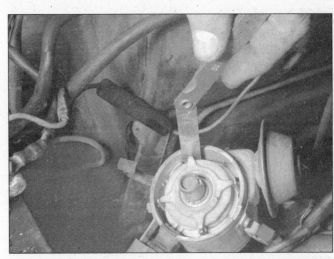

11.2 Use a feeler gauge to check the air gap (be sure the gauge rubs lightly against the trigger wheel as well as the locating pin for the correct adjustment)

Replacement

Ignition control unit

12 Make sure the ignition key is OFF.
13 Disconnect the electrical connector(s) from the control unit.
14 Remove the mounting screws from the control unit and lift it from the engine compartment.
15 Installation is the reverse of removal. **Note:** *On Bosch control units, a special dielectric grease is used between the heat sink and the back side of the control unit. In the event the two are separated (replacement or testing) the old grease must be removed and the heat sink cleaned off using 180-grit sandpaper. Apply Curil K2 (Bosch part number 81 22 9 243). A silicon dielectric compound can be used as a substitute. This treatment is very important for the long life of these expensive ignition parts.*

Impulse generator

16 Disconnect the cable from the negative battery terminal. **Caution:** *If the radio in your vehicle is equipped with an anti-theft system, make sure you have the correct activation code before disconnecting the battery, Refer to the information on page 0-7 at the front of this manual before detaching the cable.* **Note:** *If, after connecting the battery, the wrong language appears on the instrument panel display, refer to page 0-7 for the language resetting procedure.*
17 Remove the distributor from the engine (see Section 8).
18 Using a pair of snap-ring pliers, remove the snap-ring retaining the trigger wheel **(see illustration)**.
19 Use two flat-bladed screwdrivers positioned at opposite sides of the trigger wheel and carefully pry it up **(see illustration)**. **Note:** *Push the screwdrivers in as far as possible without bending the trigger wheel. Pry only on the strongest, center portion of the trigger wheel. In the event the trigger wheel is bent, it must be replaced with a new one.* **Note:** *Be sure not to lose the roll pin when lifting out the trigger wheel.*
20 Remove the mounting screws from the impulse generator electrical connector, the vacuum diaphragm and the base plate.
21 Remove the two screws from the vacuum advance unit and separate it from the distributor by moving the assembly down while unhooking it from the base plate pin.
22 Use snap-ring pliers to remove the snap-ring that retains the impulse generator and the base plate assembly.
23 Carefully remove the impulse generator and the base plate assembly as a single unit.
24 Remove the three screws and separate the base plate assembly from the impulse generator.
25 Installation is the reverse of removal. **Note:** *Be sure to position the insulating ring between the coil and the base plate. It must be centered before tightening the mounting screws. Also, it is necessary to adjust the air gap once the trigger wheel has been removed or tampered with to the point that the clearance is incorrect (see Section 11).*

11 Air gap (318i models) – check and adjustment

Refer to illustration 11.2

1 Disconnect the cable from the negative terminal on the battery. **Caution:** *If the radio in your vehicle is equipped with an anti-theft system, make sure you have the correct activation code before disconnecting the battery, Refer to the information on page 0-7 at the front of this manual before detaching the cable.* **Note:** *If, after connecting the battery, the wrong language appears on the instrument panel display, refer to page 0-7 for the language resetting procedure.*
2 Insert a brass feeler gauge between the trigger wheel tab and the impulse generator **(see illustration)**. Slide the feeler gauge up and down – you should feel a slight drag on the feeler gauge as it is moved if the gap is correct. The gap must be between 0.012 to 0.028 inch.
3 To adjust the gap, it is necessary to remove the impulse genera-

Chapter 5 Engine electrical systems

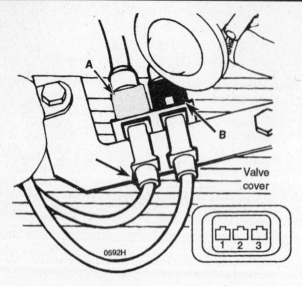

12.1a Location of the reference sensor (gray connector) (A) and the speed sensor (black connector) (B) on Motronic systems (1982 through 1985 528e and 325e(es) models)

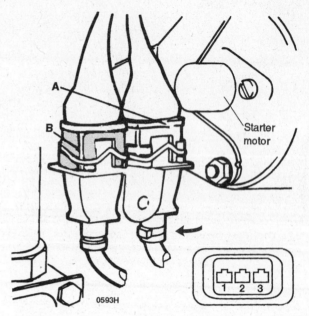

12.1b Location of the reference sensor (gray connector) (A) and the speed sensor (black connector) (B) on Motronic systems (1986 through March 1987 528e and 325e(es) models)

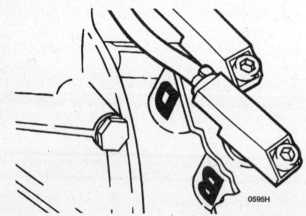

12.5 Location of the reference sensor (B) and speed sensor (D) on the bellhousing on all Motronic systems – do not interchange the sensors or the engine will not start

tor and the base plate assembly from the distributor **(see illustration 10.1)**.
4   Follow steps 17 through 24 in Section 10 and loosen the screws that retain the impulse generator to the base plate assembly.
5   Carefully insert the feeler gauge and tighten the screws.
6   Install the assembly back into the distributor and recheck the adjustment.

12 Ignition speed sensors – check and replacement (Motronic ignition systems)

Note: *Some models are equipped with a TDC sensor mounted on the front of the engine. This sensor is strictly for the BMW service test unit and it is not part of the Motronic ignition system.*

Reference and speed sensors (Motronic systems)

Refer to illustrations 12.1a, 12.1b and 12.5

Check

1   Locate the two electrical connectors for the sensors **(see illustrations)**. The gray connector is for the reference sensor and the white connector is for the speed sensor.
2   Using an ohmmeter, check resistance between terminal 1 (yellow wire) and terminal 2 (black wire) on the sensor side of each connector. The resistance should be 860 to 1,060 ohms.
3   Also check the resistance between terminal 3 and either terminal 1 or terminal 2. The resistance should be approximately 100,000 ohms.
4   If the reading(s) are incorrect, replace the sensor with a new one.

Replacement

5   Be sure the connectors are not interchanged when changing the sensor. The bellhousing is marked with a B for the reference sensor (gray connector) and D for the speed sensor (black connector) **(see illustration)**. **Note:** *It is a good idea to check the condition of the raised pin on the flywheel while the sensors are out of the sockets. The pin is visible through the timing check hole in the bellhousing. If the pin is missing or damaged, the sensor will not be able to accurately determine speed and crankshaft position.*
6   Tighten the sensor mounting screw securely but be careful not to overtighten it.

Pulse sensor (Motronic 1.1 systems)

Refer to illustrations 12.7a, 12.7b, 12.8, 12.10 and 12.11

Check

7   Locate the two electrical connectors for the sensor **(see illustrations)**. Disconnect the electrical connector from the front.
8   Using an ohmmeter, check resistance between terminal 1 (yellow wire) and terminal 2 (black wire) on the sensor side of each connector. The resistance should be 500 to 600 ohms.
9   If the reading is incorrect, replace the sensor.

Replacement

10 Remove the pulse sensor mounting bolt using a 5 mm hex wrench **(see illustration)**.
11 Use a brass feeler gauge to position the tip of the sensor the correct distance from the pulse wheel **(see illustration)**.
12 Tighten the mounting screw but be careful not to overtighten it.

Chapter 5 Engine electrical systems

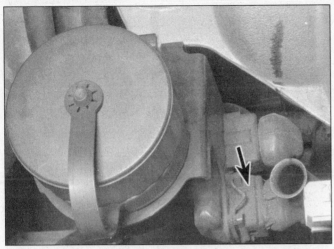

12.7a On 325is models, the pulse sensor connector is located next to the 20 pin diagnostic connector (arrow) (Motronic 1.1 system)

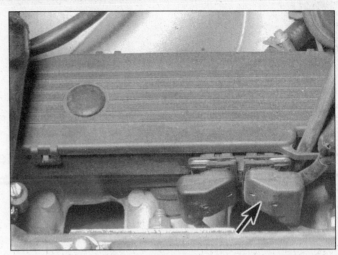

12.7b On 525i models, the pulse sensor connector is located next to the valve cover (arrow) (Motronic 1.1 system)

12.8 The resistance of the pulse sensor should be 500 to 600 ohms (Motronic 1.1 system)

12.10 The pulse sensor (arrow) is located on the timing belt cover off to the side of the pulley on all Motronic 1.1 systems

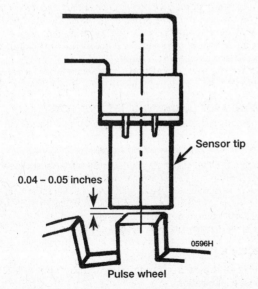

12.11 The sensor tip should be set at 0.040 to 0.050 inch from the pulse wheel

13 Charging system – general information and precautions

This alternator has an output amperage rating between 12 and 90 amps depending on the load and the engine rpm. There are two different types of alternators installed on these models; Bosch and Motorola. Also, there are three different amperage ratings available; 65A, 80A or 90A. It is important to remember that the measured amperage output at the battery with the engine running will be 10 to 15 amps lower than the specified output because of power requirements for various electrical units (air conditioning, ABS, DME, power door locks, etc) incorporated in the vehicle. A stamped serial number on the rear of the alternator will identify the type and amperage rating. Perform the charging system checks (see Section 14) to diagnose any problems with the alternator.

The purpose of the voltage regulator is to limit the alternator's voltage to a preset value. This prevents power surges, circuit overloads, etc., during peak voltage output.

The voltage regulator and the alternator brushes are mounted as a single assembly. On Bosch alternators, this unit can be removed from the alternator (see Section 16) and the components serviced individually.

The alternator on all models is mounted on the left front of the engine and utilizes a V-belt and pulley drive system. Drivebelt tension and battery service are the two primary maintenance requirements for these systems. See Chapter 1 for the procedures regarding engine drivebelt checking and battery servicing. Other than that, the charging system doesn't ordinarily require periodic maintenance.

The dashboard warning light should come on when the ignition key is turned to Start, then go off immediately. If it remains on, there is a malfunction in the charging system (see Section 14). Some vehicles are also equipped with a voltmeter. If the voltmeter indicates abnormally high or low voltage, check the charging system (see Section 14). **Note:** *On 1982 through 1986 models, a burned-out dashboard warning light will prevent the alternator from charging. After 1987 a resistor is wired in parallel with the warning light in order to allow current to bypass the light in the event of a broken circuit (burned-out light).*

Be very careful when making electrical circuit connections to a vehicle equipped with an alternator and note the following:

a) When reconnecting wires to the alternator from the battery, be sure to note the polarity.
b) Before using arc welding equipment to repair any part of the vehicle, disconnect the wires from the alternator and the battery terminals. **Caution:** *If the radio in your vehicle is equipped with an anti-theft system, make sure you have the correct activation code before disconnecting the battery, Refer to the information on page 0-7 at the front of this manual before detaching the cable.* **Note:** *If, after connecting the battery, the wrong language appears on the instrument panel display, refer to page 0-7 for the language resetting procedure.*
c) Never start the engine with a battery charger connected.
d) Always disconnect both battery cables before using a battery charger.
e) The alternator is turned by an engine drivebelt which could cause serious injury if your hands, hair or clothes become entangled in it with the engine running.
f) Because the alternator is connected directly to the battery, it could arc or cause a fire if overloaded or shorted out.
g) Wrap a plastic bag over the alternator and secure it with rubber bands before steam cleaning the engine.

14 Charging system – check

1   If a malfunction occurs in the charging circuit, don't automatically assume that the alternator is causing the problem. First check the following items:

a) Check the drivebelt tension and condition (see Chapter 1). Replace it if it's worn or deteriorated.
b) Make sure the alternator mounting and adjustment bolts are tight.
c) Inspect the alternator wiring harness and the connectors at the alternator and voltage regulator. They must be in good condition and tight.
d) Check the fuses.
e) Start the engine and check the alternator for abnormal noises (a shrieking or squealing sound indicates a bad bearing).
f) Check the specific gravity of the battery electrolyte. If it's low, charge the battery (doesn't apply to maintenance-free batteries).
g) Make sure the battery is fully charged (one bad cell in a battery can cause overcharging by the alternator).
h) Disconnect the battery cables (negative first, then positive). Inspect the battery posts and the cable clamps for corrosion. Clean them thoroughly if necessary (see Chapter 1). **Caution:** *If the radio in your vehicle is equipped with an anti-theft system, make sure you have the correct activation code before disconnecting the battery, Refer to the information on page 0-7 at the front of this manual before detaching the cable.* **Note:** *If, after connecting the battery, the wrong language appears on the instrument panel display, refer to page 0-7 for the language resetting procedure.* Reconnect the cable to the positive terminal.
i) With the key off, connect a test light between the negative battery post and the disconnected negative cable clamp.
   1) If the test light does not come on, reattach the clamp and proceed to Step 3.

15.2 Depending on how many accessories the vehicle has, sometimes it's easier to remove the alternator from the brackets first and then turn it sideways to gain access to the rear of the alternator body

   2) If the test light comes on, there is a short (drain) in the electrical system of the vehicle. The short must be repaired before the charging system can be checked. **Note:** *Accessories which are always on (such as the clock) must be disconnected before preforming this check.*
   3) Disconnect the alternator wiring harness.
      a) If the light goes out, the alternator is bad.
      b) If the light stays on, pull each fuse until the light goes out (this will tell you which component is shorted).

2   Using a voltmeter, check the battery voltage with the engine off. It should be approximately 12-volts.

3   Start the engine and check the battery voltage again. It should now be approximately 14-to-15 volts.

4   Turn on the headlights. The voltage should drop, and then come back up, if the charging system is working properly.

5   If the voltage reading is more than the specified charging voltage, replace the voltage regulator (refer to Section 16). If the voltage is less, the alternator diode(s), stator or rotor may be bad or the voltage regulator may be malfunctioning.

6   If the battery is constantly discharging, the alternator drivebelt is loose (see Chapter 1), the alternator brushes are worn, dirty or disconnected (see Section 17), the voltage regulator is malfunctioning (see Section 16) or the diodes, stator coil or rotor coil is defective. Repairing or replacing the diodes, stator coil or rotor coil is beyond the scope of the home mechanic. Replace the alternator. **Note:** *On 1982 through 1986 models, a burned-out dashboard warning light will prevent the alternator from charging the battery. After 1987 a resistor is wired in parallel with the warning light in order to allow current to bypass the light in the event of a broken circuit (burned out light).*

15 Alternator – removal and installation

Refer to illustration 15.2

1   Detach the cable from the negative terminal of the battery. **Caution:** *If the radio in your vehicle is equipped with an anti-theft system, make sure you have the correct activation code before disconnecting the battery, Refer to the information on page 0-7 at the front of this manual before detaching the cable.* **Note:** *If, after connecting the battery, the wrong language appears on the instrument panel display, refer to page 0-7 for the language resetting procedure.*

2   Detach the electrical connectors from the alternator **(see illustration)**. **Note:** *On some models, it may be necessary to remove the air*

Chapter 5 Engine electrical systems

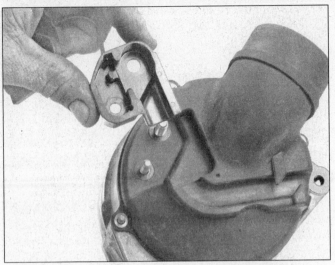

16.4a Remove the nuts and lift over the small terminal protector from the alternator cover, then remove the nuts and the cover

16.4b The regulator pulls straight out on Bosch alternators

16.5 Use a flashlight and check the condition of the slip rings for scoring or deep grooves

cleaner assembly and airflow meter to gain access to the alternator.
3 Loosen the alternator adjustment and pivot bolts and detach the drivebelt (see Chapter 1).
4 Remove the adjustment and pivot bolts and separate the alternator from the engine.
5 If you are replacing the alternator, take the old one with you when purchasing a replacement unit. Make sure the new/rebuilt unit looks identical to the old alternator. Look at the terminals – they should be the same in number, size and location as the terminals on the old alternator. Finally, look at the identification numbers – they will be stamped into the housing or printed on a tag attached to the housing. Make sure the numbers are the same on both alternators.
6 Many new/rebuilt alternators DO NOT have a pulley installed, so you may have to switch the pulley from the old unit to the new/rebuilt one. When buying an alternator, find out the shop's policy regarding pulleys – some shops will perform this service free of charge.
7 Installation is the reverse of removal.
8 After the alternator is installed, adjust the drivebelt tension (see Chapter 1).
9 Check the charging voltage to verify proper operation of the alternator (see Section 14).

16 Voltage regulator – replacement

1 The voltage regulator controls the charging system voltage by limiting the alternator output. The regulator is a sealed unit and isn't adjustable.
2 If the ammeter fails to register a charge rate or the red warning light on the dash comes on and the alternator, battery, drivebelt tension and electrical connections seem to be fine, have the regulator checked by a dealer service department or an automotive electrical repair shop.
3 Disconnect the cable from the negative battery terminal. **Caution:** *If the radio in your vehicle is equipped with an anti-theft system, make sure you have the correct activation code before disconnecting the battery, Refer to the information on page 0-7 at the front of this manual before detaching the cable.* **Note:** *If, after connecting the battery, the wrong language appears on the instrument panel display, refer to page 0-7 for the language resetting procedure.*

Bosch alternator
Refer to illustrations 16.4a, 16.4b and 16.5
4 The voltage regulator is located on the exterior of the alternator housing. To replace the regulator, remove the mounting screws **(see illustration)** and lift it off the alternator body **(see illustration)**. **Note:** *Some Bosch alternators incorporate an integral voltage regulator which is part of the brush assembly.*
5 Installation is the reverse of removal. **Note:** *Before installing the regulator, check the condition of the slip rings* **(see illustration)**. *Use a flashlight and check for any scoring or deep wear grooves. Replace the alternator if necessary.*

Motorola alternator
Refer to illustration 16.7
6 Remove the alternator from the engine compartment (see Section 15).
7 Remove the rear cover and diode carrier, remove the voltage regulator mounting screws **(see illustration)** and lift the regulator off the alternator body.
8 Installation is the reverse of removal.

17 Alternator brushes – check and replacement

1 Disconnect the negative cable from the battery. **Caution:** *If the radio in your vehicle is equipped with an anti-theft system, make sure you have the correct activation code before disconnecting the battery, Refer to the information on page 0-7 at the front of this manual before detaching the cable.* **Note:** *If, after connecting the battery, the wrong*

5-14 Chapter 5 Engine electrical systems

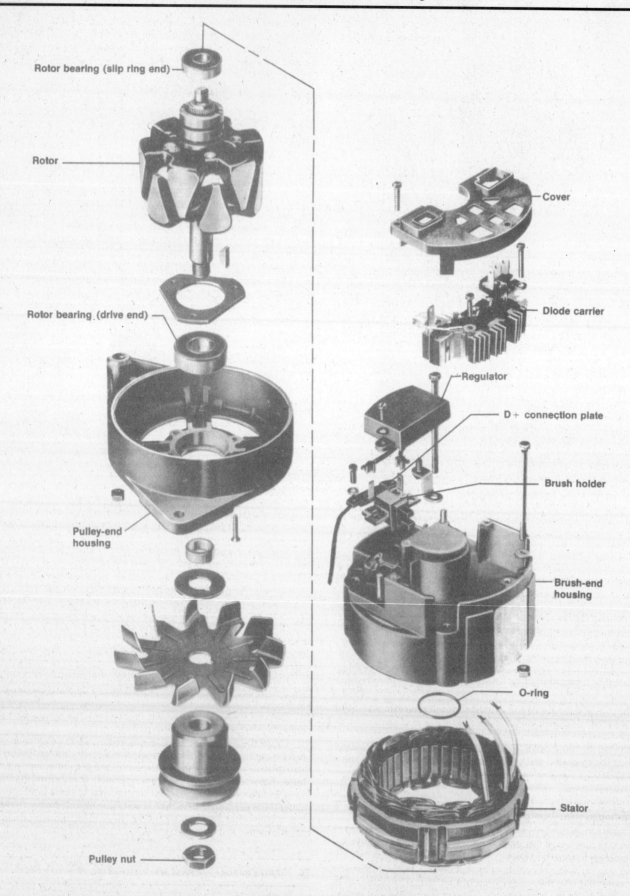

16.7 Exploded view of the Motorola alternator

Chapter 5 Engine electrical systems

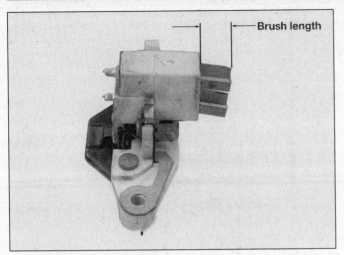

17.3 Check the brush length in a normal rest position (spring uncoiled)

language appears on the instrument panel display, refer to page 0-7 for the language resetting procedure. Label the wires and detach them from the alternator terminals.

Bosch alternator

Refer to illustration 17.3

2 Remove the voltage regulator from the back of the alternator (see Section 16).
3 Measure the length of the brushes **(see illustration)**. They should not be less than 7/32-inch. If they are worn past this point, replace them with a new set.
4 Also, check for excessively worn slip rings **(see illustration 16.5)**.
5 The brushes are retained by either set screws or by solder. **Note:** *Be careful not to apply heat to the solder joint for more than 5 seconds. If necessary, install a heat sink to capture the excess heat. This can be accomplished by clamping a pair of needle-nose pliers next to the solder joint.*
6 On the screw type, hold the assembly in place and install the screws. Tighten them evenly, a little at a time, so the holder isn't distorted.
7 Install the regulator assembly onto the alternator.
8 Reconnect the negative battery cable.

Motorola alternator

9 Remove the alternator.
10 The brushes are mounted under the regulator on the rear of the alternator **(see illustration 16.7)**.
11 Remove the mounting screws and insulating washers and separate the voltage regulator and brush holder from the brush end housing.
12 Measure the length of the brushes **(see illustration 17.3)**. If they're less then 7/32-inch long, replace them.
13 Make sure the brushes move smoothly in the holder.
14 Reinstall the brush holder/regulator. Tighten the screws securely. Make sure the brushes aren't grounded.
15 Installation is the reverse of removal.

18 Starting system – general information and precautions

The sole function of the starting system is to turn over the engine quickly enough to allow it to start.
The starting system consists of the battery, the starter motor, the starter solenoid and the wires connecting them. The solenoid is mounted directly on the starter motor. The starter/solenoid motor assembly is installed on the lower part of the engine, next to the transmission bellhousing.

When the ignition key is turned to the Start position, the starter solenoid is actuated through the starter control circuit. The starter solenoid then connects the battery to the starter. The battery supplies the electrical energy to the starter motor, which does the actual work of cranking the engine.
The starter motor on a vehicle equipped with a manual transmission can only be operated when the clutch pedal is depressed; the starter on a vehicle equipped with an automatic transmission can only be operated when the transmission selector lever is in Park or Neutral.
Always observe the following precautions when working on the starting system:
a) Excessive cranking of the starter motor can overheat it and cause serious damage. Never operate the starter motor for more than 10-seconds at a time without pausing to allow it to cool for at least two minutes.
b) The starter is connected directly to the battery and could arc or cause a fire if mishandled, overloaded or shorted out.
c) Always detach the cable from the negative terminal of the battery before working on the starting system.

19 Starter motor – in-vehicle check

Note: *Before diagnosing starter problems, make sure the battery is fully charged.*

1 If the starter motor does not turn at all when the switch is operated, make sure that the shift lever is in Neutral or Park (automatic transmission) or that the clutch pedal is depressed (manual transmission).
2 Make sure that the battery is charged and that all cables, both at the battery and starter solenoid terminals, are clean and secure.
3 If the starter motor spins but the engine is not cranking, the over-running clutch in the starter motor is slipping and the starter motor must be replaced.
4 If, when the switch is actuated, the starter motor does not operate at all but the solenoid clicks, then the problem lies with either the battery, the main solenoid contacts or the starter motor itself (or the engine is seized).
5 If the solenoid plunger cannot be heard when the switch is actuated, the battery is bad, the fusible link is burned (the circuit is open) or the solenoid itself is defective.
6 To check the solenoid, connect a jumper lead between the battery (+) and the ignition switch wire terminal (the small terminal) on the solenoid. If the starter motor now operates, the solenoid is OK and the problem is in the ignition switch, neutral start switch or the wiring.
7 If the starter motor still does not operate, remove the starter/solenoid assembly for disassembly, testing and repair.
8 If the starter motor cranks the engine at an abnormally slow speed, first make sure that the battery is charged and that all terminal connections are tight. If the engine is partially seized, or has the wrong viscosity oil in it, it will crank slowly.
9 Run the engine until normal operating temperature is reached, then disconnect the coil wire from the distributor cap and ground it on the engine.
10 Connect a voltmeter positive lead to the positive battery post and connect the negative lead to the negative post.
11 Crank the engine and take the voltmeter readings as soon as a steady figure is indicated. Do not allow the starter motor to turn for more than 10 seconds at a time. A reading of 9 volts or more, with the starter motor turning at normal cranking speed, is normal. If the reading is 9 volts or more but the cranking speed is slow, the motor is faulty. If the reading is less than 9 volts and the cranking speed is slow, the solenoid contacts are probably burned, the starter motor is bad, the battery is discharged or there is a bad connection.

20 Starter motor – removal and installation

Refer to illustration 20.4

1 Detach the cable from the negative terminal of the battery. **Cau-**

20.4 Working under the vehicle, remove the starter lower mounting bolt and nut (arrows)

tion: *If the radio in your vehicle is equipped with an anti-theft system, make sure you have the correct activation code before disconnecting the battery, Refer to the information on page 0-7 at the front of this manual before detaching the cable.* **Note***: If, after connecting the battery, the wrong language appears on the instrument panel display, refer to page 0-7 for the language resetting procedure.*

2 Raise the vehicle and support it securely on jackstands.
3 Clearly label, then disconnect the wires from the terminals on the starter motor and solenoid. **Note:** *On some models it may be necessary to remove the coolant expansion tank (see Chapter 3) and the heater hoses to gain access to the top of the starter. Carefully label any hoses or components that need to be removed from the engine compartment to avoid confusion when reassembling.*
4 Remove the mounting bolts **(see illustration)** and detach the starter.
5 Installation is the reverse of removal.

21 Starter solenoid – removal and installation

1 Disconnect the cable from the negative terminal of the battery. **Caution:** *If the radio in your vehicle is equipped with an anti-theft system, make sure you have the correct activation code before disconnecting the battery, Refer to the information on page 0-7 at the front of this manual before detaching the cable.* **Note:** *If, after connecting the battery, the wrong language appears on the instrument panel display, refer to page 0-7 for the language resetting procedure.*
2 Remove the starter motor (see Section 20).
3 Disconnect the strap from the solenoid to the starter motor terminal.
4 Remove the screws which secure the solenoid to the starter motor.
5 Detach the solenoid from the starter body.
6 Remove the plunger and plunger spring.
7 Installation is the reverse of removal.

Chapter 6 Emissions and engine control systems

Contents

Catalytic converter	7
EFI system self diagnosis capability (Motronic 1.1 systems only)	2
Electronic Control Unit (ECU) - general information and replacement	3
Evaporative emissions control (EVAP) system	6
Evaporative emissions control system check	See Chapter 1
General information	1
Information sensors	4
Positive crankcase ventilation (PCV) system	5

1 General information

Refer to illustration 1.6

To prevent pollution of the atmosphere from incompletely burned or evaporating gases, and to maintain good driveability and fuel economy, a number of emission control systems are used on these vehicles. They include the:

Catalytic converter (CAT) system
Evaporative emission control (EVAP) system
Positive crankcase ventilation (PCV) system
Electronic engine controls

The Sections in this Chapter include general descriptions and checking procedures within the scope of the home mechanic and component replacement procedures (when possible) for each of the systems listed above.

Before assuming that an emissions control system is malfunctioning, check the fuel and ignition systems carefully. The diagnosis of some emission control devices requires specialized tools, equipment and training. If checking and servicing become too difficult or if a procedure is beyond your ability, consult a dealer service department or other repair shop. Remember - the most frequent cause of emissions problems is simply a loose or broken vacuum hose or wire, so always check the hose and wiring connections first.

This doesn't mean, however, that emission control systems are particularly difficult to maintain and repair. You can quickly and easily perform many checks and do most of the regular maintenance at home with common tune-up and hand tools. **Note**: *Because of a Federally* mandated extended warranty which covers the emission control system components, check with your dealer about warranty coverage before working on any emissions-related systems. Once the warranty has expired, you may wish to perform some of the component checks and/or replacement procedures in this Chapter to save money.

Pay close attention to any special precautions outlined in this Chapter. It should be noted that the illustrations of the various systems may not exactly match the system installed on your vehicle because of changes made by the manufacturer during production or from year to year.

A Vehicle Emissions Control Information (VECI) label is located in the engine compartment (see illustration). This label contains important emissions specifications and adjustment information. When servicing the engine or emissions systems, the VECI label in your particular vehicle should always be checked for up-to-date information. You will also find the accompanying vacuum hose diagrams helpful when you troubleshoot emissions problems.

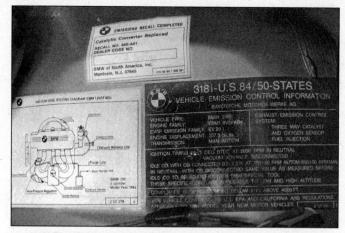

1.6 The Vehicle Emission Control Information (VECI) label contains such essential information as the types of emission control systems installed on the engine and the idle speed and ignition timing specifications (also notice that there is an additional sticker indicating a catalytic converter recall that was performed on the exhaust system)

2 EFI system self-diagnosis capability (Motronic 1.1 systems only)

The EFI system control unit (computer) has a built-in self-diagnosis system which detects malfunctions in the system sensors and alerts the driver by illuminating a CHECK ENGINE warning light in the instrument panel. The computer stores the failure code until the diagnostic system is cleared by removing the negative battery cable for a period of five seconds or longer. The warning light goes out automatically (after five engine starts) when the malfunction is repaired. **Caution:** *If the radio in your vehicle is equipped with an anti-theft system, make sure you have the correct activation code before disconnecting the battery. Refer to the information on page 0-7 at the front of this manual before detaching the cable.* **Note**: *If, after connecting the battery, the wrong language appears on the instrument panel display, refer to page 0-7 for the language resetting procedure.*

The CHECK ENGINE warning light should come on when the ignition switch is placed in the On position. When the engine is started, the warning light should go out. The light will remain on (with the engine running) once the diagnostic system has detected a malfunction or abnormality in the system. In order to read the codes on 3-Series models (1989 and later only), it is necessary to turn the key to the ON position (engine not running) and wait (approximately three-Seconds) for any stored codes to be displayed. In order to read the codes on 5-Series models (1989 and later only), it is necessary to turn the key to the ON position (engine not running), depress the accelerator pedal 5 times (make sure the pedal reaches wide open throttle [WOT] each time) and wait for any stored codes to be displayed.

The diagnostic code is the number of flashes indicated on the CHECK ENGINE light. If any malfunction has been detected, the light will blink the digit(s) of the code. For example on 3-Series models, code 3 (coolant temperature sensor malfunction) will blink three flashes. There will be a pause (3 seconds) and then any other codes that are stored will b flashed. On 5-Series models, code 1223 (coolant temperature sensor malfunction) will flash the first digit and then pause, flash the second digit (2 flashes) pause, flash the third digit (2 flashes) pause and finally flash the fourth digit (3 flashes). There will be another pause and the ECU will start the next stored trouble code (if any) or it will repeat the code 1223. Once all the codes have been displayed, the CHECK ENGINE light will remain on. In order to re-check the codes, simply turn the ignition key OFF and then back ON (repeat procedure) and the codes will be repeated.

The accompanying tables explain the code that will be flashed for each of the malfunctions. The accompanying charts indicate the diagnostic code - in blinks - along with the system, diagnosis and specific areas. **Note**: *Diagnostic codes that are not emissions or engine control related (electronic transmission, ABS, etc.) will not be accessed by this system. Take the vehicle to a dealer service department for the diagnostic procedures.*

1989 AND LATER 3-SERIES DIAGNOSTIC CODES

Code number and sensor/circuit	Malfunction	Remedy
Code 1 Airflow meter	Airflow meter flap binding Airflow meter potentiometer faulty	Check the airflow meter (see Chapter 4)
Code 2 Oxygen sensor	Oxygen sensor faulty Fuel mixture too lean or too rich Faulty fuel or ignition system causing imbalance	Test the oxygen sensor (see Section 4) Have exhaust gas analyzed
Code 3 Coolant temperature sensor	Faulty coolant temperature sensor Faulty cooling system causing overheating	Test the coolant temperature sensor (see Section 4)
Code 4 Throttle position sensor	Faulty TPS Adjustment incorrect	Test the TPS (see Section 4)

Chapter 6 Emissions and engine control systems

1989 AND LATER 5-SERIES DIAGNOSTIC CODES

CODE	PROBABLE CAUSE
1000	End of diagnosis
1211	Electronic Control Unit (ECU)
1215	Airflow sensor
1221	Oxygen sensor
1222	Oxygen sensor regulation
1223	Coolant temperature sensor
1224	Air temperature sensor
1231	Battery voltage out of range
1232	Idle switch
1233	Full throttle switch
1251	Fuel injectors (final stage 1)
1252	Fuel injectors (final stage 2)
1261	Fuel pump relay
1262	Idle speed controller
1263	Tank vent
1264	Oxygen sensor heating relay
1444	No faults in memory

Check the indicated system or component or take the vehicle to a dealer service department to have the malfunction repaired. After repairs have been made, the diagnostic code can be canceled by disconnecting the negative battery cable for 5 seconds or longer. After cancellation, perform a road test and make sure the warning light does not come on. If desired, the check can be repeated. **Caution:** *If the radio in your vehicle is equipped with an anti-theft system, make sure you have the correct activation code before disconnecting the battery, Refer to the information on page 0-7 at the front of this manual before detaching the cable.* **Note**: *If, after connecting the battery, the wrong language appears on the instrument panel display, refer to page 0-7 for the language resetting procedure.*

3 Electronic control unit (ECU) - general information and replacement

1 The Electronic Control Unit (ECU) is located either inside the passenger compartment under the dashboard (right side) or in the engine compartment (see Chapter 4).
2 Disconnect the negative battery cable from the battery. **Caution:** *If the radio in your vehicle is equipped with an anti-theft system, make sure you have the correct activation code before disconnecting the battery, Refer to the information on page 0-7 at the front of this manual before detaching the cable.* **Note:** *If, after connecting the battery, the wrong language appears on the instrument panel display, refer to page 0-7 for the language resetting procedure.*
3 First remove the access cover from the right side of the engine compartment to locate the ECU on models equipped with the ECU (see Chapter 4) up front (usually late models with Motronic 1.1).
4 If the ECU is located under the dashboard on the passenger side (see Chapter 4), remove the lower trim panel.
5 Unplug the electrical connectors from the ECU.
6 Remove the retaining nut from the ECU bracket.
7 Carefully remove the ECU. **Note**: *Avoid any static electricity damage to the computer by using gloves and a special anti-Static pad to store the ECU on once it is removed.*

4 Information sensors

Note: *Refer to Chapters 4 and 5 for additional information on the location and diagnosis of the information sensors that are not directly covered in this section.*

Coolant temperature sensor

Refer to illustrations 4.1 and 4.2

General description

1 The coolant temperature sensor **(see illustration)** is a thermistor (a resistor which varies its resistance value in accordance with temperature changes). The change in the resistance value regulates the amount of voltage that can pass through the sensor. As the sensor temperature DECREASES, the resistance values will INCREASE. As the sensor temperature INCREASES, the resistance values will DECREASE. A failure in this sensor circuit should set a Code 3 (3-Series models) or a code 1223 (5-Series models). This code indicates a failure in the coolant temperature sensor circuit, so in most cases the appropriate solution to the problem will be either repair of a wire or replacement of the sensor.

Check

2 To check the sensor, check the resistance value **(see illustration)** of the coolant temperature sensor while it is completely cold (50 to 80-degrees F = 2,100 to 2,900 ohms). Next, start the engine and warm it up until it reaches operating temperature. The resistance should be lower (180 to 200 degrees F = 270 to 400 ohms). **Note**: *If the lack of access to the coolant temperature sensor makes it difficult to position electrical probes on the terminals, remove the sensor and perform the tests in a pan of heated water to simulate the conditions.* **Note**: *In the case of extremely cold temperatures (about 15-degrees F) the sensor should read 8,000 to 11,000 ohms.*

4.1 The coolant temperature sensor (arrow) is usually located next to the temperature sending unit near the fuel pressure regulator

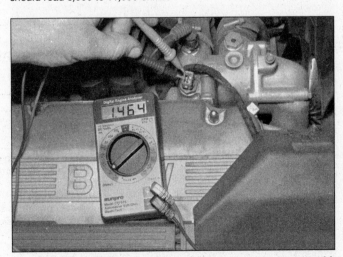

4.2 Check the resistance of the coolant temperature sensor cold and warm

Chapter 6 Emissions and engine control systems

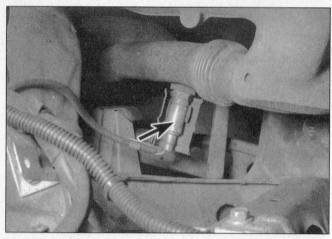

4.6 The oxygen sensor (arrow) is usually located in the exhaust pipe, downstream from the exhaust manifold

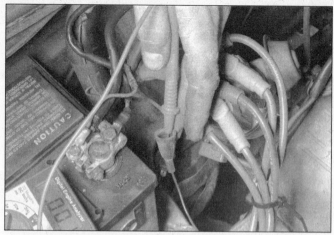

4.12a The oxygen sensor, once it is warmed up (600-degrees F), puts out a very small voltage signal. To verify it is working, check for voltage with a digital voltmeter (the voltage signals usually range from 0.02 to 0.85-volts)

Replacement

Warning: *Wait until the engine is completely cool before beginning this procedure.*

3 Before installing the new sensor, wrap the threads with Teflon tape to prevent leakage and thread corrosion.
4 To remove the sensor, depress the spring lock, unplug the electrical connector, then carefully unscrew the sensor. Be prepared for some coolant spillage and install the new sensor as quickly as possible. **Caution:** *Handle the coolant sensor with care. Damage to this sensor will affect the operation of the entire fuel injection system.* **Note:** *It may be necessary to drain a small amount of the coolant from the radiator before removing the sensor.*
5 Installation is the reverse of removal.

Oxygen sensor

General description

Refer to illustration 4.6

Note: *Most oxygen sensors are located in the exhaust pipe, downstream from the exhaust manifold. On 535 models, the oxygen sensor is mounted in the catalytic converter. The sensor's electrical connector is located near the firewall (left side) for easy access.*

6 The oxygen sensor, which is located in the exhaust system **(see illustration)**, monitors the oxygen content of the exhaust gas stream. The oxygen content in the exhaust reacts with the oxygen sensor to produce a voltage output which varies from 0.1-volt (high oxygen, lean mixture) to 0.9-volts (low oxygen, rich mixture). The ECU constantly monitors this variable voltage output to determine the ratio of oxygen to fuel in the mixture. The ECU alters the air/fuel mixture ratio by controlling the pulse width (open time) of the fuel injectors. A mixture ratio of 14.7 parts air to 1 part fuel is the ideal mixture ratio for minimizing exhaust emissions, thus allowing the catalytic converter to operate at maximum efficiency. It is this ratio of 14.7 to 1 which the ECU and the oxygen sensor attempt to maintain at all times.
7 The oxygen sensor produces no voltage when it is below its normal operating temperature of about 600-degrees F. During this initial period before warm-up, the ECU operates in open loop mode.
8 If the engine reaches normal operating temperature and/or has been running for two or more minutes, and if the oxygen sensor is producing a steady signal voltage below 0.45-volts at 1,500 rpm or greater, the ECU will set a Code 2 (3-Series models) or a code 1221, 1222 or 1264 (5-Series models).
9 When there is a problem with the oxygen sensor or its circuit, the ECU operates in the open loop mode - that is, it controls fuel delivery in accordance with a programmed default value instead of feedback information from the oxygen sensor.
10 The proper operation of the oxygen sensor depends on four conditions:
 a) Electrical - The low voltages generated by the sensor depend

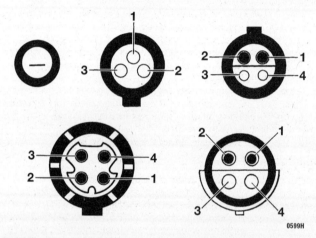

4.12b These oxygen sensor terminal designations are for the harness side only. Use the corresponding terminals on the sensor side for the testing procedures (be sure to note there are three different four-wire oxygen sensor connectors available - don't get them mixed up)

upon good, clean connections which should be checked whenever a malfunction of the sensor is suspected or indicated.
 b) Outside air supply - The sensor is designed to allow air circulation to the internal portion of the sensor. Whenever the sensor is removed and installed or replaced, make sure the air passages are not restricted.
 c) Proper operating temperature - The ECU will not react to the sensor signal until the sensor reaches approximately 600-degrees F. This factor must be taken into consideration when evaluating the performance of the sensor.
 d) Unleaded fuel - The use of unleaded fuel is essential for proper operation of the sensor. Make sure the fuel you are using is of this type.
11 In addition to observing the above conditions, special care must be taken whenever the sensor is serviced.
 a) The oxygen sensor has a permanently attached pigtail and electrical connector which should not be removed from the sensor. Damage or removal of the pigtail or electrical connector can adversely affect operation of the sensor.
 b) Grease, dirt and other contaminants should be kept away from the electrical connector and the louvered end of the sensor.
 c) Do not use cleaning solvents of any kind on the oxygen sensor.
 d) Do not drop or roughly handle the sensor.

Chapter 6 Emissions and engine control systems

e) The silicone boot must be installed in the correct position to prevent the boot from being melted and to allow the sensor to operate properly.

Check

Refer to illustration 4.12a and 4.12b

12 Warm up the engine and let it run at idle. Disconnect the oxygen sensor electrical connector and connect the positive probe of a voltmeter to the oxygen sensor connector terminal (output signal) (see chart) and the negative probe to ground **(see illustrations)**.
Note: *Most oxygen sensor electrical connectors are located at the rear of the engine, near the firewall. Look for a large rubber boot attached to a thick wire harness. On 1985 and 1986 535i models, the connector for the oxygen sensor heater circuit is directly under the driver's seat, under the vehicle. Look for a small protective cover. These models should have the updated oxygen sensor installed to make the access similar to other models. Consult your dealer service department for additional information.*

13 Increase and then decrease the engine speed and monitor the voltage.
14 When the speed is increased, the voltage should increase to 0.5 to 1.0 volts. When the speed is decreased, the voltage should decrease to about 0 to 0.4 volts.
15 Also inspect the oxygen sensor heater (models with multi-wire sensors). With the ignition key ON, disconnect the oxygen sensor electrical connector and connect a voltmeter between the designated terminal in the chart (see below). There should be battery voltage (approximately 12 volts). **Note:** *Early 318i models are not equipped with an oxygen sensor heater.*
16 If the reading is not correct, check the oxygen sensor heater relay (see Chapter 12). If the information is not available, check your owner's manual for the exact location of the oxygen sensor heater relay. The relay should receive battery voltage.
17 If the oxygen sensor fails any of these tests, replace it.

Replacement

Note: *Because it is installed in the exhaust manifold, converter or pipe, which contracts when cool, the oxygen sensor may be very difficult to loosen when the engine is cold. Rather than risk damage to the sensor (assuming you are planning to reuse it in another manifold or pipe), start and run the engine for a minute or two, then shut it off. Be careful not to burn yourself during the following procedure.*

18 Disconnect the cable from the negative terminal of the battery.
Caution: *If the radio in your vehicle is equipped with an anti-theft system, make sure you have the correct activation code before disconnecting the battery. Refer to the information on page 0-7 at the front of this manual before detaching the cable.* **Note:** *If, after connecting the battery, the wrong language appears on the instrument panel display, refer to page 0-7 for the language resetting procedure.*
19 Raise the vehicle and place it securely on jackstands.
20 Disconnect the electrical connector from the sensor.
21 Carefully unscrew the sensor. **Caution:** *Excessive force may damage the threads.*
22 Anti-Seize compound must be used on the threads of the sensor to facilitate future removal. The threads of new sensors will already be coated with this compound, but if an old sensor is removed and reinstalled, recoat the threads.
23 Install the sensor and tighten it securely.
24 Reconnect the electrical connector of the pigtail lead to the main engine wiring harness.

4.28a The TPS on L-Jetronic systems (1984 and 1985 318i only) is located under the intake manifold (arrow)

25 Lower the vehicle and reconnect the cable to the negative terminal of the battery.

Throttle Position Sensor (TPS)

General description

26 The Throttle Position Sensor (TPS) is located on the end of the throttle shaft on the throttle body. By monitoring the output voltage from the TPS, the ECU can determine fuel delivery based on throttle valve angle (driver demand). In this system the TPS acts as a switch rather than a potentiometer. One set of throttle valve switch contacts is closed (continuity) only at idle. A second set of contacts closes as the engine approaches full throttle. Both sets of contacts are open (no continuity) between these positions. A broken or loose TPS can cause intermittent bursts of fuel from the injector and an unstable idle because the ECU thinks the throttle is moving. On Motronic 1.1 systems, the self diagnosis system will set a code 4 in the event of TPS malfunction.
27 All models combine the idle and full throttle switch except for the 535i models (1986 through 1988) equipped with an automatic transmission. These systems use a separate idle position switch for the closed throttle position while the TPS sensor is used for the full throttle position. 535i models equipped with an automatic transmission, wire the throttle position sensor (TPS) directly to the automatic transmission control unit. With the throttle fully open, the A/T control unit sends the signal to the Motronic control unit that the accelerator pedal is fully open.

All 1986 through 1988 models except 535i with automatic transmission

Check

Refer to illustrations 4.28a, 4.28b and 4.28c

28 Remove the electrical connector from the TPS and install the probes of an ohmmeter onto terminal nos. 2 and 18 **(see illustrations)**

Oxygen sensor type	Sensor output signal	Heated power supply (12V)
Unheated (single wire)	black wire (+)	Not applicable
Heated (three wire)	terminal 1 (+)	terminals 3 (+) and 2 (-)
Heated (four wire)	terminal 2 (+)	terminals 4 (+) and 3 (-)

6-6   Chapter 6 Emissions and engine control systems

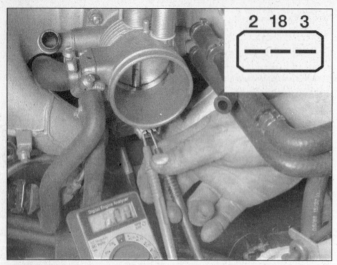

4.28b First check for continuity at the closed throttle position on terminal nos. 2 and 18 (Motronic 1.1 system shown) . . .

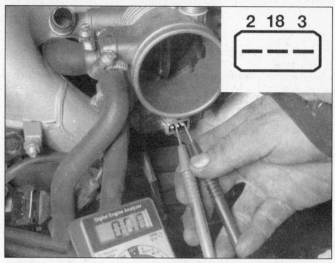

4.28c . . . then check for continuity at the full throttle position on terminal nos. 3 and 18

and using your hand, open the throttle partially. Allow the throttle to slowly release until it reaches 0.008 to 0.024 inches from the throttle stop. There should be continuity.

29 Also, check the resistance between terminal nos. 3 and 18 and open the throttle. There should be continuity when the throttle switch is within 8 to 12 degrees of fully open. If the readings are incorrect, adjust the TPS sensor.

30 If all the resistance readings are correct and the TPS is properly adjusted, check for power (5-volts) at the sensor and if necessary trace down any wiring circuit problems between the sensor and computer (see Chapter 12).

Adjustment

Note: *Adjust the TPS when installing a new one or when the original TPS was removed or diagnosed to be out of adjustment.*

31 If the adjustment is not as specified (steps 28 through 30), loosen the screws on the TPS and rotate the TPS sensor into the correct adjustment. Follow the procedure for checking the TPS and tighten the screws when the correct resistance coincides with the distance from the throttle stop.

32 Recheck the TPS sensor adjustment and if the readings are correct, reconnect the TPS harness connector.

535i models (1986 through 1988)

Check

Refer to illustration 4.34

33 First test the continuity of the TPS. Follow steps 28 through 30 and check for continuity. Use the same terminal designations.

34 Next, test the idle position switch **(see illustration)**. Unplug the electrical connector in the idle position switch harness and install the probes of an ohmmeter onto terminal no. 1 and no. 2. There should be continuity. Now, partially open the throttle and measure the resistance. There should be NO continuity.

35 Next, check for the correct voltage signals from the TPS (throttle closed). Probe the backside of the TPS connector with a voltmeter and with the ignition key ON, check for voltage at terminal no. 3 (black wire) and ground. There should be 5-volts present. Also, probe terminal no. 3 (black wire) and terminal no. 1 (brown wire). There should also be 5-volts present.

36 Check for voltage at terminal no. 2 (yellow wire) and terminal no. 1 (brown wire) and slowly open the throttle. Voltage should increase steadily from 0.7 volts to 4.8-volts (throttle fully open).

Adjustment

Note: *Adjust the TPS when installing a new one or when the original TPS was removed or diagnosed to be out of adjustment.*

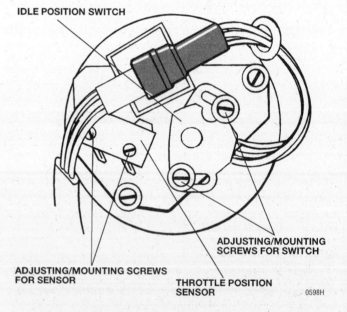

4.34 Idle position switch and TPS on 1986 through 1988 535i models with an automatic transmission

37 First measure the stabilized voltage. With the ignition ON and the throttle fully closed, measure the voltage across terminal no. 3 (black wire) and terminal no. 1 (brown wire). It should be about 5-volts.

38 Next, loosen the sensor mounting screws and connect a voltmeter to terminal no. 2 (yellow wire) and terminal no. 3 (black wire). With the throttle fully open, rotate the switch until there is 0.20 to 0.24 volts less than the stabilized voltage. **Note**: *You will need a digital voltmeter to measure these small voltage amounts.*

39 Recheck the TPS sensor adjustment and if the readings are correct, reconnect the TPS electrical connector. It is a good idea to lock the TPS screws with paint or thread locking compound.

Airflow Meter

General description

40 The airflow meter is located on the air intake duct. The airflow meter measures the amount of air entering the engine. The ECU uses this information to control fuel delivery. A large volume of air indicates ac-

Chapter 6 Emissions and engine control systems

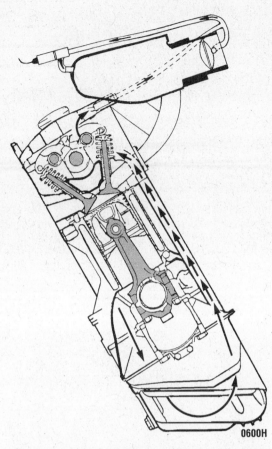

5.1 Schematic of the PCV system on a 325e model (others similar)

5.2 PCV hose being removed from the valve cover

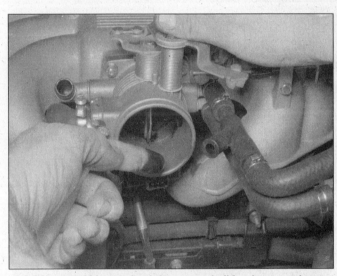

5.3 It's a good idea to check for excess buildup from crankcase vapors and residue circulating in the hoses and ports - this can eventually clog the system and cause pressure increase in the engine block

celeration, while a small volume of air indicates deceleration or idle. Refer to Chapter 4 for all the diagnostic checks and replacement procedures for the airflow meter.

Ignition timing sensors

41   Ignition timing is electronically controlled and is not adjustable on Motronic systems. During starting, a crankshaft position sensor (reference sensor) relays the crankshaft position to the ECU and an initial baseline ignition point is determined. Once the engine is running, the ignition point is continually changing based on the various input signals to the ECU. Engine speed is signaled by a speed sensor. Early Motronic systems are equipped with the reference sensor and the speed sensor mounted on the bellhousing over the flywheel on the driver's side while later Motronic 1.1 systems are equipped with a single sensor (pulse sensor) mounted up-front over the crankshaft pulley. This sensor functions as a speed sensor as well as a reference sensor (angle). Refer to Chapter 5 for checking and replacing the ignition sensors. **Note:** *Some models are equipped with a TDC sensor mounted on the front of the engine. This sensor is strictly for the BMW service test unit and it is not part of the Motronic ignition system.*

5 Positive crankcase ventilation (PCV) system

Refer to illustrations 5.1, 5.2 and 5.3

1   The Positive Crankcase Ventilation (PCV) system **(see illustration)** reduces hydrocarbon emissions by scavenging crankcase vapors. It does this by circulating blow-by gases and then rerouting them to the intake manifold by way of the air cleaner.
2   This PCV system is a sealed system. The crankcase blow-by vapors are routed directly to the air cleaner or air collector with crankcase pressure behind them. The vapor is not purged with fresh air on most models or filtered with a flame trap like most conventional systems. There are no conventional PCV valves installed on these systems - just a hose **(see illustration)**.
3   The main components of the PCV system are the hoses **(see illustration)** that connect the valve cover to the throttle body or air cleaner. If abnormal operating conditions (such as piston ring problems) arise, the system is designed to allow excessive amounts of blow-by gases to flow back through the crankcase vent tube into the intake system to be consumed by normal combustion. **Note:** *Since these models don't use a filtering element, it's a good idea to check the PCV system passageways for clogging from sludge and combustion residue* **(see illustration)**.

6 Evaporative emissions control (EVAP) system

General description

Refer to illustration 6.1

1   The evaporative emissions control system **(see illustration)**

6-8 Chapter 6 Emissions and engine control systems

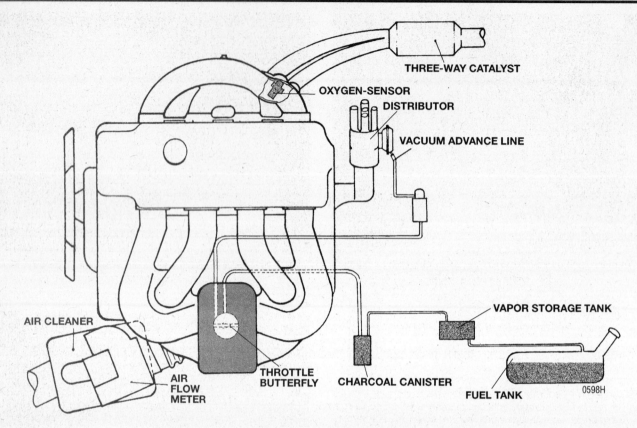

6.1 Schematic of the EVAP system on a 318i engine (others similar)

stores fuel vapors generated in the fuel tank in a charcoal canister when the engine isn't running. When the engine is started, the fuel vapors are drawn into the intake manifold and burned. The evaporative emission control system works like this: When the engine is cruising, the purge control valve (bypass valve) is opened slightly and a small amount of fuel vapor is drawn into the intake manifold and burned. When the engine is starting cold or idling, the bypass valve prevents any vapors from entering the intake manifold causing excessively rich fuel mixture.

2 Two types of purge valves or bypass valves are used on these models; an electrically operated valve or a vacuum-operated valve. To find out which type is on your vehicle, follow the hose from the charcoal canister until you locate the purge valve. Some are located on the intake manifold and others near the charcoal canister. Look for an electrical connector to the purge valve (electrically operated) or vacuum lines (vacuum-operated). **Note:** *325i models produced in December 1986 may be equipped with faulty charcoal canisters that can cause hard starting and rough idle during warm temperatures. Check for a white dot on top of the canister next to the production stamp. If a white dot is not present, contact a dealer service department for the VIN numbers of the models that are defective. These canisters have loose charcoal particles that can be drawn into the bypass valve and cause it to stick open.* **Note:** *1982 through 1984 528e models may have had the purge valve replaced as a dealer recall to prevent a rough idle condition. Look for a special BMW sticker that describes the retrofit EVAP kit and the date. Also, 528e models built before March 1987 may have a small restrictor valve installed in the hose between the charcoal canister and the throttle body. This valve eliminates poor throttle response during warm engine acceleration. Check the dealer service department for the necessary information on this BMW recall.*

3 A faulty EVAP system affects the engine driveability only when the temperatures are warm. The EVAP system is not usually the cause of hard cold starting or any other cold running problems.

Check
Vacuum-operated purge valve

4 Remove the vacuum lines from the purge valve and blow into the larger port of the valve. It should be closed and not pass any air. **Note:** *Some models are equipped with a thermo-vacuum valve that prevents canister purge until the coolant temperature reaches approximately 115-degrees F. Check this valve to make sure that vacuum is controlled at the proper temperatures. The valve is usually located in the intake manifold, near the thermo-time switch and the coolant temperature sensor.*

5 Disconnect the small vacuum hose from the purge valve and apply vacuum with a hand-held vacuum pump. The purge valve should be open and air should be able to pass through.

6 If the test results are incorrect, replace the purge valve with a new part.

Electrically operated purge valve

Refer to illustration 6.8 and 6.9

7 Disconnect any lines from the purge valve and without disconnecting the electrical connector, place it in a convenient spot for testing.

8 Check that the valve makes a "click" sound as the ignition key is turned to the On position **(see illustration)**.

9 If the valve does not "click", disconnect the valve connector and check for power to the valve using a test light or a voltmeter **(see illustration)**.

10 If there is battery voltage, replace the purge valve. If there is no voltage present, check the Motronic control unit and the wiring harness for any shorts or faulty components.

Canister

11 Label, then detach all hoses to the canister.
12 Slide the canister out of its mounting clip.

Chapter 6 Emissions and engine control systems

6.8 When the ignition key is turned ON, there should be a distinct "click" from the purge valve

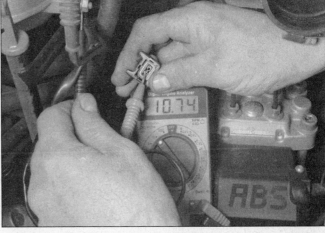

6.9 Check for battery voltage at the electrical connector to the purge valve

13 Visually examine the canister for leakage or damage.
14 Replace the canister if you find evidence of damage or leakage.

7 Catalytic converter

Note: *Because of a Federally mandated extended warranty which covers emissions-related components such as the catalytic converter, check with a dealer service department before replacing the converter at your own expense.*

General description

Refer to illustration 7.1

1 To reduce hydrocarbons (HC), carbon monoxide (CO) and oxides of nitrogen (NOx), the vehicles covered by this manual are equipped with a catalytic converter **(see illustration)**.

Check

2 Visually examine the converters for cracks or damage. Make sure all fasteners are tight.
3 Inspect the insulation covers (if equipped) welded onto the converters - they should be tight. **Caution:** *If an insulation cover is touching a converter housing, excessive heat at the floor may result.*
4 Start the engine and run it at idle speed.
5 Check for exhaust gas leakage from the converter flanges. Check the body of each converter for holes.

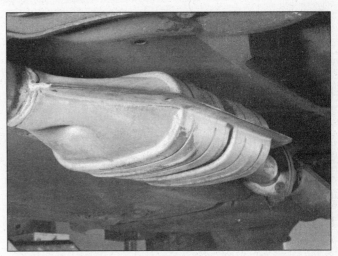

7.1 Typical catalytic converter (1984 318i shown, others similar)

Component replacement

6 See Chapter 4 for removal and installation procedures.

Notes

Chapter 7 Part A Manual transmission

Contents

General information... 1	Shift lever - removal and installation... 2
Manual transmission lubricant change........................... See Chapter 1	Transmission - removal and installation.. 5
Manual transmission lubricant level check..................... See Chapter 1	Transmission mounts - check and replacement............................ 4
Oil seals - replacement... 3	Transmission overhaul - general information................................ 6

Specifications

Torque Specifications

Ft-lbs

Output shaft flange nut
   3-Series
      ZF transmission .... 74 to 89
      Getrag 240 and 260 transmissions
         First stage.. 125
         Second stage 89
   5-Series .... See Getrag above
Transmission-to-engine hex-head bolts
   M8 .... 16 to 20
   M10 .... 35 to 38
   M12 .... 49 to 60
Transmission-to-engine Torx-head bolts
   M8 .... 15 to 18
   M10 .... 28 to 35
   M12 .... 47 to 59
Rear transmission support nut .... 16 to 18
Rubber mount nut (to transmission or support) .... 32 to 35
Transmission drain plug/fill plug..................................... 30 to 44

7A-2 Chapter 7 Part A Manual transmission

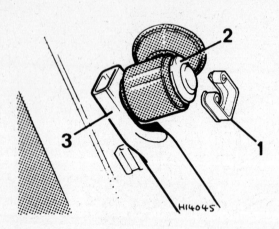

2.5 Working from under the vehicle, remove the retaining clip (1) to disconnect the shift lever (2) from the shift rod (3)

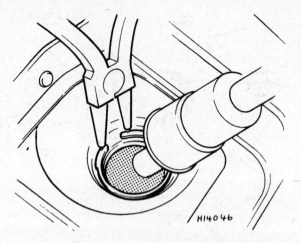

2.6 On models with a sheet-metal shift console, remove the snap-ring from the top of the shift lever bearing and lift out the shift lever

1 General information

The manual transmission used in the vehicles covered by this manual is a five-speed Getrag or ZF. To identify a transmission, look for the manufacturer's stamp and code numbers and letters. The manufacturer's stamp is on the case, just in front of the mounting for the clutch release cylinders; the identification code and letters are located on top of the bellhousing.

2 Shift lever - removal and installation

Refer to illustrations 2.5, 2.6 and 2.7

1 Put the transmission in reverse.
2 Pull the shift lever knob up and off the shift lever.
3 Remove the shift boot and the sound deadening material underneath.
4 Remove the rubber dust cover. On models with an aluminum shift console, unplug the connector for the backup light switch before removing the dust cover.
5 Working from under the vehicle, disconnect the shift lever from the shift rod by removing the retaining clip **(see illustration)**.
6 On models with a sheet-metal shift console, remove the snap-ring from the top of the shift lever bearing **(see illustration)** and lift out the shift lever.
7 On models with an aluminum shift console, use two screwdrivers to engage the slots of the bearing retaining ring from below the lower shift lever bearing. To unlock the ring, turn it counterclockwise 1/4-turn **(see illustration)** and lift out the shift lever.
8 Installation is the reverse of the removal procedure.

3 Oil seals - replacement

1 Worn or damaged transmission oil seals can cause lubricant leaks. Insufficient lubricant in the transmission can cause hard shifting, jumping out of gear and transmission noise.
2 All three transmission seals can be replaced without disassembling the transmission; two of them can be replaced without even removing the transmission.
3 If you suspect a leaking seal, get underneath the transmission and inspect the seals closely. Raise the front of the vehicle and support it securely on jackstands. Be sure to block the rear wheels to keep the vehicle from rolling.
4 If transmission lubricant is coating the front of the driveshaft, either the selector shaft or the output shaft seal is leaking. If you find oil

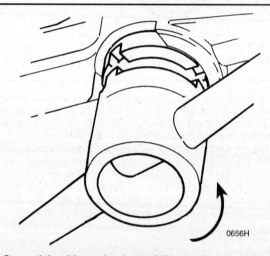

2.7 On models with an aluminum shift console, turn the bearing retaining ring counterclockwise 1/4-turn and lift out the shift lever (special tool shown here, but two screwdrivers, with tips engaged in opposite slots, can be used)

in the bottom of the bellhousing, the input shaft seal may be leaking (of course, the crankshaft rear seal could also cause oil to collect in the bottom of the bellhousing). Whenever you decide to replace a seal, always check the vent on top of the transmission housing first. A clogged or damaged vent can cause pressure inside the transmission to rise enough to pump lubricant past the seals. If you simply replace a seal without cleaning out a clogged vent, the new seal will quickly fail.

Output shaft oil seal

Refer to illustration 3.9

5 Raise the front of the vehicle and support it securely on jackstands. Be sure to block the rear wheels to keep the vehicle from rolling.
6 Remove the exhaust system (see Chapter 4).
7 Remove the driveshaft (see Chapter 8).
8 Bend the lockplate tabs out of their respective grooves and remove the lockplate. Immobilize the flange and remove the flange nut with a 30 mm thin-walled box wrench. Remove the flange from the output shaft. Use a puller, if necessary.
9 Using a seal removal tool or a small screwdriver, carefully pry out the old oil seal **(see illustration)**. Make sure you don't damage the seal bore when you pry out the seal.

Chapter 7 Part A Manual transmission

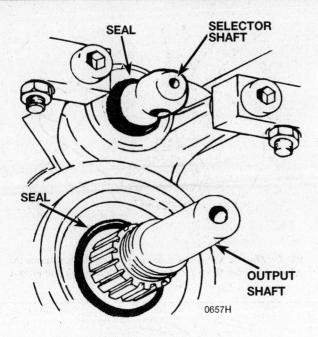

3.9 Using a seal removal tool or a small screwdriver, carefully pry out the old output shaft oil seal or selector shaft seal

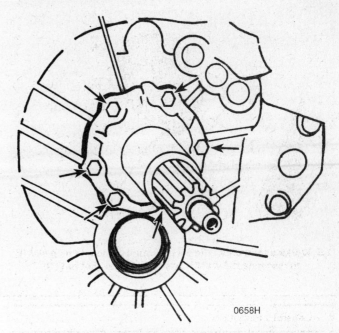

3.22a Locations of the guide sleeve bolts on a 3-Series model (Getrag 240 transmission shown)

10   Apply a light coat of oil to the lip of the new seal, clean off the end of the output shaft, slide the seal onto the shaft and carefully drive it into place using a short section of pipe with an outside diameter slightly smaller than the outside diameter of the seal.

11   Install the flange. Coat the side of the nut that faces toward the flange with sealant to prevent oil leaks. Tighten the flange nut to the torque listed in this Chapter's Specifications. Note that the nut must be tightened in two stages on Getrag transmissions. Install the lockplate and bend the tabs into their respective grooves.

12   The remainder of installation is the reverse of removal.

Selector shaft oil seal

13   Follow steps 5 through 8 above.
14   To disconnect the shift rod from the selector shaft on 3-Series models, remove the bushing lockring, then use a small drift punch to drive out the pin.
15   To disconnect the shift rod from the selector shaft on 5-Series models, put the transmission in 3rd gear, pull back on the spring sleeve and drive out the pin.
16   Pry out the selector shaft seal **(see illustration 3.9)** with a seal removal tool or a small screwdriver. Make sure you don't damage the seal bore while prying out the seal.
17   Coat the new seal lip with oil, slide the seal onto the end of the selector shaft and drive the seal into place with a socket or a piece of pipe of suitable dimensions.
18   Reattach the shift rod to the selector shaft.
19   Install the flange (see Step 11 above). The remainder of installation is the reverse of removal.

Input shaft seal

Refer to illustrations 3.22a and 3.22b
20   Remove the transmission (see Section 5).
21   Remove the clutch release bearing and release lever (see Chapter 8).
22   Remove the bolts for the clutch release guide sleeve **(see illustrations)**. The bolts are not all the same size on some models, so be sure to note where each bolt goes. Remove the sleeve and any spacers behind it.
23   Using a seal removal tool or a screwdriver, pry out the old seal. Make sure you don't damage the seal bore.
24   Lubricate the lip of the new seal with oil and drive it into place

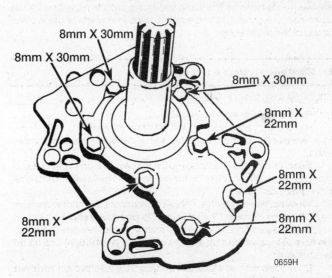

3.22b Locations of the guide sleeve bolts on a 5-Series model (Getrag 265 transmission shown)

with a section of pipe of suitable dimensions.

25   Clean the mounting bolts and sealing surfaces of the guide sleeve, and the bolt threads in the case. On 3-Series models with ZF transmissions, use a new gasket without sealant. On all other 3-Series and on all 5-Series models (Getrag transmissions), apply sealant (Loctite 573 or equivalent) to the guide sleeve sealing surface. Install the guide sleeve and spacer(s). **Caution:** *On Getrag 260 transmissions (3-Series models only), make sure the groove in the sleeve is aligned with the oil bore in the transmission case.*

26   Coat the bolt shoulders with sealant, install the bolts and tighten them to the torque listed in this Chapter's Specifications.
27   The remainder of installation is the reverse of removal.
28   Check the transmission lubricant level and add the recommended lubricant (see Chapter 1) as necessary.
29   Lower the vehicle, test drive it and check for leaks.

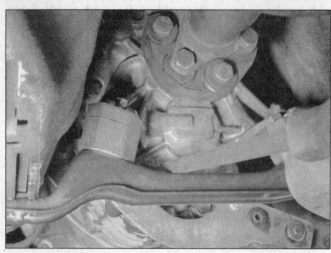

4.1 To check a transmission mount for wear, insert a large screwdriver or prybar over the crossmember and under the end of the transmission and try to pry the transmission upward - if the transmission lifts easily, replace the mount(s)

4 Transmission mounts - check and replacement

Refer to illustration 4.1

1   Raise the vehicle and support it securely on jackstands. Insert a large screwdriver or prybar over the crossmember and under the rear of the transmission housing **(see illustration)**. Try to pry the transmission up slightly.
2   The transmission should move very little. If it moves easily, the mount is probably cracked or torn. Replace it.
3   To replace the mount, support the transmission with a floor jack, remove the nuts and bolts attaching the mount to the crossmember and to the transmission, raise the transmission slightly and remove the mount.
4   Installation is the reverse of removal.

5 Transmission - removal and installation

1   Disconnect the cable from the negative battery terminal. **Caution:** *If the radio in your vehicle is equipped with an anti-theft system, make sure you have the correct activation code before disconnecting the battery, Refer to the information on page 0-7 at the front of this manual before detaching the cable.* **Note**: *If, after connecting the battery, the wrong language appears on the instrument panel display, refer to page 0-7 for the language resetting procedure.*
2   Remove the exhaust system (see Chapter 4).
3   Remove the driveshaft (see Chapter 8).
4   Remove the shift lever console (see Section 2).
5   Remove the shift rod (see Section 3).
6   Remove the TDC and reference sensors, if equipped, from the bellhousing (see Chapter 5).
7   Remove the clutch release cylinder (see Chapter 8). Do NOT disconnect the clutch hydraulic fluid hose. Hang the release cylinder with a piece of wire. **Caution**: *Do NOT operate the clutch while the release cylinder is removed from the transmission. The release cylinder could be damaged if the pushrod is forced out of the cylinder.*
8   Support the transmission with a transmission jack or a floor jack and remove the transmission support mounting nuts.
9   Carefully lower the transmission/engine assembly until it's resting on the front crossmember. **Caution**: *Never allow the weight of the transmission to be supported by the transmission input shaft. Such a load will damage the clutch and the transmission components.*
10 Remove the bolts that attach the transmission to the engine. Some bolts may have a Torx head; you'll need to use a special Torx socket for these bolts.
11 Carefully separate the transmission from the engine. Make sure you don't place any strain on the transmission input shaft. Lower the jack until there's enough clearance to pull the transmission out from under the vehicle.
12 Installation is basically the reverse of removal. Be sure to lightly lubricate the input shaft splines with Microlube GL 261 or its equivalent, then carefully position the transmission on the engine. The splines of the input shaft and the splines of the clutch disc may not quite line up; if they don't, turn the crankshaft pulley with a wrench to rotate the crankshaft slightly. **Caution**: *BMW recommends using only Microlube GL 261 on the clutch disc and transmission input shaft splines. Otherwise, the clutch disc could bind on the input shaft and cause hard shifting. Microlube GL 261 is available at BMW dealers (BMW Part No. 81 22 9 407 436).* **Note**: *If any of the transmission bolts are the Torx-head type, be sure to use washers when installing the bolts; installing bolts without washers can cause the bolts to seal so tightly against the transmission housing that they're extremely difficult to loosen.*
13 If the transmission was drained, be sure to refill it with the correct lubricant once the installation is complete (see Chapter 1).

6 Transmission overhaul - general information

The overhaul of a manual transmission is a somewhat difficult job for the home mechanic. It involves the disassembly and reassembly of many small parts. Numerous clearances must be precisely measured and, if necessary, adjusted with selective fit spacers and snap-rings. For this reason, we strongly recommend that the home mechanic limit himself to the removal and installation of the transmission and leave overhaul to a transmission repair shop. Exchange transmissions are available at reasonable prices for this vehicle and the time and money involved in the at home overhaul is almost sure to exceed the cost of a rebuilt unit.

Chapter 7 Part B Automatic transmission

Contents

Automatic transmission - removal and installation 6	Shift lever linkage - adjustment.. 3
Automatic transmission fluid level check .... See Chapter 1	Throttle Valve (TV) cable and kickdown - adjustment.................... 4
Diagnosis - general 2	Transmission oil seals - replacement 7
General information... 1	Transmission mount - replacement .... See Chapter 7A
Neutral start switch - check and adjustment 5	

Specifications

Transmission Throttle Valve (TV) cable adjustment
Distance between lead cable plug and end of
   cable housing (at idle) .... 0.020 ± 0.010 inch
With accelerator pedal at kickdown position .... 1.5 inches

Torque Specifications
                                                                                                        Ft-lbs

Transmission-to-engine hex-head bolts
   M8 18
   M10 33
   M12 .... 58 to 63
Transmission-to-engine Torx-head bolts
   M8 15
   M12 .... 46 to 53
Rear transmission support-to-body .... 16 to 18
Torque converter-to-driveplate
   M8 .... 18 to 20
   M10 .... 35 to 38
Transmission reinforcement plate .... 16 to 18
ATF cooler lines
   3-Series
      Cooler lines-to-transmission case 26
   5-Series
      Cooler line coupling nuts 13
      Cooler line hollow bolt 26
      Cooler line adapter coupling 26
ATF filler tube-to-ATF sump
   3 HP 22 .... 74 to 85
   4 HP 22 H/EH 72
ATF sump drain plug .... 11 to 13

Chapter 7 Part B Automatic transmission

1.4 The automatic transmission identification plate is located on the left side of the transmission housing, just behind the manual lever

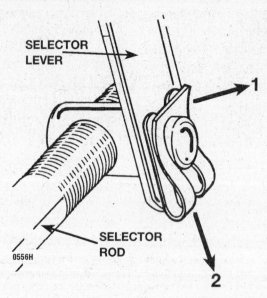

3.2 To disconnect the selector rod from the lower shift lever on a ZF 3 HP 22 automatic transmission, pull the top of the clip out (1), then slide the clip off the pin (2)

1 General information

Refer to illustration 1.4

Earlier models use a three-speed ZF 3 HP 22 automatic transmission. Later models use a four-speed unit, the ZF 4 HP 22. Current models use a ZF 4 HP EH with electronic shifting control.

Because of the complexity of the clutch mechanisms and the hydraulic control systems, and because of the special tools and expertise needed to rebuild an automatic transmission, this work is usually beyond the scope of the home mechanic. The procedures in this Chapter are therefore limited to general diagnosis, routine maintenance, adjustments and transmission removal and installation.

Should the transmission require major repair work, take it to a dealer service department or a transmission repair shop. You can, however, save the expense of removing and installing the transmission by doing it yourself.

If you have to replace the transmission with a new or rebuilt unit, make sure you purchase the right unit. The transmission identification plate is located on the left side of the housing, just behind the manual valve lever **(see illustration)**.

2 Diagnosis - general

Note: *Automatic transmission malfunctions may be caused by five general conditions: poor engine performance, improper adjustments, hydraulic malfunctions, mechanical malfunctions or malfunctions in the computer or its signal network. Diagnosis of these problems should always begin with a check of the easily repaired items: fluid level and condition (see Chapter 1), shift linkage adjustment and throttle linkage adjustment. Next, perform a road test to determine if the problem has been corrected or if more diagnosis is necessary. If the problem persists after the preliminary tests and corrections are completed, additional diagnosis should be done by a dealer service department or transmission repair shop. Refer to the Troubleshooting section at the front of this manual for information on symptoms of transmission problems.*

Preliminary checks

1 Drive the vehicle to warm the transmission to normal operating temperature.
2 Check the fluid level as described in Chapter 1:
  a) If the fluid level is unusually low, add enough fluid to bring the level within the designated area of the dipstick, then check for external leaks (see below).
  b) If the fluid level is abnormally high, drain off the excess, then check the drained fluid for contamination by coolant. The presence of engine coolant in the automatic transmission fluid indicates that a failure has occurred in the internal radiator walls that separate the coolant from the transmission fluid (see Chapter 3).
  c) If the fluid is foaming, drain it and refill the transmission, then check for coolant in the fluid or a high fluid level.
3 Check the engine idle speed. **Note:** *If the engine is malfunctioning, do not proceed with the preliminary checks until it has been repaired and runs normally.*
4 Check the Throttle Valve (TV) cable for freedom of movement. Adjust it if necessary (see Section 3). **Note:** *The TV cable may function properly when the engine is shut off and cold, but it may malfunction once the engine is hot. Check it cold and at normal engine operating temperature.*
5 Inspect the shift control linkage (see Section 5). Make sure that it's properly adjusted and that the linkage operates smoothly.

Fluid leak diagnosis

6 Most fluid leaks are easy to locate visually. Repair usually consists of replacing a seal or gasket. If a leak is difficult to find, the following procedure may help.
7 Identify the fluid. Make sure it's transmission fluid and not engine oil or brake fluid (automatic transmission fluid is a deep red color).
8 Try to pinpoint the source of the leak. Drive the vehicle several miles, then park it over a large sheet of cardboard. After a minute or two, you should be able to locate the leak by determining the source of the fluid dripping onto the cardboard.
9 Make a careful visual inspection of the suspected component and the area immediately around it. Pay particular attention to gasket mating surfaces. A mirror is often helpful for finding leaks in areas that are hard to see.
10 If the leak still cannot be found, clean the suspected area thoroughly with a degreaser or solvent, then dry it.
11 Drive the vehicle for several miles at normal operating temperature and varying speeds. After driving the vehicle, visually inspect the suspected component again.
12 Once the leak has been located, the cause must be determined before it can be properly repaired. If a gasket is replaced but the sealing flange is bent, the new gasket will not stop the leak. The bent flange must be straightened.
13 Before attempting to repair a leak, check to make sure that the following conditions are corrected or they may cause another leak.

Chapter 7 Part B Automatic transmission

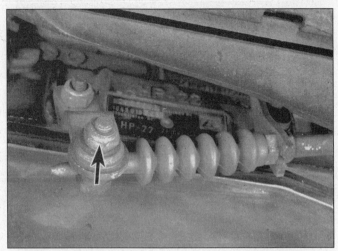

3.8 To adjust the shift linkage on a ZF 4 HP 22 automatic transmission, loosen this nut (arrow), push the manual valve lever forward (toward the engine), pull the selector cable rod back and tighten the nut securely. Note: *The linkage shown here (on a later model 5-Series) is typical; however, the lever on some older units may face up instead of down and utilize a clevis arrangement between the rod and the lever instead of a nut).*

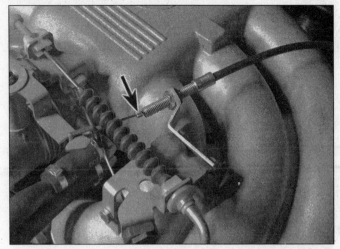

4.2 With the throttle at the idle position, check the distance between the lead plug (arrow) and the end of the cable sleeve (the correct dimensions are listed in this Chapter's Specifications); if the distance is incorrect, loosen the locknuts, adjust the position of the cable sleeve and tighten the nuts

Note: *Some of the following conditions cannot be fixed without highly specialized tools and expertise. Such problems must be referred to a transmission repair shop or a dealer service department.*

Gasket leaks

14 Check the pan periodically. Make sure the bolts are tight, no bolts are missing, the gasket is in good condition and the pan is flat (dents in the pan may indicate damage to the valve body inside).
15 If the pan gasket is leaking, the fluid level or the fluid pressure may be too high, the vent may be plugged, the pan bolts may be too tight, the pan sealing flange may be warped, the sealing surface of the transmission housing may be damaged, the gasket may be damaged or the transmission casting may be cracked or porous. If sealant instead of gasket material has been used to form a seal between the pan and the transmission housing, it may be the wrong sealant.

Seal leaks

16 If a transmission seal is leaking, the fluid level or pressure may be too high, the vent may be plugged, the seal bore may be damaged, the seal itself may be damaged or improperly installed, the surface of the shaft protruding through the seal may be damaged or a loose bearing may be causing excessive shaft movement.
17 Make sure the dipstick/filler tube nut is tight. Periodically check the area around the speedometer gear or sensor for leakage. If transmission fluid is evident, check the O-ring for damage.

Case leaks

18 If the case itself appears to be leaking, the casting is porous and will have to be repaired or replaced.
19 Make sure the oil cooler hose fittings are tight and in good condition.

Fluid comes out vent pipe or fill tube

20 If this condition occurs, the transmission is overfilled, there is coolant in the fluid, the case is porous, the dipstick is incorrect, the vent is plugged or the drain back holes are plugged.

3 Shift lever linkage - adjustment

3 HP 22 transmissions

Refer to illustration 3.2

1 Raise the vehicle and support it securely on jackstands. On some models, it may be necessary to remove the exhaust system (see Chapter 4) and the heat shield to reach the lower end of the shift lever.
2 Remove the retaining clip **(see illustration)** from the selector pin, pull out the pin and disconnect the selector rod from the bottom of the selector lever.
3 Pull the transmission manual valve lever back as far as it will go, then push it forward two clicks. The transmission should now be in Neutral.
4 Working from inside the vehicle, move the shift lever to N, then push it forward against the Neutral position stop in the shift gate.
5 Hold the shift lever against the gate, turn the selector pin on the selector rod until the pin lines up with the hole on the lower shift lever and shorten the linkage by turning the pin an extra one or two turns in a clockwise direction.
6 Install the selector pin and verify that the linkage is correctly adjusted by starting the engine and moving the shift lever though all shift positions.

4 HP 22 transmissions

Refer to illustration 3.8

7 Place the shift lever in the Park (P) position.
8 Loosen the nut **(see illustration)** which attaches the cable to the manual valve lever.
9 Push the manual valve lever forward (toward the engine) and pull the selector cable rod backward.
10 Reattach the cable rod at the manual valve and tighten the nut securely.

4 Throttle Valve (TV) cable and kickdown - adjustment

Refer to illustrations 4.2 and 4.3

1 Before making the following adjustment, check the full-throttle adjustment of the accelerator cable and, if necessary, adjust the accelerator cable (see Chapter 4).
2 With the throttle at the idle position, use a feeler gauge to check the distance between the lead plug and the end of the cable sleeve **(see illustration)** and compare your measurement with the adjustment dimensions listed in this Chapter's Specifications. If the distance is incorrect, loosen the locknuts, adjust the position of the cable sleeve and tighten the nuts.
3 Locate the kickdown switch locknut under the accelerator pedal

Chapter 7 Part B Automatic transmission

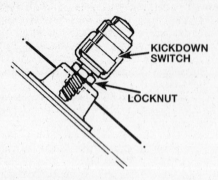

4.3 To adjust the kickdown switch, loosen the locknut, screw in the switch until it's no longer touching the underside of the accelerator pedal, depress the pedal to the point at which you can feel transmission resistance, hold it in this position, back out the switch until it contacts the underside of the pedal and tighten the locknut

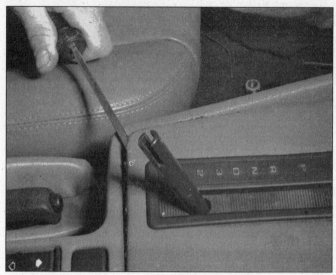

5.2 Pry out the trim panel around the base of the shift lever (on some 5-Series models, you'll also need to remove the trim bezel screws and the trim bezel)

5 Neutral start switch - check and replacement

Refer to illustration 5.2

1 If the engine can be started with the shift lever in any position other than Park or Neutral, or if the back-up lights do not come on when the lever is moved to Reverse, check the neutral start switch.

2 Remove the shift lever knob retaining screw and pull off the shift lever knob. Pry out the trim panel around the base of the shift lever **(see illustration)**. Remove the trim bezel screws and the trim bezel if necessary.

3-Series models

3 Position the shift lever in Neutral, detach the wiring connector to the switch, then connect an ohmmeter between the terminals for the blue/white wire and the green/black wire and verify that there's continuity. Put the shift lever in Park and try the same test again. There should still be continuity. Place the shift lever in Drive, 1 and 2, and verify there's no continuity when you hook up the ohmmeter between the blue/white wire and the green/yellow wire terminals AND when you put the ohmmeter probes between the terminals for the blue/white wire and the green/black wire. Finally, put the shift lever in Reverse and verify that there's continuity between the terminals for blue/white wire and the green/yellow wire.

5.4 Neutral start switch wiring connector on earlier 5-Series models

(see illustration) and screw it in far enough so that it's not touching the underside of the accelerator pedal.

4 Depress the accelerator pedal to the point at which you can feel transmission resistance, hold the pedal in this position, back out the kickdown switch so it just contacts the pedal and tighten the locknut.

5 Check your adjustment by depressing the accelerator pedal to the kickdown position and noting the position of the lead plug (see Step 2 above). If the cable can't be adjusted to the correct specifications, recheck and, if necessary readjust, the full-throttle adjustment of the accelerator cable. Then try adjusting the throttle valve cable one more time.

5-Series models

Refer to illustrations 5.4, 5.5a and 5.5b

Note: *If the connector terminal pattern for the neutral start switch on your vehicle doesn't match either of the two types shown here, special checking procedures may be required to determine if the switch is op-*

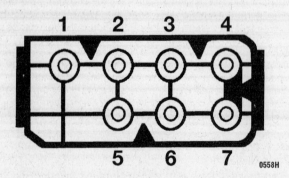

5.5a Neutral start switch wiring connector on later 5-Series models

Shift lever position	Continuity between terminals
Park	1 and 3 and 7
Reverse	1 and 4 and 7; 5 and 6
Neutral	1 and 2 and 3
Drive	1 and 4
3	1 and 2
2	1 and 7
1	1 and 2 and 7

5.5b Neutral start switch continuity table for later 5-Series models

Chapter 7 Part B Automatic transmission

erating properly. Have the switch checked by a dealer service department or other repair shop.

4 On earlier 5-Series models, unplug the C1 and C2 connectors **(see illustration)**, then:
   a) Put the shift lever in the Park position and verify that there's continuity between terminals 3 and 4.
   b) Put the lever in the Reverse position and verify that there's continuity between terminals 3 and 4.
   c) Put the lever in the Neutral position and verify that there's continuity between terminals 3 and 6.
   d) Put the lever in the Drive position and verify that there's continuity between terminals 3 and 7.
   e) Put the lever in the "1" position and verify that there's continuity between terminals 3 and 9.
   f) Put the lever in the "2" position and verify that there's continuity between terminals 3 and 8.
   g) Put the lever in the "3" position and verify that there's continuity between terminal 3 in the C1 connector and terminal 1 in the C2 connector

5 On later 5-Series models, unplug the connector **(see illustration)** and check the continuity of the indicated terminals with the shift lever in each position **(see illustration)**.

6 Automatic transmission - removal and installation

Refer to illustration 6.8

1 Disconnect the cable from the negative battery terminal. **Caution:** *If the radio in your vehicle is equipped with an anti-theft system, make sure you have the correct activation code before disconnecting the battery. Refer to the information on page 0-7 at the front of this manual before detaching the cable.* **Note:** *If, after connecting the battery, the wrong language appears on the instrument panel display, refer to page 0-7 for the language resetting procedure.*

2 Disconnect the Throttle Valve (TV) cable from the throttle body and detach it from the bracket on the intake manifold (see Chapter 4).

3 On 318i models, remove the distributor cap to prevent damage when the engine tilts back during transmission removal.

4 Raise the vehicle and support it securely on jackstands.

5 Remove the exhaust system (see Chapter 4).

6 Disconnect the selector rod from the bottom of the shift lever (see Section 3).

7 Remove the driveshaft (see Chapter 8).

8 Drain the transmission (see Chapter 1). Detach the filler tube and disconnect the fluid cooler lines **(see illustration)**. Plug the lines to prevent contaminants from entering the transmission.

9 On 3 HP 22 transmissions, remove the transmission cover at the bottom of the bellhousing; on other units, remove the reinforcement plate and the transmission bump stop at the bottom of the bellhousing (see *Oil pan - removal and installation* in Chapter 2A).

10 Reach through the opening in the bellhousing and remove the bolts that attach the torque converter to the driveplate (see *Engine - removal and installation* in Chapter 2B). On 3 HP 22 transmissions, there are four bolts; on 4 HP 22 units, there are three.

11 On vehicles with a Motronic fuel injection system, remove the engine rpm sensor and the reference sensor mounted in the bellhousing (see Chapter 5).

12 On vehicles with the 4 HP 22 EH transmission, unplug the control unit electrical connector from the left side of the transmission, just above the ATF sump

13 Support the transmission from underneath with a transmission jack or a floor jack. If you use a floor jack, place a block of wood between the jack head and the oil pan to protect the pan. Remove the rear transmission support and, if equipped, the reinforcement bar by removing the bolts that hold them to the body. Lower the engine and transmission onto the front axle carrier. **Caution:** *Do NOT allow the torque converter shaft to support the weight of the transmission or you will damage the torque converter or transmission.*

14 Remove the bolts that attach the transmission bellhousing to the engine.

15 To gain access to the torque converter, remove the inspection grille from the side of the bellhousing. While using a prybar to ensure that the torque converter stays firmly mounted to the transmission, pull the transmission off the engine.

16 If you're replacing the transmission, remove the torque converter and install it on the new/rebuilt unit. To get the torque converter off the old transmission, install two long bolts halfway into the converter mounting holes and pull evenly on both bolts. To install the converter, see Section 7, Step 7.

17 Installation is basically the reverse of removal, but note the following:
   a) Tighten the torque converter bolts to the torque listed in this Chapter's Specifications.
   b) Before reattaching the ATF cooler lines and cooler, blow them out with compressed air, then flush them with clean ATF fluid, to remove any friction lining particles that could clog the new or rebuilt transmission.
   c) Use a new gasket on the ATF sump drain plug and a new O-ring on the ATF filler tube connection to the ATF sump.
   d) When you plug in the control unit connector on 4 HP 22 EH transmissions, make sure the markings are aligned.

18 Refill the transmission with clean ATF (see Chapter 1).

19 Adjust the shift lever linkage and the throttle valve cable (see Sections 3 and 4).

7 Transmission oil seals - replacement

1 There are three seals in the automatic transmission - the torque converter seal, the manual valve seal and the transmission output shaft seal - that can cause ATF leaks. If the torque converter seal is leaking, you'll find ATF fluid in the bottom of the bellhousing. If the manual valve seal leaks, you'll see ATF fluid on the side of the transmission case and the pan. If the output shaft seal leaks, you'll find ATF fluid on the extension housing and the front of the driveshaft.

2 You can replace all of these seals without disassembling the transmission, although replacing the torque converter seal requires removal of the transmission and the torque converter.

Torque converter seal

Refer to illustration 7.5

3 Remove the transmission and the torque converter (see Section 6).

4 Inspect the bushing in the torque converter hub. If it's worn, the new seal will soon leak. Try removing any sharp edges or burrs with fine emery cloth. If the hub is too deeply scored to smooth out, the

6.8 ATF filler tube fitting (arrow) and ATF cooler lines (arrows)

7.5 Using a hooked seal removal tool or a large screwdriver, carefully pry the old torque converter seal out of the transmission

7.16 Bend back the tab on the locking plate, immobilize the flange and remove the flange nut with a 30 mm deep-well socket, then remove the flange from the output shaft, using a puller if necessary

torque converter should be replaced.

5   Using a hooked seal removal tool or a screwdriver, pry the old seal out of the transmission case **(see illustration)**. Make sure you don't gouge the surface of the seal bore when prying out the old seal.

6   Lubricate the lip of the new seal with ATF and carefully drive it into place with a socket or a section of pipe with an outside diameter slightly smaller than the outside diameter of the seal.

7   Install the converter, making sure it is completely engaged with the front pump (to do this, turn the converter as you push it into place - continue turning and pushing until you're sure it's fully seated). Install the transmission/converter assembly (see Section 6).

Manual valve seal

8   Raise the vehicle and place it securely on jackstands.

9   Remove the manual valve lever **(see illustration 3.8)** from the transmission. Do not disconnect the linkage from the lever or you'll have to readjust it.

10   Pry out the old seal with a small hooked seal removal tool or a small screwdriver. Make sure you don't gouge the surface of the seal bore when prying out the old seal.

11   Lubricate the lip of the new seal with ATF and carefully drive it into place with a socket or a section of pipe with an outside diameter slightly smaller than the outside diameter of the seal.

12   Install the manual lever and tighten the nut to the torque listed in this Chapter's Specifications. If you didn't disconnect the linkage, it's unnecessary to readjust the linkage.

Output shaft seal

Refer to illustration 7.16

13   Raise the rear of the vehicle and support it securely on jackstands. Be sure to block the front wheels to keep the vehicle from rolling.

14   Remove the exhaust system (see Chapter 4).

15   Remove the driveshaft (see Chapter 8).

16   Bend back the tab on the locking plate **(see illustration)**, immobilize the flange and remove the flange nut with a 30 mm deep-well socket. Remove the flange from the output shaft. Use a puller, if necessary.

17   Using a seal removal tool or a small screwdriver, carefully pry out the old oil seal. Make sure you don't damage the seal bore when you pry out the seal.

18   Apply a light coat of oil to the lip of the new seal, clean off the end of the output shaft, slide the seal onto the shaft and carefully drive it into place using a short section of pipe with an outside diameter slightly smaller than the outside diameter of the seal.

19   Install the flange. Coat the side of the nut that faces toward the flange with sealant to prevent oil leaks. Tighten the flange nut to the torque listed in this Chapter's Specifications. Install a new lockplate and bend the tab into the slot.

20   The remainder of installation is the reverse of removal.

Chapter 8 Clutch and driveline

Contents

Center bearing - check and replacement	11
Check and boot replacement	15
Clutch - description and check	2
Clutch components - removal, inspection and installation	8
Clutch hydraulic system - bleeding	7
Clutch master cylinder - removal and installation	3
Clutch pedal - adjustment	4
Clutch slave cylinder - removal and installation	6
Clutch start switch - check and replacement	5
Constant velocity (CV) joints and boots - check and replacement	16
Differential housing - removal and installation	18
Differential housing side seals - replacement	17
Differential lubricant change	See Chapter 1
Differential lubricant level check	See Chapter 1
Driveaxles - removal and installation	15
Driveshaft constant velocity (CV) joint - replacement	14
Driveshaft - removal and installation	9
Flexible coupling - replacement	10
Fluid level checks	See Chapter 1
Front centering guide - check and replacement	13
General information	1
Pinion oil seal - replacement	19
Rear driveaxle boot check	See Chapter 1
Universal joints - check and replacement	12

Specifications

Clutch release cylinder pushrod travel	
3-Series	3/4-inch
5-Series	9/10-inch
Clutch pedal-to-firewall distance	About 10 inches
Maximum allowable clutch disc runout	0.020 inch
Minimum allowable clutch disc thickness	19/64-inch

Torque specifications

Ft-lbs (unless otherwise indicated)

Clutch master cylinder mounting bolts	80 to 84 in-lbs
Clutch release cylinder bolts	18
Driveshaft	
Rear U-joint-to-final drive flange	53
Flexible coupling-to-transmission flange	
M10 (Grade 8.8) bolts	34
M10 (Grade 10.9) bolts	53
M12 bolts	90 to 91
Center bearing mounting bolts	16
Splined coupling clamping sleeve	13
Rear transmission bracket (models with CV joints)	
M10 bolts	32 to 35
M8 bolts	16 to 17
CV joint bolts-to-transmission flange	24
Driveaxles	
CV joint Allen-head bolts	42 to 46
Driveaxle nut (3-Series)	144 to 155
Pinion shaft flange nut	
318i, 5-Series models with M20 bolts	111, or until marks line up
Other 3-Series models, 5-Series models with M22 bolts	229, or until marks line up

Chapter 8 Clutch and driveline

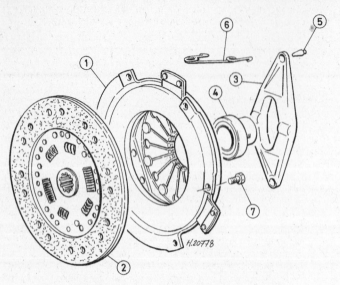

2.1 An exploded view of a typical clutch assembly

1. Pressure plate
2. Clutch disc
3. Release lever
4. Release bearing
5. Ball stud
6. Lever retaining spring
7. Bolt

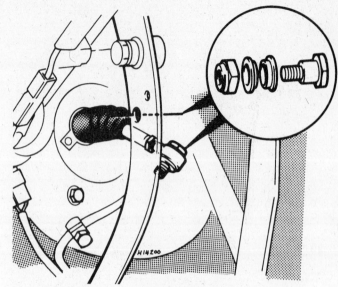

3.4 Typical clutch master cylinder pushrod-to-clutch pedal installation

1 General information

Clutch

Models with manual transmissions are equipped with a hydraulically-actuated diaphragm-type clutch. The clutch disc is clamped by spring pressure between the diaphragm-type pressure plate assembly and the engine flywheel. The splined hub of the clutch rides on the transmission input shaft; the pressure plate is bolted to the engine flywheel.

When the clutch pedal is depressed, the master cylinder generates hydraulic pressure, which forces a pushrod out of the clutch slave (release) cylinder, moving the release lever and release bearing. As the release bearing pushes against the pressure plate release tangs, clamping pressure on the clutch disc is eased, disconnecting the engine from the transmission.

Two different types of clutch discs are used on the models covered by this manual, depending on the type of flywheel. On models with a conventional plate-type flywheel, the clutch disc has integral cushion springs and dampening springs. The cushion springs - located between the friction surfaces of the disc - help reduce the shock when the clutch is engaged; the dampening springs - visible in the center of the hub - help absorb the rotating power pulses of the engine. On models with dual-mass flywheels, the dampening springs are integrated into the flywheel itself (see Chapter 2A).

Driveshaft

Power is transmitted from the transmission to the rear axle by a two-piece driveshaft joined behind the center bearing by a "slip joint," a sliding, splined coupling. The slip joint allows slight fore-and-aft movement of the driveshaft. The forward end of the driveshaft is attached to the output flange of the transmission by a flexible rubber coupling. This coupling, sometimes referred to as a "Guibo" coupling or a "flex-disc," isolates the driveshaft and differential from the sudden torque forces of the engine. On some models, a vibration damper is mounted between the transmission output flange and the flexible coupling. The middle of the driveshaft is supported by the center bearing, which is rubber-mounted to isolate driveshaft vibrations. The bearing housing is bolted to the vehicle body. The forward end of the driveshaft is aligned with the transmission by a centering guide recessed into the driveshaft; the centering guide engages a centering pin on the transmission output flange. Universal joints are located at the center bearing and at the rear end of the driveshaft to compensate for flexing of the chassis.

Differential assembly

The rear-mounted differential assembly includes the drive pinion, the ring gear, the differential and the output flanges. The drive pinion, which drives the ring gear, is also known as the differential input shaft. It's connected to the driveshaft via an input flange. The differential is bolted to the ring gear and drives the rear wheels through a pair of output flanges bolted to driveaxles with constant velocity (CV) joints at either end. The differential allows the wheels to turn at different speeds when cornering (the outside wheel must travel farther - and therefore faster - than the inside wheel in the same period of time). The differential assembly is bolted to the rear axle carrier and is attached to the body by flexible rubber bushings.

Major repair work on the differential assembly components (drive pinion, ring-and-pinion and differential) requires many special tools and a high degree of expertise, and therefore should not be attempted by the home mechanic. If major repairs become necessary, we recommend that they be performed by a dealer service department or other repair shop.

Driveaxles and CV joints

The driveaxles deliver power from the differential output flanges to the rear wheels. The driveaxles are equipped with Constant Velocity (CV) joints at each end. A CV joint is similar in function to a standard universal joint, but can accept more radical driveaxle angles than the U-joint. CV joints allow the driveaxles to deliver power to the rear wheels while moving up and down with the rear suspension, even though the differential assembly, the driveaxles and the wheels are never in perfect alignment.

The inner CV joints on all models are bolted to the differential flanges. The outer CV joints on 3-Series models engage the splines of the wheel hubs and are secured by an axle nut. The outer joints on 5-Series models are identical to the inner joints. Each joint is packed with a special CV joint lubricant and sealed by a rubber boot. Inspect the boots regularly to make sure they're in good condition. A ripped or torn boot will allow dirt to contaminate the CV joint. The CV joints are not rebuildable on the vehicles covered by this manual but can be disassembled for cleaning during boot replacement (except for the outer joint on 3-Series models). The inner CV joint can be replaced individually on all models, as well as the outer joint on 5-Series models. On 3-Series models the outer CV joint and the axleshaft must be replaced as a complete assembly.

Chapter 8 Clutch and driveline

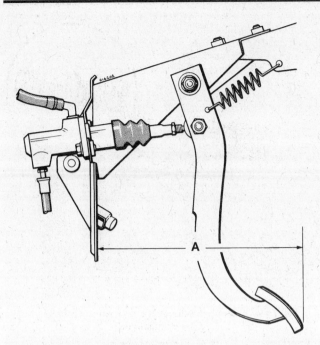

4.1 To check the clutch pedal adjustment, measure the distance from the lower rear edge of the pedal to the firewall (dimension A)

2 Clutch - description and check

Refer to illustration 2.1

1   Models equipped with a manual transmission feature a single dry plate, diaphragm spring-type clutch **(see illustration)**. The actuation is through a hydraulic system.
2   When the clutch pedal is depressed, hydraulic fluid (under pressure from the clutch master cylinder) transfers force to the slave cylinder. Because the slave cylinder is connected to the clutch release fork, the fork moves the release bearing into contact with the clutch cover/pressure plate release fingers (tangs), disengaging the clutch disc.
3   The hydraulic system adjusts the clutch automatically, so no adjustment of the linkage or pedal is required, unless you replace the clutch master cylinder and/or pushrod.
4   Terminology can be a problem regarding clutch components because manufacturers, parts departments and the automotive aftermarket use different names for the same parts. For example, the driven plate is also called the clutch plate or disc, the clutch release bearing is sometimes called a throwout bearing, the slave cylinder is sometimes called the operating or release cylinder, the pressure plate is also called a clutch cover, etc.
5   Components with obvious damage must be replaced, but there are some preliminary checks you can make to diagnose a clutch system failure caused by less obvious problems.

  a) The first check should be of the fluid level in the clutch master cylinder. If the fluid level is low, add fluid as necessary and retest. If the master cylinder runs dry, or if any of the hydraulic components are serviced, bleed the hydraulic system (see Section 7).
  b) To check "clutch spin-down time," run the engine at normal idle speed with the transmission in Neutral (clutch pedal up). Disengage the clutch (pedal down), wait nine seconds and shift the transmission into Reverse. You should hear no grinding noises. A grinding noise indicates component failure in the clutch cover assembly or the clutch disc.
  c) To check for complete clutch release, run the engine (with the brake on to prevent movement) and hold the clutch pedal approximately 1/2-inch from the floor mat. Shift the transmission between First gear and Reverse several times. If the shift is not smooth, component failure is indicated.
  d) Visually inspect the clutch pedal bushing at the top of the clutch pedal to make sure there is no sticking or excessive wear.

3 Clutch master cylinder - removal and installation

Removal

Refer to illustration 3.4

1   Remove the brake fluid reservoir cap and remove the brake fluid in the reservoir with a syringe until the fluid level is below the brake line connection for the master cylinder.
2   Disconnect the threaded fitting for the clutch hydraulic line from the front of the master cylinder, which protrudes through the firewall into the engine compartment. Have a shop rag handy to wipe up any spilled fluid.
3   Working under the dash, remove the lower left dash trim panel (see Chapter 11).
4   Remove the bolt that attaches the clutch master cylinder pushrod to the clutch pedal **(see illustration)**.
5   Remove the two bolts that hold the master cylinder to the clutch pedal bracket, detach the threaded fitting for the fluid line on top of the master cylinder and remove the master cylinder. Have a shop rag handy to wipe up any spilled fluid.

Installation

6   Installation is basically the reverse of removal. Be sure to tighten the clutch master cylinder mounting bolts and the pushrod-to-pedal bolt to the torque listed in this Chapter's Specifications. Tighten the threaded fitting for the clutch fluid line securely.
7   Check the clutch pedal adjustment (see Section 4) and the clutch start switch adjustment (see Section 5).
8   Fill the clutch master cylinder reservoir with brake fluid conforming to DOT 3 specifications and bleed the clutch system (see Section 7).

4 Clutch pedal - adjustment

Refer to illustration 4.1
Note: *Hydraulic clutch systems are self-adjusting, so a correctly adjusted clutch pedal should never get out of adjustment. However, clutch pedal height should be checked when installing a new clutch master cylinder to ensure correct initial adjustment.*

1   Measure the distance from the lower rear edge (driver's side) of the pedal to the firewall **(see illustration)** and compare your measurement to the dimension listed in this Chapter's Specifications.
2   If the distance is incorrect, loosen the pushrod nut and bolt. The bolt is eccentric; rotating it changes the distance from the pedal to the firewall. When the distance is correct, tighten the nut and bolt.
3   Some models also have an adjustable clutch master cylinder pushrod. If the pedal can't be adjusted by rotating the eccentric bolt, loosen the pushrod locknut and turn the pushrod until the distance from the pedal to the firewall is correct, then tighten the locknut securely. **Caution:** *Don't screw the pushrod all the way in. This could cause the locknut to jam against the clutch pedal during operation, which could break the pushrod. If you do change the pushrod length, make sure you check the locknut clearance before operating the clutch pedal. Also, don't overtighten the pushrod locknut. If the threads are stripped, the master cylinder may jam, causing clutch failure.*

5 Clutch start switch - check and replacement

1   The clutch start switch, located at the top of the clutch pedal, prevents the engine from being started unless the clutch pedal is depressed. The switch, which is part of the starting circuit, is normally open. When the clutch pedal is depressed, it depresses a small pushrod in the switch, closing the circuit and allowing the starter to operate.

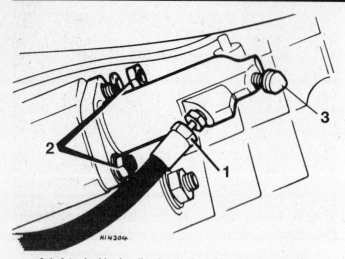

6.4 A typical hydraulic clutch slave (or release) cylinder

1. Hydraulic line
2. Mounting nuts
3. Bleeder screw

2 If the vehicle starts in gear, either the switch is malfunctioning, or is incorrectly adjusted. To check its operation, unplug the electrical connector and hook up an ohmmeter across the terminals of the connector. Verify that there's no continuity when the pedal is released (pedal up) and continuity when the pedal is depressed. If the switch doesn't operate as described, try adjusting it. If it still doesn't operate correctly, replace it.

6 Clutch slave cylinder - removal and installation

Refer to illustration 6.4

Note: *Before beginning this procedure, contact local parts stores and dealers concerning the purchase of a rebuild kit or a new slave cylinder. Availability and cost of the necessary parts may dictate whether the cylinder is rebuilt or replaced with a new one. If you decide to rebuild the cylinder, follow the instructions contained in the rebuild kit.*

1 Remove the brake fluid reservoir cap and remove the brake fluid in the reservoir with a syringe until the fluid level is below the brake line connection for the master cylinder.
2 Raise the vehicle and support it securely on jackstands.
3 Wipe the area around the threaded fitting for the hydraulic fluid line and disconnect the fluid line. Use a flare nut wrench to avoid damaging the fitting. Have a shop rag handy to mop up any spilled fluid. Plug or tape the end of the line to contain the remaining fluid in the line and to keep dirt out.
4 Remove the slave cylinder mounting nuts **(see illustration)** and remove the cylinder from the transmission.
5 Inspect the plastic tip of the pushrod and the release lever for wear. The pushrod should be worn only on the tip. The release lever should show wear only in the recess for the pushrod. Wear on the sides of the pushrod tip, or wear on the release lever anywhere besides the recess, indicates that the pushrod is misaligned.
6 Inspect the slave cylinder for leaks. If the pushrod seal is leaking, replace the slave cylinder. If you have to replace the slave cylinder, make sure you obtain the correct replacement unit. Models with a conventional flywheel use a slave cylinder with a diameter of 0.813-inch; models with a dual-mass flywheel use a slave cylinder with an 0.874-inch diameter.
7 Installation is the reverse of removal. Be sure to apply a light coat of moly-based grease to the plastic pushrod tip and make sure the tip engages the recess in the release lever properly. Tighten the slave cylinder mounting nuts to the torque listed in this Chapter's Specifications.
8 Refill the brake fluid reservoir and bleed the clutch hydraulic system (see Section 7).

7 Clutch hydraulic system - bleeding

1 If air gets into the clutch hydraulic system, the clutch may not release completely when the pedal is depressed. Air can enter the system whenever any part of it is dismantled or if the fluid level in the master cylinder reservoir runs low. Air can also leak into the system through a hole too small to allow fluid to leak out. In this case, it indicates that a general overhaul of the system is required.
2 To bleed the air out, you will need an assistant to pump the clutch pedal, a supply of new brake fluid of the recommended type, a clear plastic container, a section of clear, flexible plastic tubing which will fit over the bleeder screw and a wrench for the bleeder screw.
3 Check the fluid level in the master cylinder reservoir. Add fluid, if necessary, to bring the level up to the full mark. Use only the recommended brake fluid and do not mix different types. Never use fluid from a container that has been standing uncapped. You will have to check the fluid level in the master cylinder reservoir often during the bleeding procedure. If the level drops too far, air will enter the system through the master cylinder.
4 Raise the front of the vehicle and support it securely on jackstands. Apply the parking brake.
5 Remove the bleeder screw cap from the bleeder screw on the slave cylinder **(see illustration 6.4)**.
6 Attach one end of the plastic tube to the bleeder screw fitting and place the other end in the container, submerged in clean brake fluid.
7 Loosen the bleeder screw slightly, then tighten it to the point where it is snug yet easily loosened.
8 Have your assistant pump the pedal several times and hold it in the fully depressed position.
9 With pressure on the pedal, open the bleeder screw approximately one-half turn. As the fluid stops flowing through the tube and into the jar, tighten the bleeder screw. Again, pump the pedal, hold it in the fully depressed position and loosen the bleeder screw momentarily. Do not allow the pedal to be released with the bleeder screw in the open position.
10 Repeat the procedure until no air bubbles are visible in the fluid flowing through the tube. Be sure to check the fluid level in the master cylinder reservoir while performing the bleeding operation.
11 Tighten the bleeder screw completely, remove the tube and install the bleeder screw cap.
12 Check the clutch fluid level in the master cylinder to make sure it is adequate, then test drive the vehicle and check for proper clutch operation.

8 Clutch components - removal, inspection and installation

Removal

Refer to illustrations 8.3, 8.4 and 8.5.

1 Access to the clutch components is normally accomplished by removing the transmission, leaving the engine in the vehicle. If, of course, the engine is being removed for major overhaul, then the opportunity should always be taken to check the clutch for wear and replace worn components as necessary. The following procedures will assume that the engine stays in place.
2 Referring to Chapter 7A, remove the transmission from the vehicle.
3 To support the clutch disc during removal, install a clutch alignment tool or an old transmission input shaft through the middle of the clutch **(see illustration)**.
4 Carefully inspect the flywheel and pressure plate for indexing marks. If no marks are visible, use a center punch, scribe or paint to mark the pressure plate and flywheel so they can be reinstalled in the same relative positions **(see illustration)**.
5 Turning each bolt only a little at a time, loosen the pressure plate-to-flywheel bolts. Work in a criss-cross pattern, until all spring pressure is relieved **(see illustration)**. Grasp the pressure plate and completely remove the bolts, then separate the pressure plate and clutch disc from the flywheel.

Chapter 8 Clutch and driveline

8.3 Use a clutch alignment tool to support the clutch disc during pressure plate removal and to center the disc during installation

8.4 Be sure to mark the pressure plate and flywheel to ensure proper alignment during installation (this is only necessary if the same pressure plate will be used)

8.5 Loosen the pressure plate bolts (arrows) a little at a time until the spring tension is released

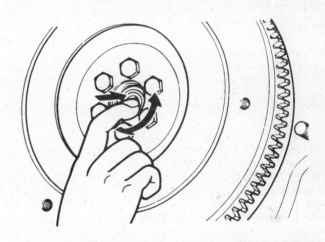

8.8 Turn the pilot bearing by hand while pressing on it - if it's rough or noisy, install a new one

8.9a You'll need a slide hammer puller with a special attachment to extract the pilot bearing

8.9b Using a hammer and a socket, carefully drive the new bearing into place

Inspection

Refer to illustrations 8.8, 8.9a, 8.9b, 8.10, 8.11 and 8.12

6   Ordinarily, when a problem develops in the clutch, it can be attributed to wear of the clutch driven plate assembly (clutch disc). However, all components should be inspected at this time.
7   Inspect the flywheel (see Chapter 2A).
8   Check the pilot bearing in the end of the crankshaft. Make sure that it turns smoothly and quietly **(see illustration)**. If the transmission input shaft contact surface is worn or damaged, replace the bearing with a new one.
9   Use a puller to remove the pilot bearing **(see illustration)**. Install a new bearing with a socket and hammer **(see illustration)**.

8-6   Chapter 8   Clutch and driveline

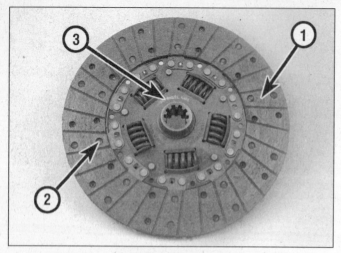

8.10 The clutch plate

1   *Lining* - This will wear down in use
2   *Rivets* - These secure the lining and will damage the flywheel or pressure plate if allowed to contact the surfaces
3   *Markings* - "Flywheel side" or something similar

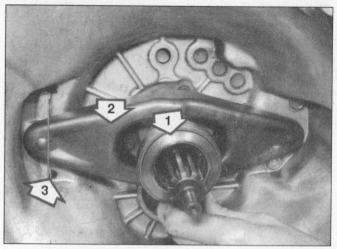

8.11 To remove the release bearing (1), simply slide it off the input shaft; to remove the release arm (2), pop off the return spring (3)

10   Inspect the lining on the clutch disc **(see illustration)**. Look for oil contamination and general wear. If the disc lining looks contaminated or worn, replace the disc. Check the thickness and runout of the clutch disc and compare your measurements to the figures listed in this Chapter's Specifications. If the clutch is too thin or is warped, replace it. Check for loose rivets, distortion, cracks, broken springs and any other obvious damage. As mentioned above, the clutch disc is normally replaced each time it is removed, so if there is any doubt about its condition, install a new one.

11   Ordinarily, the release or throwout bearing is also replaced along with the clutch disc. Pull the bearing off the input shaft **(see illustration)**. It's also a good idea to pop loose the return spring, remove the release arm and inspect the recess for the tip of the slave cylinder pushrod. Replacing the release arm and bearing is simply the reverse of this step.

12   Check the machined surface of the pressure plate. If it's cracked, grooved, scored, discolored by heat or oil contamination or otherwise damaged, it must be be replaced. Lay a straightedge across the surface and verify that it's flat. Look for loose rivets and bent or misaligned release fingers **(see illustration)**. If the pressure plate shows any signs of the above, replace it with a new or factory rebuilt unit.

Installation

13   Before installation, carefully clean the flywheel and pressure plate machined surfaces with brake system cleaner. Handle the parts only with clean hands. DO NOT get oil or grease on the clutch friction surfaces.

14   Position the clutch disc and pressure plate on the flywheel with the clutch disc held in place with the alignment tool. Make sure the disc is installed properly (replacement discs are usually marked with "flywheel side" or some similar identification). If the same pressure plate is being used, make sure the previously applied match marks are in alignment.

15   Tighten the pressure plate-to-flywheel bolts finger tight only, working around the pressure plate.

16   Insert an alignment tool, (if not already in place) through the middle of the clutch disc **(see illustration 8.3)**. Move the disc until it is exactly in the center so the transmission input shaft will pass easily through the disc and into the pilot bearing.

17   Tighten the pressure plate-to-flywheel bolts a little at a time, working in a diagonal pattern to prevent distorting the pressure plate, to the torque listed in this Chapter's Specifications.

18   Using high-temperature grease, lubricate the entire inner surface of the release bearing. Make sure the groove inside is completely filled. Also place grease on the ball socket and the fork fingers.

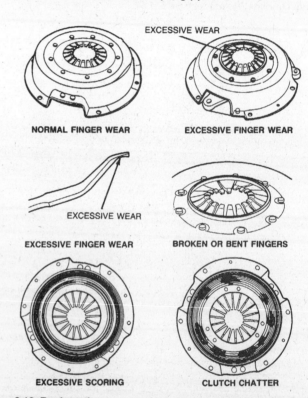

8.12 Replace the pressure plate if any of these conditions are present

19   Install the transmission, slave cylinder and all components removed previously.

9   Driveshaft - removal and installation

Removal

Refer to illustrations 9.6a, 9.6b, 9.10 and 9.11

1   Raise the rear of the vehicle and support it securely on jackstands. Block the front tires to keep the vehicle from rolling.

2   On 3-Series models, remove the rear reinforcement bar and, if equipped, the exhaust and fuel tank heat shields. **Note:** *On some models, the holder for the oxygen sensor plug is attached to the body*

Chapter 8 Clutch and driveline

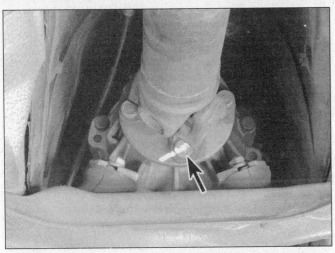

9.6a To insure that the driveshaft is reinstalled in the same position, paint or scribe alignment marks across the transmission output flange and front U-joint flange, as shown on this 5-Series model (most other models use a flexible coupling; a few 5-Series models use a CV joint) . . .

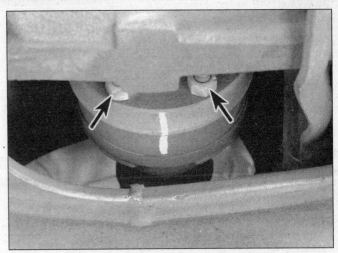

9.6b . . . and across the rear U-joint and differential input flange

9.10 Support the weight of the driveshaft at the center bearing and remove the bolts (arrows) from the center bearing assembly

9.11 Bend the driveshaft down at the center bearing, then slide it off the centering pin on the transmission output flange - slide the splined coupling together to make enough room if you have to, but do NOT pull the two driveshaft sections apart

by one of the reinforcement bar bolts. On these models, make sure you don't damage the sensor leads during removal of the bar.
3   Remove the exhaust system (see Chapter 4).
4   On 5-Series models, remove the exhaust heat shield.
5   At the sliding coupling (slip joint), loosen the clamping sleeve (if equipped) several turns, but don't remove it. **Note:** *Later 5-Series models don't use a clamping sleeve.*
6   Using paint, a punch or a scribe, make alignment marks across the transmission output flange and flexible coupling (or front U-joint, on some models), and across the rear U-joint and differential input flange, to insure that the transmission is reinstalled in the same position **(see illustrations)**.
7   Remove the nuts and bolts which attach the flexible coupling **(see illustration 10.3)** or U-joint **(see illustration 9.6a)** to the transmission output flange. But don't separate the driveshaft from the flange yet.
8   On (some 5-Series) models with a front CV joint, use a suitable transmission jack to support the transmission, loosen the rear transmission support mounting nuts and bolts (see Chapter 7), slide the mount toward the rear and remove the six CV joint-to-transmission mounting nuts and bolts.
9   Remove the nuts and bolts that attach the U-joint flange of the rear section of the driveshaft to the differential input flange (see illus-

tration 9.6b).
10 Support the driveshaft at the center and remove the center bearing bolts **(see illustration)**.
11 Bend the driveshaft down at the center bearing, then slide it off the centering pin on the transmission output flange **(see illustration)**. You may have to slide the splined coupling together to make enough room. Do NOT pull the two driveshaft sections apart. **Note:** *On models with a vibration damper, rotate the damper assembly 60-degrees before pulling it off the centering pin on the transmission output flange.*
12 If you're planning to clean the splined coupling, or replace the rubber bushing, mark the two sections of the driveshaft before pulling the coupling apart. After cleaning the splines, lubricate them with moly-based grease before reassembly. If you pull the two sections apart without marking them, reassemble the splined coupling so that the U-joints are on the same plane. If the driveshaft vibrates after reinstallation, the two sections are reassembled 180-degrees out of balance. You'll have to remove the driveshaft, separate the two sections at the splined coupling, rotate one section 180-degrees and reinstall it.

Installation
Refer to illustration 9.18
13 Installation is basically the reverse of removal. However the fol-

Chapter 8 Clutch and driveline

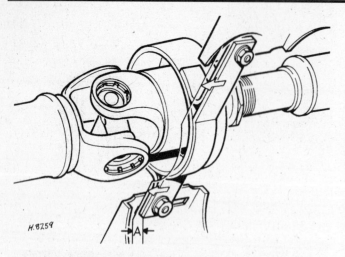

9.18 Preload the center bearing by sliding it forward 5/32 to 15/64-inch from its unloaded position and tighten the bolts to the torque listed in this Chapter's Specifications

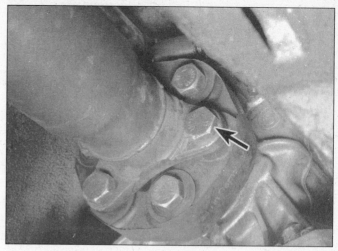

10.3 Remove the flexible coupling nuts and bolts

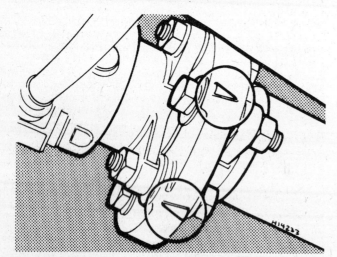

10.4 If the coupling has directional arrows on it, the arrows must face toward the flange arms, not toward the bolt heads

lowing points should be noted:

14 Apply a light coat of moly-based grease to the centering pin on the transmission output flange.

15 Position the driveshaft against the differential input flange first, slide it onto the centering pin at the transmission output flange, then position the center bearing up against the pan and loosely install the mounting bolts, but don't tighten them yet.

16 Align the marks you made across the rear U-joint flange and the differential input flange prior to removal. Reattach the rear driveshaft section to the differential input flange with NEW self-locking nuts and tighten the nuts to the torque listed in this Chapter's Specifications. **Caution:** *Don't re-use self-locking nuts. Use new nuts.*

17 Align the marks you made at the front end of the driveshaft prior to removal and reconnect the flexible coupling, U-joint or CV joint to the transmission output shaft with new self-locking nuts. Tighten the bolts to the torque listed in this Chapter's Specifications. **Note 1:** *Several grades of bolts are used and different grades must be tightened to different torques; you'll find the bolt grade on the bolt head.* **Note 2:** *To avoid stressing the flexible coupling when tightening the bolts, hold the bolt with a wrench and tighten the nut on the flange side.* **Note 3:** *On flexible couplings, it's easier to install the bolts if you place a large hose clamp around the coupling and tighten it slightly to compress the coupling.*

18 Preload the center bearing by sliding it forward 5/32 to 15/64-inch from its unloaded position **(see illustration)**, then tighten the mounting bolts to the torque listed in this Chapter's Specifications.

19 On (5-Series) models with CV joints, slide the rear transmission bracket forward and tighten the mounting bolts to the torque listed in this Chapter's Specifications. Remove the transmission jack.

20 Tighten the clamping sleeve on the splined coupling and tighten it to the torque listed in this Chapter's Specifications.

21 The remainder of installation is the reverse of removal.

10 Flexible coupling - replacement

Refer to illustrations 10.3 and 10.4

1 Although BMW specifies no maintenance interval for the flexible coupling between the front section of the driveshaft and the transmission, you should inspect it whenever you're working under the vehicle. The coupling is made of rubber, so it's vulnerable to road grit, heat generated by the exhaust system, and the twisting torque of the driveshaft. Inspect the coupling for cracks, tears, missing pieces and distortion. If it's cracked or generally worn, replace it.

2 Remove the driveshaft (see Section 9).

3 Remove the flexible coupling nuts and bolts **(see illustration)** and remove the coupling from the driveshaft. **Note:** *To facilitate removal and installation of the bolts, place a large hose clamp around the flexible coupling and tighten the clamp slightly to compress the coupling.*

4 Install the new coupling with NEW self-locking nuts. If the coupling has directional arrows on it, the arrows must face toward the flange arms, NOT toward the bolt heads **(see illustration)**. Tighten the nuts to the torque listed in this Chapter's Specifications.

5 Install the driveshaft (see Section 9).

11 Center bearing - check and replacement

Check

1 The center bearing consists of a ball-bearing assembly encircled by a rubber mount. The bearing assembly is pressed onto the rear end of the front section of the driveshaft and is secured by a snap-ring. The center bearing is protected by dust caps on either side, but water and dirt can still enter the bearing and ruin it. Driveshaft vibration can also lead to bearing failure. To inspect the center bearing properly, you'll need to remove the driveshaft (see Section 9). Conversely, whenever the driveshaft is removed for any reason, always use the opportunity to inspect the bearing. Rotate the bearing and note whether

Chapter 8 Clutch and driveline

11.5 Remove the center bearing snap-ring and dust guard

1 Snap-ring   2 Dust guard

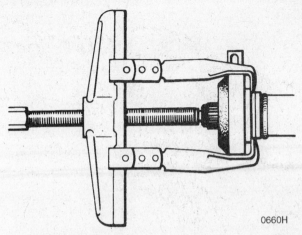

11.6 When you install a puller to remove the center bearing assembly, make sure it's pulling on the hub; pulling on the outer ring of the mount may tear the rubber, necessitating the replacement of the entire bearing/mount assembly

it turns smoothly; if it's difficult to turn, or if it has a gritty feeling, replace it. Also inspect the rubber portion. If it's cracked or deteriorated, replace it.

Replacement

Refer to illustration 11.5 and 11.6

2   Make matching marks on the front and rear sections of the driveshaft.
3   Unscrew and remove the clamping sleeve, then pull the two sections of the driveshaft apart. Remove the rubber bushing, washer and clamping sleeve from the front section of the driveshaft.
4   Inspect the condition of the rubber bushing for the splined coupling. If it's torn or cracked, replace it.
5   Remove the center bearing snap-ring and dust guard **(see illustration)**.
6   Using a puller, remove the bearing mount **(see illustration)**. **Caution:** *Install the puller so that it's pulling on the bearing hub. Pulling on the outer ring of the mount may tear the rubber, necessitating the replacement of the entire bearing/mount assembly.*
7   If you're replacing the bearing or the mount, take the assembly to an automotive machine shop and have the old bearing pressed out of the bearing mount and a new bearing pressed in (or have the old bearing pressed out of the old mount and pressed into a new mount).
8   Make sure the dust guard is still on the driveshaft, then press the center mount assembly onto the driveshaft so that it's flush with the dust guard.
9   On models with a sliding driveshaft, install the clamping sleeve, washer and rubber bushing on the front driveshaft section. Lubricate the splines with moly-based grease, then reassemble the driveshaft. Make sure the matching marks align. Don't retighten the clamping sleeve until the driveshaft is installed.
10 Reinstall the driveshaft (see Section 9).
11 On models without a sliding driveshaft, install the driveshaft yoke mounting bolt with Loctite and tighten it to the torque listed in this Chapter's Specifications.
12 On models with a sliding driveshaft, tighten the clamping sleeve to the torque listed in this Chapter's Specifications.

12 Universal joints - check and replacement

Check

1   Wear in the universal joints is characterized by vibration in the transmission, noise during acceleration, and metallic squeaking and grating sounds as the bearings disintegrate.
2   It's easy to check the U-joints for wear with the driveshaft installed. To check the rear U-joint, try to turn the driveshaft while immobilizing the differential input flange. To check the front U-joint, try to turn the driveshaft while holding the transmission output flange. Freeplay between the driveshaft and the front or rear flanges indicates excessive wear.
3   If driveshaft is already removed, you can check the universal joints by holding the shaft in one hand and turning the yoke or flange with the other. If the axial movement is excessive, replace the driveshaft.

Replacement

4   The universal joints are not serviceable. Therefore, when the joints become worn, the driveshaft and U-joints must be replaced as a single unit.
5   Removal and installation of the driveshaft is covered in Section 9.

13 Front centering guide - check and replacement

Check

Refer to illustration 13.1

1   The front centering guide **(see illustration)** positions the driveshaft precisely in relation to the transmission output flange. The guide, which is press-fitted into a cavity in the front of the driveshaft, engages a guide pin on the transmission output flange. A worn centering guide will allow the front of the driveshaft to wobble and vibrate.
2   To check the centering guide, remove the nuts and bolts that attach the flexible coupling to the transmission output flange, grasp the front end of the driveshaft and try to move it up and down and side to side. It should fit snugly over the pin on the transmission output flange. If there's excessive freeplay, replace the guide. **Note:** *Some driveshafts have a dust cover on the front end, over the centering guide. This cover is sometimes bent or distorted during removal of the driveshaft. Damage to this cover shouldn't affect the operation of the centering guide and shouldn't be confused with guide wear.*

Replacement

Refer to illustration 13.5

3   Remove the driveshaft (see Section 9).
4   Pack the cavity behind the centering guide with a heavy grease until the grease is flush with the lower edge of the guide. Insert a 1/2-inch diameter metal rod into the guide, cover the guide with a rag and strike the bar with a hammer (don't try to use a rod that's too small - it won't work; the grease will escape around the sides). The pressure of the rod against the grease should push the centering guide out of the driveshaft.

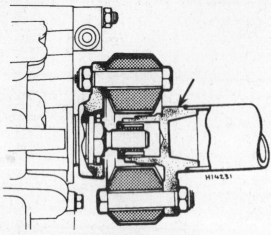

13.1 A cutaway of the centering guide (arrow)

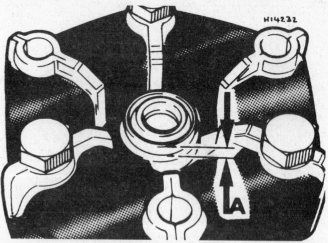

13.5 Make sure the sealing lip of the new centering guide protrudes 11/64-inch (dimension A) from the front of the driveshaft

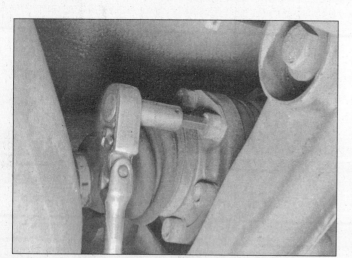

15.5 On all models, remove the Allen-head bolts from the inner CV joint and detach the joint from the differential flange (5-Series model shown, 3-Series models similar)

15.6 On 5-Series models, remove the Allen-head bolts from the outer CV joint and detach the joint from the drive flange

5 Lubricate the new centering guide with moly-based grease and drive it into the driveshaft. Make sure the sealing lip of the guide faces out and protrudes from the face of the driveshaft by about 11/64-inch **(see illustration)**.
6 Install the driveshaft (see Section 9).

14 Driveshaft constant velocity (CV) joint - replacement

1 The constant velocity (CV) joint on the front of the driveshaft on some models bolts directly to the transmission output flange. Replacing it requires removal of the driveshaft. Make sure you replace the clamps, snap-rings, boots, gaskets and covers.
2 Remove the driveshaft (see Section 9).
3 Remove the boot clamp from the back of the CV joint.
4 At the front of the joint, remove the snap-ring that attaches the CV joint to the shaft. Pull the joint off the shaft with a suitable puller (the joint is cemented to the shaft with a thread locking agent and you can't break it free without a puller).
5 Remove the mounting bolts from the CV joint. The bolt shoulders are knurled, so you'll have to drive them out of the CV joint.
6 Pack the new CV joint with 2 ounces of CV joint grease. Work the grease down into the cage. **Caution:** *Do NOT swivel the CV joint inner race while packing it. The balls can fall out of the cage.*
7 Install a new gasket and dust cover.

8 Tap the bolts and retainers through the joint holes. Clean the splines on the driveshaft and apply a small amount of bolt cement to the splines.
9 Place the boot on the shaft. Install the joint onto the shaft and tap it on with a wooden dowel.
10 Install a new snap-ring on the shaft and a install a new boot clamp on the boot.
11 Install the driveshaft (see Section 9).

15 Driveaxles - removal and installation

Refer to illustrations 15.5 and 15.6

1 Set the parking brake firmly. Loosen the rear wheel lug bolts, but don't remove them.
2 On 3-Series models, pry off the dust cap in the center of the wheel hub, lift out the lockplate around the nut and loosen - but don't remove - the nut.
3 Raise the rear of the vehicle and support it securely with jackstands. Remove the wheel(s).
4 Wipe off the inner and outer CV joints to prevent dirt from contaminating them when they're opened.
5 Remove the six Allen-head bolts **(see illustration)** from the inner CV joint and detach it from the inner flange.
6 On 5-Series models, remove the six Allen-head bolts from the

Chapter 8 Clutch and driveline

16.5a Pry open the large boot clamp . . .

16.5b . . . and the smaller boot clamp and discard both of them

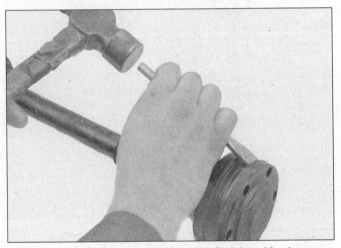

16.6 Tap off the protective cap from the CV joint with a hammer and chisel

16.7 Remove the large snap-ring with a pair of snap-ring pliers

outer CV joint **(see illustration)** and detach it from the outer flange.
7 On 3-Series models, remove the axle nut and press the stub axle out of the wheel hub with a puller. If necessary, prevent the hub from turning by reinstalling two lug bolts and wedging a large screwdriver or prybar between them.
8 On 5-Series models, remove the driveaxle.
9 Installation is the reverse of removal. Be sure to tighten the CV joint Allen bolts and (on 3-Series models) the axle nut, to the torque listed in this Chapter's Specifications.

16 Constant velocity (CV) joints and boots - check and replacement

Note 1: *Some auto parts stores carry "split" type replacement boots, which can be installed without removing the driveaxle from the vehicle. This is a convenient alternative; however, we recommend that the driveaxle be removed, the CV joint cleaned to ensure that it's free from contaminants such as moisture and dirt (which will accelerate CV joint wear) and a new one-piece boot installed.*
Note 2: *If the outer boot on 3-Series models requires replacement, you'll have to remove the inner CV joint to install it.*
Note 3: *If the CV joints exhibit signs of wear (usually caused by torn boots), they must be replaced. On 3-Series models the inner CV joint can be replaced separately; if the outer CV joint is damaged, you must replace the axleshaft as well - they can't be separated. On 5-Series models, both joints are replaceable.*

Note 4: *Complete rebuilt driveaxles are available on an exchange basis, which eliminates much time and work. Whichever route you choose to take, check on the cost and availability of parts before disassembling your vehicle.*

Check

1 The CV joints and boots should be inspected periodically and whenever the vehicle is raised.
2 Raise the vehicle and support it securely on jackstands. Inspect the CV joint boots for cracks, leaks and broken retaining bands. If lubricant leaks out through a hole or crack in a boot, the CV joint will wear prematurely and require replacement. Replace any damaged boots immediately.
3 Grasp each driveaxle, rotate it in both directions and move it in-and-out to check for excessive movement, indicating worn splines or loose CV joints.

CV joint and boot replacement

Note: *If you're replacing the outer driveaxle boot on a 3-Series model, the inner CV joint and boot must be removed first.*
Refer to illustrations 16.5a, 16.5b, 16.6, 16.7, 16.8, 16.13, 16.14 16.15a, 16.15b, 16.16a and 16.16b

4 Remove the driveaxle (see Section 15).
5 Pry open and remove the boot clamps **(see illustrations)** and discard them.
6 Remove the protective cap from the CV joint **(see illustration)**.
7 Remove the large snap-ring that retains the CV joint on the axleshaft **(see illustration)**.

16.8 Using a soft-faced hammer, carefully tap on the the end of the axleshaft (NOT the CV races) and separate the shaft from the joint

16.13 Cover the splines on the end of the axleshaft with tape to protect the new boot(s) from damage during installation

16.14 Carefully tap the CV joint onto the axleshaft - use a soft-faced hammer so you don't damage the races

16.15a Fill the cavity in the backside of the CV joint with CV joint grease . . .

8   Carefully tap on the the end of the axleshaft (NOT the CV races) and separate the shaft from the joint **(see illustration)**.
9   Pull the old boot off the axleshaft.
10   If you're working on a 3-Series model and are also replacing the outer boot, cut the inner and outer boot clamps and pull that boot off too.
11   Thoroughly clean the CV joint(s) with solvent. Wash out all the old grease and blow the solvent out with compressed air, if available. **Warning:** *Wear safety goggles when using compressed air.*
12   Inspect the CV joint(s) for scoring, pitting or other signs of excessive wear. If a CV joint is damaged, replace it. **Note:** *If the outer CV joint on a 3-Series model is damaged, the axleshaft and outer CV joint must be replaced as an assembly.*
13   Wrap the splines on the end of the axleshaft with tape to protect the boot(s) from damage and install the new boot(s) **(see illustration)**. Remove the tape.
14   Carefully tap the CV joint onto the axleshaft **(see illustration)**. Be extremely careful not to damage the races. Use a brass or rubber hammer. Install the snap-ring.
15   Fill the cavity in the backside of the joint with CV joint grease and work it into the joint **(see illustrations)**.
16   Seat the boot into the groove on the axleshaft and install the small boot clamp **(see illustrations)**.

16.15b . . . and work it into the joint

Chapter 8 Clutch and driveline

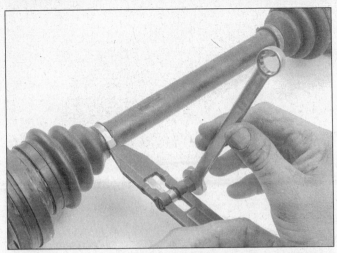

16.16a Slide the boot into place, install the smaller boot clamp and tighten it with a band tightening tool (available at most auto parts stores)

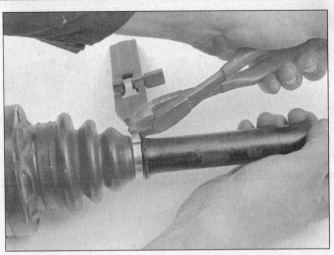

16.16b Bend the end of the clamp back and cut off the excess

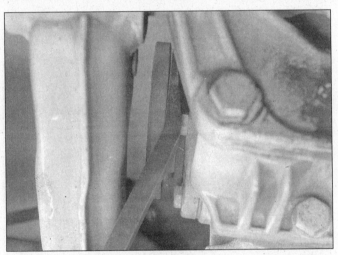

17.4 Using a prybar or a big screwdriver, pry the differential flange out of the differential - be ready to catch the flange as it comes out

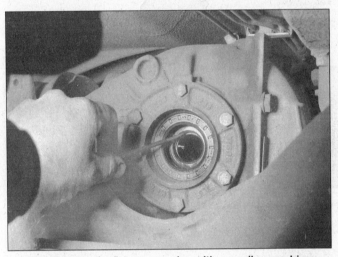

17.5 Remove the flange snap-ring with a small screwdriver

17.7 Pry out the old seal with a seal removal tool (shown) or a large screwdriver

17 Fill the rest of the CV joint with CV joint grease **(see illustrations 16.15a and 16.15b)**, work it into the joint and install the protective end cover.

18 Install the large boot clamp and tighten it **(see illustrations 16.16a and 16.16b)**.

17 Differential housing side seals - replacement

Refer to illustrations 17.4, 17.5, 17.7 and 17.8

1 Raise the rear of the vehicle and support it on jackstands.
2 Drain the differential lubricant (see Chapter 1).
3 Separate the driveaxle from the differential (see Section 15) and suspend it out of the way with a piece of wire.
4 Using a prybar or a big screwdriver, pry the differential flange out of the differential **(see illustration)**. Be ready to catch the flange as it comes out.
5 Remove the flange snap-ring **(see illustration)**.
6 Inspect the stub axle of the flange. If there's a groove worn in the shaft where it contacts the oil seal, replace the flange.
7 Pry out the old seal with a seal removal tool or a large screwdriver **(see illustration)**. Make sure you don't nick or gouge the seal bore in the differential housing.

8-14 Chapter 8 Clutch and driveline

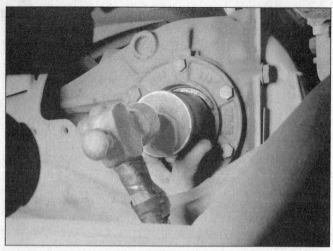

17.8 Coat the new seal with differential lubricant, then use a large socket or section of pipe and a hammer to install the seal

18.4 Disconnect the wires for the speedometer pulse sender, remove the two bolts that attach the sender and remove it (3-Series model shown, 5-Series models similar)

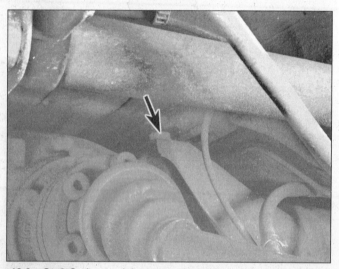

18.6a On 3-Series models, remove the differential rear mounting bolt (arrow)

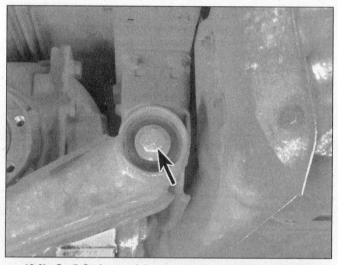

18.6b On 5-Series models, remove the two differential rear mounting bolts (arrow) on the top of the rear axle carrier (left side shown)

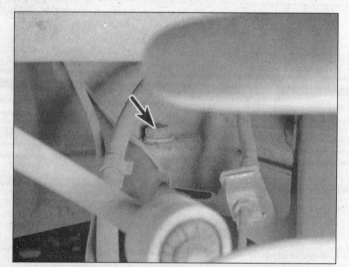

18.7a On 3-Series models, support the differential with a floor jack, then remove the front mounting bolts (arrow) (right side shown)

8 Coat the new seal with differential lubricant, then use a large socket or section of pipe and a hammer to install the seal **(see illustration)**.

9 Install a new snap-ring in the groove. Make sure both ends of the snap-ring are fully seated in the groove.

10 Coat the lip of the seal with grease, then install the differential flange. Push the flange in until you feel the snap-ring engage.

11 The rest of installation is the reverse of removal.

18 Differential housing - removal and installation

Removal

Refer to illustrations 18.4, 18.6a, 18.6b, 18.7a, 18.7b and 18.8

1 Raise the rear of the vehicle and place it securely on jackstands. Drain the differential lubricant (see Chapter 1).

2 Disconnect the driveshaft from the pinion flange (see Section 9). Support the end of the driveshaft with a piece of wire so the center bearing isn't damaged.

3 Unbolt the inner ends of the driveaxles from the differential flanges (see Section 15).

4 Unplug the electrical connector from the speedometer pulse

Chapter 8 Clutch and driveline

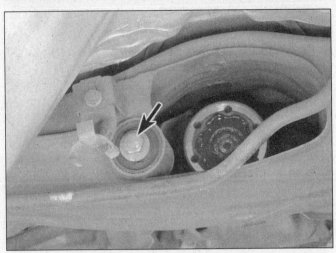

18.7b On 5-Series models, remove the differential front mounting bolt (arrow)

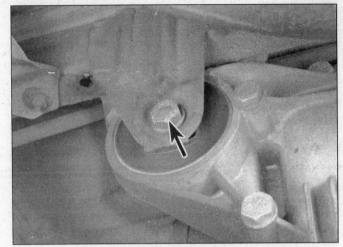

18.8 Remove the nut and bolt (arrow) for the rubber mounting bushing

Installation

10 Position the differential housing in place against its mounts and loosely install the bolts (don't tighten any of them until they are all installed), then tighten all of the fasteners securely.
11 The remainder of installation is the reverse of removal.

19 Pinion oil seal - replacement

Refer to illustration 19.3

1 Raise the rear of the vehicle and place it securely on jackstands.
2 Remove the differential housing (see Section 18).
3 Make matching marks on the pinion shaft, the flange nut and the flange **(see illustration)**.
4 Remove the lockplate, then hold the input flange stationary and remove the flange nut.
5 Using a puller, remove the input flange.
6 Pry out the old seal with a seal removal tool or a screwdriver. Make sure you don't nick or gouge the seal bore.
7 Dip the new seal in differential lubricant and drive it into position with a large socket or a section of pipe.
8 Lightly lubricate the input shaft and press the input flange back on, but don't bottom it out. Install the flange nut and, using a torque wrench, slowly tighten the nut, noting the torque at which the marks line up - it should be near the torque value listed in this Chapter's Specifications. If the nut is tightened and the marks are in alignment, but the torque required to align the marks isn't fairly close to the specified torque, the pinion shaft sleeve probably needs to be replaced. Due to the special tools required and the critical nature of the job, replacement of the pinion shaft sleeve should be left to a dealer service department or other qualified repair shop.
9 If the specified torque is reached *before* the match marks line up, don't continue to turn the nut until they do. **Caution:** *Don't tighten the nut past the match marks and then back it off, as this could over-compress the pinion shaft sleeve.*

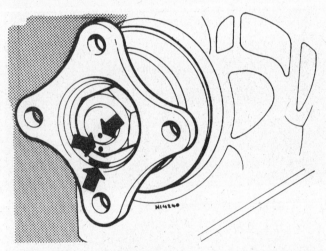

19.3 Scribe or paint alignment marks on the flange, the pinion shaft and the pinion shaft nut to insure proper reassembly

sender **(see illustration)**. Remove the two bolts that attach the sender and remove the sender.
5 Support the differential with a floor jack.
6 On 3-Series models, remove the differential rear mounting bolt **(see illustration)**; on 5-Series models, remove the two differential rear mounting bolts on the top of the rear axle carrier **(see illustration)**.
7 Remove the two front mounting bolts (3-Series models) or the single front mounting bolt (5-Series models) **(see illustrations)**.
8 On 3-Series models, remove the nut and bolt for the rubber mounting bushing **(see illustration)**.
9 Lower the differential and remove it from under the vehicle.

Notes

Chapter 9 Brakes

Contents

Anti-lock brake system (ABS) - general information	2
Brake booster - check, removal and installation	8
Brake check	See Chapter 1
Brake disc - inspection, removal and installation	5
Brake fluid level check	See Chapter 1
Brake hoses and lines - inspection and replacement	15
Brake hydraulic system - bleeding	16
Brake light switch - check, adjustment and replacement	14
Brake pedal - adjustment	13
Disc brake caliper - removal, overhaul and installation	4
Disc brake pads - replacement	3
Drum brake shoes (318i models) - replacement	6
General information	1
Hydraulic brake booster (533i and 535i models) - description, removal and installation	9
Master cylinder - removal and installation	7
Parking brake assembly - check, removal and installation	12
Parking brake cable - adjustment	11
Parking brake cable(s) - replacement	10
Parking brake assembly - check, removal and installation	12

Specifications

General
Brake fluid type .... See Chapter 1

Disc brakes
Minimum brake pad thickness .... See Chapter 1
Brake disc minimum permissible thickness*
  Front
    3-Series
      Solid discs .... 0.421 inch
      Ventilated discs .... 0.787 inch

5-Series
    528e, 533i .... 0.787 inch
    535i .... 0.905 inch
Rear .... 0.315 inch
Brake disc minimum thickness after machining
  Front
    3-Series
      Solid discs .... 0.437 inch
      Ventilated discs .... 0.803 inch
    5-Series
      528i, 533i .... 0.803 inch
      535i .... 0.921 inch
  Rear .... 0.331 inch
Parallelism (difference between any two measurements) .... 0.0008 inch
Maximum disc runout .... 0.008 inch

* Refer to marks cast into the disc (they supersede information printed here)

Brake pedal adjustments

Brake pedal pushrod adjustment (3-Series) .... 4-59/64 inches
Brake pedal height (pedal-to-firewall distance)
  3-Series 9-1/4-inches
  5-Series .... 9-7/64 inches
Brake light switch adjustment 13/64-inch

Parking brakes

Parking brake shoe lining minimum thickness 1/32-inch
Parking brake lever travel .... 5 to 8 clicks

Torque specifications   **Ft-lbs** (unless otherwise indicated)

Front disc brake caliper
  Caliper guide (mounting) bolts .... 22 to 26
  Carrier-to-strut housing bolts .... 80 to 89
Rear disc brake caliper
  Caliper guide (mounting) bolts .... 22 to 25
  Carrier-to-trailing arm bolts .... 44 to 49
Brake hose-to-caliper fitting .... 120 to 144 in-lbs
Master cylinder-to-brake booster nuts
  3-Series .... 16 to 17
  5-Series
    528e .... 18 to 21
    533i, 535i .... 19 to 24
Brake booster mounting nuts .... 16 to 17
Hydraulic line-to-hydraulic brake booster threaded fittings .... 18 to 26
Wheel lug bolts .... See Chapter 1

1 General information

All models covered by this manual, except for the 318i, are equipped with hydraulically operated front and rear disc brake systems. The 318i is equipped with disc brakes in the front and drum brakes at the rear. Both front and rear brakes are self adjusting.

Hydraulic system

The hydraulic system consists of two separate circuits. The master cylinder has separate reservoirs for the two circuits, and, in the event of a leak or failure in one hydraulic circuit, the other circuit will remain operative. Some later models are equipped with an Anti-lock Braking System (ABS).

Power brake booster

The power brake booster, utilizing engine manifold vacuum and atmospheric pressure to provide assistance to the hydraulically operated brakes, is mounted on the firewall in the engine compartment.

A hydraulic brake booster system is used on 533i and 535i models. This system uses hydraulic pressure from the power steering pump to assist braking.

Parking brake

The parking brake lever operates the rear brakes through cable actuation. It's activated by a lever mounted in the center console. The parking brake assembly on 318i models is part of the rear drum brake assembly. On all other models, it uses a pair of brake shoes located inside the center portion of the rear brake disc.

Brake pad wear warning system

The brake pad wear warning system turns on a red light in the instrument cluster when the brake pads have worn down to the point at which they must be replaced. Do NOT ignore this reminder. If you don't replace the pads shortly after the brake pad wear warning light comes on, the brake discs will be damaged.

On 1984 and most 1985 3-Series models and on 1982 through (most) 1985 5-Series models, the brake pad wear warning system also includes an early warning light that comes on only when the brake pedal is depressed, letting you know in advance that the pads need to be replaced.

The wear sensor is attached to the brake pads. On 318i models, the sensor is located at the left front wheel; on all other models, there's a sensor at the left front wheel and another at the right rear wheel. The wear sensor is part of a closed circuit. Once the pads wear down to the

Chapter 9 Brakes

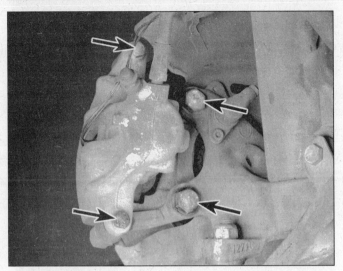

3.5a On 3-Series models, unscrew the caliper mounting bolts (left arrows); right arrows point to the caliper bracket bolts, which should only be removed if you're removing the brake disc

point at which they're flush with the sensor, the disc grinds away the side of the sensor facing the disc, the wire inside the sensor is broken, the circuit is opened and the red light on the instrument panel comes on.

Always check the sensor(s) when replacing the pads. If you change the pads before the warning light comes on, the sensor(s) may still be good; once the light has come on, replace the sensor.

Service

After completing any operation involving disassembly of any part of the brake system, always test drive the vehicle to check for proper braking performance before resuming normal driving. When testing the brakes, perform the tests on a clean, dry, flat surface. Conditions other than these can lead to inaccurate test results.

Test the brakes at various speeds with both light and heavy pedal pressure. The vehicle should stop evenly without pulling to one side or the other. Avoid locking the brakes, because this slides the tires and diminishes braking efficiency and control of the vehicle.

Tires, vehicle load and front-end alignment are factors which also affect braking performance.

2 Anti-lock Brake system (ABS) - general information

The Anti-lock Brake System is designed to maintain vehicle steerability, directional stability and optimum deceleration under severe braking conditions on most road surfaces. It does so by monitoring the rotational speed of each wheel and controlling the brake line pressure to each wheel during braking. This prevents the wheels from locking up.

The ABS system has three main components - the wheel speed sensors, the electronic control unit and the hydraulic control unit. The sensors - one at each front wheel and one at the differential - send a variable voltage signal to the control unit, which monitors these signals, compares them to its program and determines whether a wheel is about to lock up. When a wheel is about to lock up, the control unit signals the hydraulic unit to reduce hydraulic pressure (or not increase it further) at that wheel's brake caliper. Pressure modulation is handled by electrically-operated solenoid valves.

If a problem develops within the system, an "ABS" warning light will glow on the dashboard. Sometimes, a visual inspection of the ABS system can help you locate the problem. Carefully inspect the ABS wiring harness. Pay particularly close attention to the harness and connections near each wheel. Look for signs of chafing and other damage caused by incorrectly routed wires. If a wheel sensor harness is damaged, the sensor should be replaced (the harness and sensor are inte-

gral). **Warning:** *Do NOT try to repair an ABS wiring harness. The ABS system is sensitive to even the smallest changes in resistance. Repairing the harness could alter resistance values and cause the system to malfunction. If the ABS wiring harness is damaged in any way, it must be replaced.* **Caution:** *Make sure the ignition is turned off before unplugging or reattaching any electrical connections.*

Diagnosis and repair

If a dashboard warning light comes on and stays on while the vehicle is in operation, the ABS system requires attention. Although special electronic ABS diagnostic testing tools are necessary to properly diagnose the system, you can perform a few preliminary checks before taking the vehicle to a dealer service department.
 a) Check the brake fluid level in the reservoir.
 b) Verify that the computer master cylinder connectors are securely connected.
 c) Check the electrical connectors at the hydraulic control unit.
 d) Check the fuses.
 e) Follow the wiring harness to each front wheel and to the differential sensor and verify that all connections are secure and that the wiring is undamaged.

If the above preliminary checks do not rectify the problem, the vehicle should be diagnosed by a dealer service department. Due to the complex nature of this system, all actual repair work must be done by a dealer service department.

3 Disc brake pads - replacement

Refer to illustrations 3.5a through 3.5m
Warning: *Disc brake pads must be replaced on both front wheels or both rear wheels at the same time - never replace the pads on only one wheel. Also, the dust created by the brake system may contain asbestos, which is harmful to your health. Never blow it out with compressed air and don't inhale any of it. An approved filtering mask should be worn when working on the brakes. Do not, under any circumstances, use petroleum-based solvents to clean brake parts. Use brake system cleaner only! When servicing the disc brakes, use only high quality, nationally recognized brand-name pads.*
Note: *This procedure applies to both the front and rear disc brakes.*
1 Remove the cap(s) from the brake fluid reservoir and siphon off about two-thirds of the fluid from the reservoir. Failing to do this could result in the reservoir overflowing when the caliper pistons are pressed into their bores.
2 Loosen the wheel lug bolts, raise the front of the vehicle and support it securely on jackstands.
3 Remove the front wheels. Work on one brake assembly at a time, using the assembled brake for reference if necessary.
4 Inspect the brake disc carefully as outlined in Section 5. If machining is necessary, follow the information in that Section to remove the disc, at which time the pads can be removed from the calipers as well.
5 Follow the accompanying photos, beginning with **illustration 3.5a**, for the pad removal procedure. Be sure to stay in order and read the caption under each illustration. **Note 1:** *Different types of front calipers are used on 3 and 5-Series models.* **Illustrations 3.5a through 3.5e** *are for the front calipers on 3-Series models.* **Illustrations 3.5f through 3.5m** *are for the front calipers on 5-Series models. There's no photo sequence for rear calipers; although slightly different in size, they're identical in design to the front brake calipers used on 5-Series models.* **Note 2:** *Some models may have different numbers and types of anti-squeal shims and other hardware than what is shown in this Chapter. It's best to note how the hardware is installed on the vehicle before disassembly so you can duplicate it on reassembly.*
6 Be sure to inspect the wear sensors (left front wheel only on 318i models; left front and right rear on all other models). If they're okay, transfer them from the old pads to the new ones; if they're worn by abrasion, install new sensors on the new pads.
7 To install the new pads, reverse the removal procedure. When reinstalling the caliper, be sure to tighten the mounting bolts to the

Chapter 9 Brakes

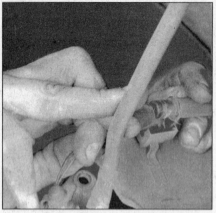

3.5b Unplug the electrical connector for the brake pad wear sensor (3-Series)

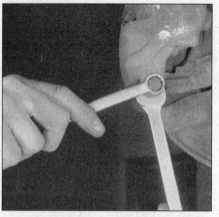

3.5c Use a back-up wrench on the guide pins while breaking loose the caliper mounting bolts (3-Series)

3.5d Remove the caliper, brake pad wear sensor and inner pad all at the same time (3-Series), then install the inner pad on the piston and bottom the piston in the bore with a C-clamp

3.5e Remove the outer brake pad (3-Series) - to install the new pads, reverse the removal procedure

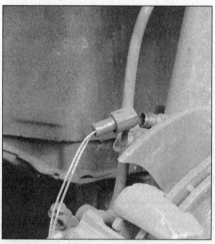

3.5f On 5-Series models, unplug the electrical connector for the brake pad wear sensor

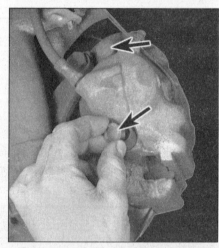

3.5g Remove the plugs for the brake caliper mounting bolts then remove the bolts (5-Series)

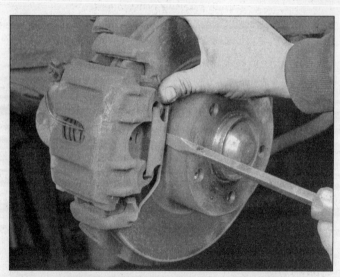

3.5h Pry off the anti-rattle spring (5-Series)

3.5i Depress the piston with a C-clamp (5-Series)

Chapter 9 Brakes

3.5j Remove the caliper and inner brake pad (5-Series)

3.5k Unclip the inner brake pad from the piston (5-Series)

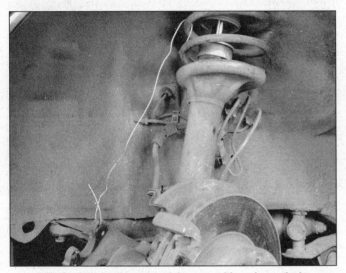

3.5l Hang the caliper out of the way with a piece of wire

3.5m Remove the outer brake pad - to install the new pads, reverse the removal procedure

torque listed in this Chapter's Specifications. **Warning:** *The manufacturer recommends replacement of the mounting bolts on 3-Series models whenever they are removed.*

8 After the job is completed, firmly depress the brake pedal a few times to bring the pads into contact with the discs. The pedal should be at normal height above the floorpan and firm. Check the level of the brake fluid, adding some if necessary. Check carefully for leaks and check the operation of the brakes before placing the vehicle into normal service.

4 Disc brake caliper - removal, overhaul and installation

Warning: *Dust created by the brake system may contain asbestos, which is harmful to your health. Never blow it out with compressed air and don't inhale any of it. An approved filtering mask should be worn when working on the brakes. Do not, under any circumstances, use petroleum-based solvents to clean brake parts. Use brake system cleaner only!*

Note: *If an overhaul is indicated (usually because of fluid leakage), explore all options before beginning the job. New and factory rebuilt calipers are available on an exchange basis, which makes this job quite easy. If you decide to rebuild the calipers, make sure that a rebuild kit is* available before proceeding. Always rebuild the calipers in pairs - never rebuild just one of them.

Removal

1 Loosen the wheel lug bolts, raise the front or rear of the vehicle and place it securely on jackstands. Remove the wheel.
2 If you're just removing the caliper for access to other components, it isn't necessary to detach the brake line. If you're removing the caliper for overhaul, disconnect the brake line from the caliper with a flare-nut wrench to protect the fitting and plug the line to keep contaminants out of the brake system and to prevent losing brake fluid unnecessarily.
3 Refer to Section 3 for the front or rear caliper removal procedure - it's part of the brake pad replacement procedure. **Note:** *The rear caliper on 3-Series models is similar in design to the calipers on 5-series models.*

Overhaul

Refer to illustrations 4.4a, 4.4b, 4.5 and 4.7

4 On all calipers except the front calipers on 3-Series models, remove the circlip for the dust boot **(see illustration),** then remove the dust boot **(see illustration).** Before you remove the piston, place a

Chapter 9 Brakes

4.4a Remove the circlip for the dust seal

wood block between the piston and caliper to prevent damage as it is removed.
5 To remove the piston from the caliper, apply compressed air to the brake fluid hose connection on the caliper body **(see illustration)**. Use only enough pressure to ease the piston out of its bore. **Warning:** *Be careful not to place your fingers between the piston and the caliper as the piston may come out with some force.* If you're working on a front caliper of a 3-Series model, remove the dust boot.
6 Inspect the mating surfaces of the piston and caliper bore wall. If there is any scoring, rust, pitting or bright areas, replace the complete caliper unit with a new one.
7 If these components are in good condition, remove the piston seal from the caliper bore using a wooden or plastic tool **(see illustration)**. Metal tools may damage the cylinder bore.
8 Remove the caliper guide pins or bolts and remove the rubber dust boots.
9 Wash all the components brake system cleaner.
10 Using the correct rebuild kit for your vehicle, reassemble the caliper as follows.

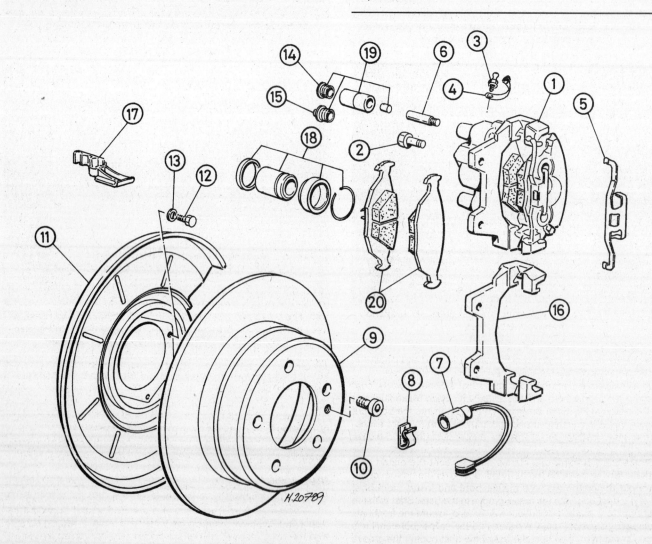

4.4b An exploded view of a typical rear caliper assembly (front calipers similar)

1	Caliper assembly	7	Brake pad wear warning light wire	11	Shield
2	Bracket mounting bolt			12	Bolt
3	Bleed screw	8	Cable clamp	13	Washer
4	Dust cap	9	Brake disc	14	Plug
5	Anti-rattle spring	10	Allen bolt	15	Plug
6	Guide bolt				

16	Caliper bracket
17	Cable clamp
18	Piston seal, piston, dust boot and circlip
19	Guide bushing repair kit
20	Brake pads

Chapter 9 Brakes

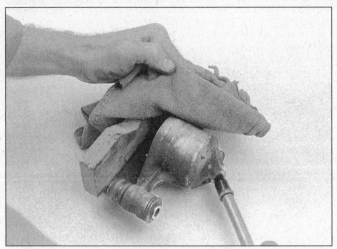

4.5 With the caliper padded to catch the piston, use compressed air to force the piston out of its bore - make sure your fingers are not between the piston and the caliper

4.7 Remove the piston seal from the caliper bore using a wooden or plastic tool (metal tools may damage the cylinder bore)

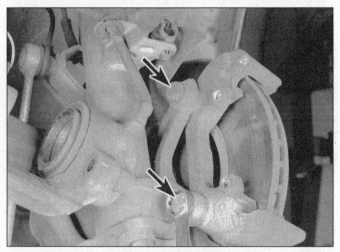

5.2 Remove the caliper mounting bracket bolts (arrows) and remove the bracket

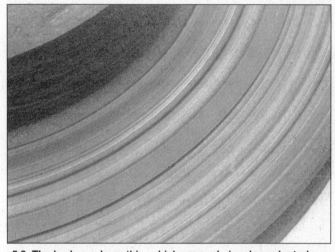

5.3 The brake pads on this vehicle were obviously neglected, as they wore down to the rivets and cut deep grooves into the disc - wear this severe means the disc must be replaced

11   Submerge the new rubber seal in clean brake fluid and install it in the lower groove in the caliper bore, making sure it isn't twisted.
12   On all calipers except the front calipers of 3-Series models, coat the walls of the caliper bore and the piston with clean brake fluid and install the piston at this time. Do not force the piston into the bore, but make sure that it is squarely in place, then apply firm (but not excessive) pressure to install it. Install the new rubber dust boot and the retaining ring.
13   On the front calipers of 3-Series models, coat the piston with clean brake fluid and stretch the new dust boot over the bottom of the piston. Hold the piston over the caliper bore and insert the rubber flange of the dust boot into the upper groove in the bore. Start with the furthest side from you and work your way around toward the front until it is completely seated. Push the piston into the caliper bore until it is bottomed in the bore, then seat the top of the dust boot in the groove in the piston.
14   Lubricate the sliding surfaces of the guide pins or bolts with silicone-based grease (usually supplied in the kit) and push them into the caliper ears. Install the dust boots.

Installation

15   Install the caliper by reversing the removal procedure (see Section 3). **Warning:** *The manufacturer recommends replacement of the mounting bolts on 3-Series models whenever they are removed.*

16   If the hose was disconnected from the caliper, bleed the brake system (see Section 16).

5   Brake disc - inspection, removal and installation

Note: *This procedure applies to both the front and rear brake discs.*

Inspection

Refer to illustrations 5.2, 5.3, 5.4a, 5.4b and 5.5

1   Loosen the wheel lug bolts, raise the vehicle and support it securely on jackstands. Remove the wheel and install three lug bolts to hold the disc in place. If the rear brake disc is being worked on, release the parking brake.
2   Remove the brake caliper as outlined in Section 4. It is not necessary to disconnect the brake hose. After removing the caliper, suspend it out of the way with a piece of wire. Remove the caliper mounting bracket **(see illustration)**.
3   Visually inspect the disc surface for scoring or damage. Light scratches and shallow grooves are normal after use and may not always be detrimental to brake operation, but deep scoring - over 0.015 inch - requires disc removal and refinishing by an automotive machine shop. Be sure to check both sides of the disc **(see illustration)**. If pul-

Chapter 9 Brakes

5.4a To check disc runout, mount a dial indicator as shown and rotate the disc

5.4b Using a swirling motion, remove the glaze from the disc surface with sandpaper or emery cloth

5.5 The disc thickness can be checked with a micrometer

5.6a Remove the disc retaining screw . . .

5.6b . . . and remove the disc from the hub

sating has been noticed during application of the brakes, suspect disc runout.

4   To check disc runout, place a dial indicator at a point about 1/2-inch from the outer edge of the disc **(see illustration)**. Set the indicator to zero and turn the disc. The indicator reading should not exceed the specified allowable runout limit. If it does, the disc should be refinished by an automotive machine shop. **Note:** *It is recommended that the discs be resurfaced regardless of the dial indicator reading, as this will impart a smooth finish and ensure a perfectly flat surface, eliminating any brake pedal pulsation or other undesirable symptoms related to questionable discs. At the very least, if you elect not to have the discs resurfaced, remove the glazing from the surface with emery cloth or sandpaper using a swirling motion* **(see illustration)**.

5   It is absolutely critical that the disc not be machined to a thickness under the specified minimum allowable thickness. The minimum wear (or discard) thickness is stamped into the hub of the disc. The disc thickness can be checked with a micrometer **(see illustration)**.

Removal

Refer to illustrations 5.6a, 5.6b, 5.6c and 5.6d

6   Remove the disc retaining screw **(see illustration)** and remove the disc from the hub **(see illustration)**. If the disc is stuck to the hub, spray a generous amount of penetrant onto the area between the hub and the disc **(see illustration)** and allow the penetrant a few minutes to loosen the rust between the two components. If a rear disc still sticks, insert a thin, flat-bladed screwdriver through the hub flange, rotate the star wheel on the parking brake adjusting screw and contract the parking brake shoes **(see illustration)**. If the front disc is stuck, thread two or three bolts into the holes provided and tighten them. Alternate between the bolts, turning them a couple of turns at a time, until the disc is free.

5.6c If the disc is stuck to the hub, spray some penetrant onto the area between the hub and the disc and give the penetrant a few minutes to dissolve the rust between the two parts

5.6d If a rear disc still sticks to the hub, insert a thin, flat-bladed screwdriver through the hub flange, rotate the star wheel on the parking brake adjusting screw and contract the parking brake shoes (hub removed for clarity)

6.2a Before trying to remove the drum, remove this retaining screw first

6.2b If the drum is stuck the hub, apply penetrant around the hub/drum area and give it a few minutes to loosen up any rust

6.2c If the brake shoes have worn a groove in the drum and it won't come off, insert a thin flat-bladed screwdriver through one of the lug bolt holes in the flange and loosen the automatic adjuster mechanism (for the sake of clarity, the hub has already been removed in this photo and the screwdriver is being inserted underneath the flange instead of though a lug bolt hole)

Installation

7   Place the disc on the hub and install the disc retaining screw. Tighten the screw securely.
8   Install the caliper mounting bracket, brake pads and caliper (see Sections 3 and 4). Tighten all fasteners to the torque listed in this Chapter's Specifications.
9   Install the wheel, then lower the vehicle to the ground. Depress the brake pedal a few times to bring the brake pads into contact with the disc.
10   Adust the parking brake shoes, if necessary.

11   Check the operation of the brakes carefully before placing the vehicle into normal service.

6   Drum brake shoes (318i models) - replacement

Refer to illustrations 6.2a, 6.2b, 6.2c, 6.3, 6.4a, 6.4b, 6.5a, 6.5b, 6.6a, 6.6b, 6.7, 6.8, 6.9, 6.10a, 6.10b, 6.10c and 6.11

Warning: *Drum brake shoes must be replaced on both wheels at the same time - never replace the shoes on only one wheel. Also, the dust created by the brake system may contain asbestos, which is harmful to your health. Never blow it out with compressed air and don't inhale any of it. Always wear an approved filtering mask when servicing the brake system. Do not, under any circumstances, use petroleum-based solvents to clean brake parts. Use brake system cleaner only.*

Caution: *Whenever the brake shoes are replaced, the retractor and hold-down springs should also be replaced. Due to the continuous heating/cooling cycle to which the springs are subjected, they lose their tension over a period of time and may allow the shoes to drag on the drum and wear at a much faster rate than normal. When replacing the rear brake shoes, use only high-quality, nationally-recognized brand name parts.*

Note: *All four rear brake shoes must be replaced at the same time, but to avoid mixing up parts, work on only one brake assembly at a time.*

1   Loosen the rear wheel lug bolts, raise the rear of the vehicle and place it securely on jackstands. Remove the rear wheels.
2   Remove the drum retaining screw **(see illustration)** and remove the drum. If the drum is stuck to the hub, spray the area between the hub and the drum with penetrant **(see illustration)**. If the drum still won't come off, the shoes have probably worn ridges into the drum and will have to be retracted. Insert a narrow flat-bladed screwdriver

6.3 The maximum allowable inside diameter of the drum is cast into the drum

6.4a Unhook the lower return spring from the front shoe . . .

6.4b . . . then unhook it from the rear shoe and remove it

6.5a Unhook the upper return spring from the front shoe . . .

6.5b . . . then unhook it from the rear shoe and remove it

6.6a Remove the front shoe hold-down spring . . .

6.6b . . . and the rear shoe hold-down spring

through one of the holes in the hub flange **(see illustration)** and back off the adjuster wheel until the drum can be removed.

3   Inspect the drum for cracks, score marks, deep scratches and hard spots, which will appear as small discolored areas. If the hard spots can't be removed with emery cloth or if any of the other conditions exist, the drum must be taken to an automotive machine shop to have the drum resurfaced. **Note:** *Professionals recommend resurfacing the drums whenever a brake job is done. Resurfacing will eliminate the possibility of out-of-round drums. If the drums are worn so much that they can't be resurfaced without exceeding the maximum allowable diameter (which is cast into the drum)* **(see illustration)**, *then new ones will be required. At the very least, if you elect not to have the drums resurfaced, remove the glazing from the surface with emery cloth or sandpaper, using a swirling motion.*

Chapter 9 Brakes

6.7 Remove the front shoe and automatic adjuster lever and spring as an assembly then remove the lever and spring and set them aside for attachment to the new shoe

6.8 Remove the automatic adjuster assembly

6.10a Before you install the new shoes, apply some white grease to the friction surfaces where the inner edge of the shoe slides on the brake backing plate - when you install the automatic adjuster mechanism, make sure each end engages properly with its respective notch in the brake shoe

6.9 To disconnect the parking brake cable from the parking brake lever, pull on the plug at the end of the cable and detach the cable from the bracket on the upper end of the lever (diagonal cutting pliers are being used here because they grip the cable well, but care must be taken not to nick the cable)

6.10c . . . then hook the lower end of the spring onto the lever as shown, stretch the spring and hook the upper end into its hole in the parking brake shoe

6.10b Install the automatic adjuster lever first - make sure it's properly engaged with the notch in the front end of the adjuster mechanism . . .

4   Unhook and remove the lower return spring (see illustrations).
5   Unhook and remove the upper return spring (see illustrations).
6   Remove the front and rear brake shoe hold-down springs (see illustrations).
7   Remove the front shoe (see illustration).

8   Remove the adjuster assembly (see illustration).
9   Disconnect the parking brake cable from the parking brake lever and remove the rear shoe (see illustration).
10 Installation is basically the reverse of removal:
  a) Be sure to apply high-temperature brake grease to the backing plate (see illustration).
  b) Make sure the adjuster assembly is properly engaged with its respective notch in the parking brake lever.
  c) When reattaching the automatic adjustment mechanism, install the lever on the shoe first (see illustration), then hook the lower end of the spring onto the lever and the upper end into its hole in the front shoe (see illustration).

9-12　Chapter 9　Brakes

6.11 When you get everything back together, this is how it should look!

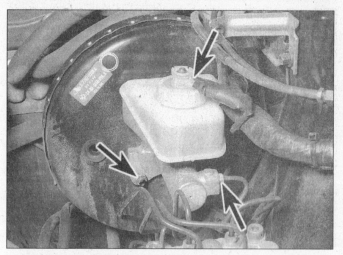

7.1 To remove the master cylinder, unplug the electrical connector (top arrow), disconnect the brake fluid hydraulic line fittings (lower right arrow, other fitting not visible in this photo) and remove the two master cylinder mounting nuts (lower left arrow, other nut not visible in this photo) (5-Series master cylinder shown, 3-Series similar)

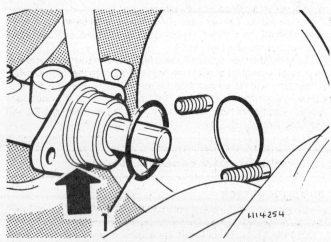

7.6 Always discard the old O-ring (arrow) between the master cylinder and the brake booster and replace it

11　When you're done, the brake assembly should look like this **(see illustration)**. Now proceed to the other brake.
12　When you're done with both brakes, install the brake drums. Inspect and adjust the parking brake lever as necessary.
13　Install the wheel and lug bolts. Lower the vehicle to the ground and tighten the lug bolts to the torque listed in the Chapter 1 Specifications. Check the operation of the brakes carefully before driving the vehicle in traffic.

7　Master cylinder - removal and installation

Warning: *On some 1984 533i models, incorrect machining of the mating surface between the master cylinder and the hydraulic booster may result in a warped or cracked master cylinder flange. A white paint mark on the booster verifies that the vehicle has been inspected for this problem. If there is no paint mark on the booster on these models, contact your BMW dealer.*
Note: *Although master cylinder parts and rebuild kits are available for most models, we recommend replacing the master cylinder with a unit that's already rebuilt. It will take you more time to rebuild the master cylinder than to replace it, and you can't even determine whether the master cylinder is in good enough condition to rebuild it until you have disassembled it. You may very well find that it can't be rebuilt because of its internal condition.*

Removal

Refer to illustrations 7.1 and 7.6
1　The master cylinder is connected to the brake booster and both are attached to the firewall, located on the driver's side of the engine compartment **(see illustration)**.
2　Remove as much fluid as you can from the reservoir with a syringe.
3　Place rags under the line fittings and prepare caps or plastic bags to cover the ends of the lines once they are disconnected. **Caution:** *Brake fluid will damage paint. Cover all body parts and be careful not to spill fluid during this procedure.*
4　Loosen the tube nuts at the ends of the brake lines where they enter the master cylinder. To prevent rounding off the flats on these nuts, a flare-nut wrench, which wraps around the nut, should be used.
5　Pull the brake lines away from the master cylinder slightly and plug the ends to prevent contamination.
6　Disconnect any electrical connectors at the master cylinder, then remove the nuts attaching the master cylinder to the power booster. Pull the master cylinder off the studs and lift it out of the engine compartment. Again, be careful not to spill fluid as this is done. Discard the old O-ring **(see illustration)** between the master cylinder and the booster unit. **Warning:** *The O-ring should always be replaced. A faulty O-ring can cause a vacuum leak, which can reduce braking performance and cause an erratic idle.*

Bench bleeding procedure

7　Before installing a new or rebuilt master cylinder it should be bench bled. Because it will be necessary to apply pressure to the master cylinder piston and, at the same time, control flow from the brake line outlets, it is recommended that the master cylinder be mounted in a vise. Use caution not to clamp the vise too tightly, or the master cylinder body might crack.
8　Insert threaded plugs into the brake line outlet holes and snug them down so that there will be no air leakage past them, but not so tight that they cannot be easily loosened.
9　Fill the reservoir with brake fluid of the recommended type (see *Recommended lubricants and fluids* in Chapter 1).
10　Remove one plug and push the piston assembly into the master cylinder bore to expel the air from the master cylinder. A large Phillips screwdriver can be used to push on the piston assembly.
11　To prevent air from being drawn back into the master cylinder, the plug must be replaced and snugged down before releasing the pressure on the piston assembly.
12　Repeat the procedure until only brake fluid is expelled from the

Chapter 9 Brakes

8.9 Disconnect the brake pedal return spring, then remove the clip and clevis pin (arrows) to disconnect the pushrod from the brake pedal

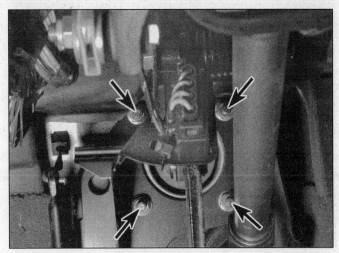

8.10 Remove the four mounting nuts (arrows) and withdraw the booster unit from the engine compartment

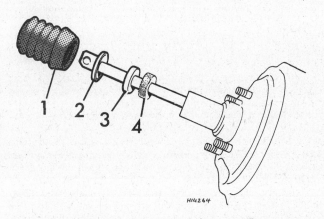

8.11 An exploded view of a typical booster pushrod assembly

1 Boot
2 Holder
3 Damper
4 Air filter

brake line outlet hole. When only brake fluid is expelled, repeat the procedure with the other outlet hole and plug. Be sure to keep the master cylinder reservoir filled with brake fluid to prevent the introduction of air into the system.

13 Since high pressure is not involved in the bench bleeding procedure, an alternative to the removal and replacement of the plugs with each stroke of the piston assembly is available. Before pushing in on the piston assembly, remove the plug as described in Step 10. Before releasing the piston, however, instead of replacing the plug, simply put your finger tightly over the hole to keep air from being drawn back into the master cylinder. Wait several seconds for brake fluid to be drawn from the reservoir into the piston bore, then depress the piston again, removing your finger as brake fluid is expelled. Be sure to put your finger back over the hole each time before releasing the piston, and when the bleeding procedure is complete for that outlet, replace the plug and snug it up before going on to the other port.

Installation

14 Install the master cylinder over the studs on the power brake booster and tighten the mounting nuts only finger tight at this time.
15 Thread the brake line fittings into the master cylinder. Since the master cylinder is still a bit loose, it can be moved slightly in order for the fittings to thread in easily. Do not strip the threads as the fittings are tightened.
16 Tighten the brake fittings securely and the mounting nuts to the torque listed in this Chapter's Specifications.
17 Fill the master cylinder reservoir with fluid, then bleed the master cylinder (only if the cylinder has not been bench bled) and the brake system as described in Section 16.
18 To bleed the cylinder on the vehicle, have an assistant pump the brake pedal several times and then hold the pedal to the floor. Loosen the fitting nut to allow air and fluid to escape, then tighten the nut. Repeat this procedure on both fittings until the fluid is clear of air bubbles. Test the operation of the brake system carefully before placing the vehicle into service.

8 Brake booster - check, removal and installation

Operating check

1 Depress the brake pedal several times with the engine off and make sure there is no change in the pedal reserve distance.
2 Depress the pedal and start the engine. If the pedal goes down slightly, operation is normal.

Airtightness check

3 Start the engine and turn it off after one or two minutes. Depress the brake pedal several times slowly. If the pedal goes down further the first time but gradually rises after the second or third depression, the booster is airtight.
4 Depress the brake pedal while the engine is running, then stop the engine with the pedal depressed. If there is no change in the pedal reserve travel after holding the pedal for 30 seconds, the booster is airtight.

Removal and installation

Refer to illustrations 8.9, 8.10, 8.11 and 8.13

5 Dismantling of the power booster requires special tools and cannot be performed by the home mechanic. If a problem develops, it is recommended that a new or factory rebuilt unit be installed.
6 Remove the master cylinder as described in Section 7.
7 Disconnect the vacuum hose where it attaches to the brake booster.
8 Working in the passenger compartment, remove the lower left trim panels above the brake pedal.
9 Disconnect the brake pedal return spring, then remove the clip and clevis pin (see illustration) to disconnect the pushrod from the brake pedal.
10 Remove the four mounting nuts (see illustration) and withdraw the booster unit from the engine compartment.
11 Inspect the small foam filter (see illustration) inside the rubber boot on the pushrod. If the filter is clogged, it may affect the booster's

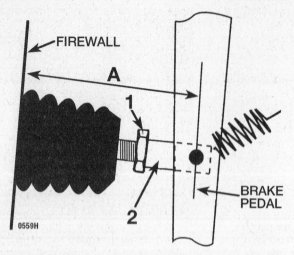

8.13 On 3-Series models, adjust dimension A (the distance between the middle of the brake lever and the firewall) by loosening the locknut (1) at the pushrod clevis (2) and turning the threaded part of the pushrod until dimension A matches the dimension listed in this Chapter's Specifications - when the basic setting is correct, tighten the locknut, then adjust the brake pedal travel and the brake light switch

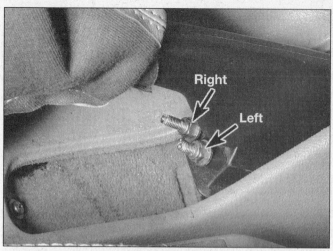

10.1 Peel back the parking brake lever boot and remove the parking brake cable adjusting nut (arrows) for the cable you're replacing

performance. To clean the filter, wash it in a mild soapy solution. If it's still dirty, replace it.

12 Installation is the reverse of the removal procedure. Tighten the brake booster mounting nuts to the torque listed in this Chapter's Specifications. Before you slide the boot into place over the booster pushrod air filter, make sure the notches in the filter offset the notches in the damper by 180-degrees.

13 On 3-Series models, adjust the basic setting of the pushrod's threaded clevis until the dimension is correct (see illustration). When the basic setting is correct, tighten the locknut, then adjust the brake pedal travel and the brake light switch (see Section 13).

14 On 5-Series models, adjust the brake pedal height and the brake light switch (see Section 13).

15 Install the master cylinder (see Section 7) and attach the vacuum hose.

16 Carefully test the operation of the brakes before placing the vehicle in normal operation.

9 Hydraulic brake booster - description, removal and installation

Description

1 On 533i and 535i models, a hydraulic brake booster system assists braking when the brake pedal is depressed. The booster unit, located between the brake pedal and the master cylinder, is operated by hydraulic pressure generated by the power steering pump. When the engine is running, the power steering pump supplies hydraulic pressure to a power flow regulator/accumulator. The regulator/accumulator stores and regulates the pressure to the hydraulic brake booster. When you depress the brake pedal, the pressure in the booster helps actuate the master cylinder, reducing pedal effort.

2 The hydraulic brake booster isn't rebuildable; if it fails, it must be replaced. Testing the system requires special tools, so troubleshooting is beyond the scope of the home mechanic. If the system fails, take it to a dealer service department or other qualified repair shop for repairs.

Removal and installation

3 With the engine off, discharge the hydraulic accumulator by depressing the brake pedal 20 times or more.

4 Remove the master cylinder (see Section 7).

5 Clean the area around the return and supply line fittings, then disconnect them and plug the lines to prevent contamination from entering the system. **Caution:** *Even a particle of dirt can damage the booster system, so be extremely careful to prevent dirt from entering the system while the lines are disconnected.*

6 Working from inside the passenger compartment, remove the lower left trim panels above the brake pedal.

7 Disconnect the brake pedal return spring, pry off the retaining clip and disconnect the pushrod from the brake pedal (see illustration 8.9).

8 Remove the four mounting nuts and remove the brake booster (see illustration 8.10).

9 Installation is the reverse of removal. Tighten the hydraulic lines to the torque listed in this Chapter's Specifications. **Note:** *Don't try to tighten these fittings without a torque wrench. If they're too loose, they can leak, which can affect system operation; if they're too tight, they can be damaged, and they'll also leak. You'll need a crowfoot-type flare-nut wrench for your torque wrench to tighten the fittings properly.*

10 When you're done, adjust the brake pedal travel and the brake light switch (see Section 13).

10 Parking brake cable(s) - replacement

Refer to illustrations 10.1, 10.4a, 10.4b and 10.6

1 Peel back the boot at the base of the parking brake lever and remove the adjusting nut (see illustration) which attaches the parking brake cable to the parking brake lever (there are two cables - one for each rear wheel - and a nut for each cable).

2 Raise the vehicle and support it securely on jackstands.

3 On 318i models, remove the rear brake drum (see Section 6). On all other models, remove the rear brake disc (see Section 5).

4 On 318i models, unhook the parking brake cable from the lever on the rear brake shoe (see Section 6). On all other models, remove the parking brake shoes and the actuator (see Section 12) and unhook the parking brake cable from the actuator (see illustrations).

5 On 318i models, pull the cable and cable conduit (tube) out of the back side of the brake backing plate then detach the cable conduit from the cable clips on the back side of the trailing arm (it's easier to pull out the old cable and install the new cable with the conduit straight instead of curved). On all other models, it's unnecessary to detach the cable conduit from the brake backing plate, but it's a good idea to detach the conduit from the clips and guides that fasten it to the trailing arm to take some of the bend out of the conduit.

6 Working from the wheel end of the cable conduit, pull the cable out of the conduit (see illustration).

10.4a To detach the parking brake cable from the parking brake actuator on models with rear disc brakes, pull on the outer cam and disconnect it from the inner cam . . .

10.4b . . . then remove the pin that attaches the cable to the inner cam and remove the inner cam

10.6 Pull the cable out of its conduit (the cable is easier to remove and install if you detach the cable conduit from the clips that attach it the trailing arm); before you install the new cable, be sure to lubricate it with multi-purpose grease

7 Lubricate the new cable with multi-purpose grease, then insert it into the cable housing and push it through until the forward end comes out at the parking brake lever.
8 Insert the cable housing through the backing plate and attach the rear end of the cable to the parking brake lever (318i models) or the actuator (all other models). Make sure you don't kink the cable while connecting it.
9 Reattach the cable housing to the clips on the back side of the trailing arm.
10 On 318i models, install the brake shoes and drum (see Section 6). On all other models, install the parking brake shoes and actuator (see Section 12) and the rear brake disc (see Section 5).
11 Lower the vehicle and install the adjusting nut at the parking brake lever. Adjust the parking brake cable (see Section 11) and reinstall the parking brake lever boot.

11 Parking brake cable - adjustment

318i models

Note: *Adjustment of the parking brake cable(s) on models with rear drum brakes should only be necessary when you replace a cable or detach if from the rear brake assembly for some reason. Failure of the parking brake system to hold the vehicle usually indicates worn brake shoes or a faulty self-adjusting mechanism.*

1 Raise the rear of the vehicle and place it securely on jackstands.
2 Fully release the parking brake lever, then apply the brakes firmly several times.
3 Pull the parking brake lever up five clicks.
4 Tighten or loosen the adjusting nuts by equal amounts until the rear brake shoes just begin to drag on the brake drum. You should feel the same amount of resistance at both wheels when you rotate them.
5 Release the parking brake lever and verify that the wheels rotate freely. If they don't, readjust them.

All other models

Note: *The parking brake system is not self-adjusting on models with rear disc brakes. The parking brakes therefore require periodic adjustment to compensate for wear. They should also be adjusted anytime either cable, brake disc or parking brake assembly is replaced or removed for some reason.*

6 Slowly apply the parking brake and count the number of clicks at the lever. If the lever can be pulled up further than the eighth click, adjust the parking brake cable as follows:
7 Peel back the parking brake lever boot and loosen the cable adjusting nut **(see illustration 10.1)**.
8 Loosen a single lug bolt in each rear wheel. Raise the vehicle and place it securely on jackstands.
9 Remove the lug bolt you loosened in each rear wheel. Turn the wheel until, using a flashlight, you can see the adjuster star wheel through the lug bolt hole.
10 Turn the adjuster - clockwise to expand the shoes, counterclockwise to retract them - until the brake shoes just contact the brake drum **(see illustration 5.6b)**. Then back off the brake shoes so the wheel spins freely (three to four teeth on the adjuster). **Note:** *If the adjuster star wheel is hard to turn, remove the wheel and brake disc, lubricate the adjuster wheel and try again.*
11 With the disc installed, set the parking brake three times to stretch and seat the cables, then slowly pull up on the parking brake lever to the fifth click. Tighten the cable adjusting nuts by equal amounts until the rear brake shoes just touch the brake drum. Verify that both wheels have the same amount of resistance.
12 Release the parking brake and verify that both rear wheels rotate freely.
13 Tighten the lug bolts to the torque listed in the Chapter 1 Specifications.

12 Parking brake assembly - check, removal and installation

Warning: *Some of the parking brake components are made of asbestos, which may cause serious harm if inhaled. When servicing these components, do not create dust by grinding or sanding the linings or by using compressed air to blow away dust. Use a water dampened cloth to wipe away the residue.*

Check

1 The parking brake system should be checked as a normal part of driving. With the vehicle parked on a hill, apply the brake, place the transmission in Neutral and check that the parking brake alone will hold the vehicle (be sure to stay in the vehicle during this check). However, every 24 months (or whenever a fault is suspected), the assembly itself should be visually inspected.
2 With the vehicle raised and supported on jackstands, remove the rear wheels.
3 On 318i models, refer to Chapter 1; checking the thickness of the brake shoes is a regular maintenance item on these models. On all other models, remove the rear discs as outlined in Section 5. Support the caliper assemblies with a coat hanger or heavy wire and do not disconnect the brake line from the caliper.
4 With the disc removed, the parking brake components are visible and can be inspected for wear and damage. The linings should last the life of the vehicle. However, they can wear down if the parking brake

12.6a Remove the lower shoe return spring (diagonal cutting pliers are being used here because they grip the spring well, but care must be taken not to cut or nick the spring)

12.6b Remove the upper shoe return spring

12.7 Remove the shoe hold-down springs

12.8 Remove the shoes

12.9 When you're done, the actuator should be properly seated between the two shoes as shown (hub removed for clarity)

system has been improperly adjusted. There is no minimum thickness specification for the parking brake shoes, but as a rule of thumb, if the shoe material is less 1/32-inch thick, you should replace them. Also check the springs and adjuster mechanism and inspect the drum for deep scratches and other damage.

Removal and installation

Refer to illustrations 12.6a, 12.6b, 12.7, 12.8 and 12.9

Note: *The following procedure applies only to models with rear disc brakes. The parking brake system on 318i models with rear drum brakes is an integral part of the brake assembly* (see Section 6).

5   Loosen the wheel lug bolts, raise the rear of the vehicle and place it securely on jackstands. Remove the rear wheels. Remove the brake discs (see Section 5) if you haven't already done so. Work on only one side at a time, so you can use the other side as a reference during reassembly.

6   Remove the shoe return springs **(see illustrations)**.
7   Remove the shoe hold-down springs **(see illustration)**.
8   Remove the shoes **(see illustration)**.
9   Installation is the reverse of removal. When you're done, the actuator should be properly seated between the two shoes as shown **(see illustration)**.
10   After installing the brake disc, adjust the parking brake shoes. Temporarily install two lug bolts, turn the adjuster **(see illustration 5.6d)** and expand the shoes until the disc locks, then back off the adjuster until the shoes don't drag (see Section 11).

13 Brake pedal - adjustment

Refer to illustrations 13.1a and 13.1b
Note: *You should always adjust brake pedal height after the master cylinder or brake booster has been removed or replaced. You should also adjust the brake light switch (see Section 14).*

1   Measure the distance between the lower edge of the brake pedal and the firewall **(see illustrations)** and compare your measurement with the dimension listed in this Chapter's Specifications. If it's not as listed, loosen the locknut on the pushrod and rotate the pushrod while holding the clevis stationary until the distance is correct.

14 Brake light switch - check, adjustment and replacement

Note: *The brake light switch on all 1984 through 1986 and most 1987 3-Series models and on all 5-Series models should be checked and, if necessary, adjusted after the master cylinder or brake booster has been removed or replaced.*

1   The brake light switch is located on a bracket at the top of the brake pedal. The switch activates the brake lights at the rear of the vehicle whenever the pedal is depressed.
2   With the brake pedal in the rest position, measure the distance between the switch contact point on the brake pedal and the switch

Chapter 9 Brakes

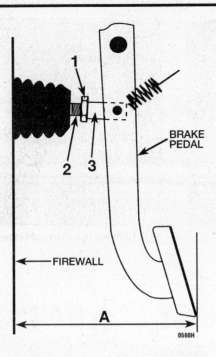

13.1a To adjust the brake pedal height on 3-Series models, loosen the locknut (1) and turn the pushrod (2) while holding the clevis (3) until dimension A (the distance between the lower edge of the brake pedal and the firewall) is within the range listed in this Chapter's Specifications

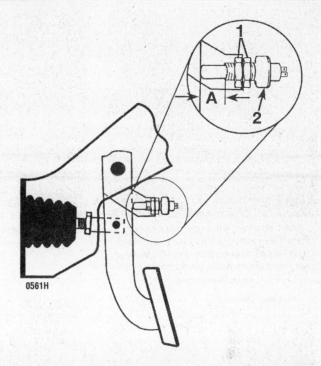

13.1b To adjust the brake pedal height on 5-Series models, follow the procedure for 3-Series models; to adjust the brake light switch on any model (3 or 5-Series), loosen the locknuts (1) and screw the switch (2) in or out until dimension A is correct

housing **(see illustration 13.1b)** and compare your measurement with the dimension listed in this Chapter's Specifications.

3 If your measurement is outside the indicated dimension, loosen the locknut, adjust the switch until the plunger is the correct dimension and retighten the locknut.

15 Brake hoses and lines - inspection and replacement

Inspection

1 About every six months, with the vehicle raised and placed securely on jackstands, the flexible hoses which connect the steel brake lines with the front and rear brake assemblies should be inspected for cracks, chafing of the outer cover, leaks, blisters and other damage. These are important and vulnerable parts of the brake system and inspection should be complete. A light and mirror will prove helpful for a thorough check. If a hose exhibits any of the above conditions, replace it with a new one.

Flexible hose replacement

Refer to illustration 15.3

2 Clean all dirt away from the ends of the hose.
3 To disconnect the hose at the frame end, put a backup wrench on the hex-shaped fitting on the end of the flexible hose and loosen the nut on the metal brake line **(see illustration)**. If the nut is stuck, soak it with penetrating oil. After the hose is disconnected from the metal line, remove the spring clip from the bracket and detach the hose from the bracket.
4 To detach the flexible hose from the caliper, simply unscrew it.
5 Installation is the reverse of the removal procedure. Make sure the brackets are in good condition and the locknuts are tightened to the torque listed in this Chapter's Specifications. Replace the spring clips if they don't fit tightly.
6 Carefully check to make sure the suspension and steering components do not make contact with the hoses. Have an assistant push

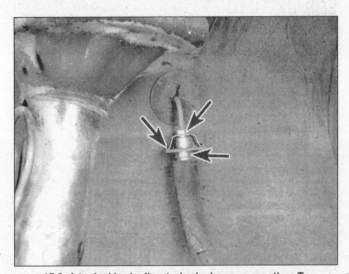

15.3 A typical brake line-to-brake hose connection: To disconnect it, put a backup wrench on the hex fitting on the end of the flexible hose (lower right arrow) and break loose the threaded fitting on the metal line with a flare-nut wrench (upper right arrow), then remove the spring clip (left arrow)

on the vehicle and also turn the steering wheel from lock-to-lock during inspection.
7 Bleed the brake system as described in Section 16.

Metal brake line replacement

8 When replacing brake lines, use the proper parts only. Do not use copper line for any brake system connections. Purchase steel brake lines from a BMW dealer or an auto parts store.
9 Genuine BMW OEM replacement brake lines are straight. You'll need a tubing bender to bend them to the proper shape.

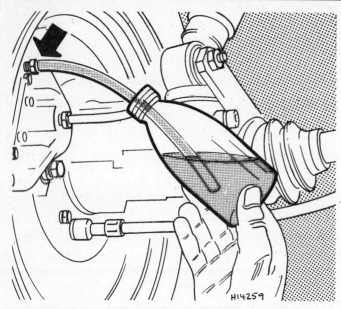

16.9 Place one end of the tubing over the bleeder screw and submerge the other end in brake fluid in the container

Note: *Bleeding the hydraulic system is necessary to remove any air which has entered the system during removal and installation of a hose, line, caliper or master cylinder.*

1 It will probably be necessary to bleed the system at all four brakes if air has entered the system due to low fluid level or if the brake lines have been disconnected at the master cylinder.
2 If a brake line was disconnected at only one wheel, then only that caliper or wheel cylinder must be bled.
3 If a brake line is disconnected at a fitting located between the master cylinder and any of the brakes, that part of the system served by the disconnected line must be bled.
4 Bleed the right rear, the left rear, the right front and the left front caliper, in that order, when the entire system is involved.
5 Remove any residual vacuum from the power brake booster and pressure in the anti-lock brake system (if equipped) by applying the brake about 30 times with the engine off.
6 Remove the master cylinder reservoir cover and fill the reservoir with brake fluid. Reinstall the cover. **Note:** *Check the fluid level often during the bleeding operation and add fluid as necessary to prevent the fluid level from falling low enough to allow air into the master cylinder.*
7 Have an assistant on hand, as well as a supply of new brake fluid, an empty clear plastic container, a length of 3/16-inch clear plastic or vinyl tubing to fit over the bleeder screws and a wrench to open and close the bleeder screws.
8 Beginning at the right rear wheel, loosen the bleeder screw slightly, then tighten it to a point where it is snug but can still be loosened quickly and easily.
9 Place one end of the tubing over the bleeder valve and submerge the other end in brake fluid in the container **(see illustration)**.
10 Have the assistant pump the brakes a few times to get pressure in the system, then hold the pedal firmly depressed. **Note:** *If the vehicle is equipped with ABS, have the assistant pump the pedal at least 12 times.*
11 While the pedal is held depressed, open the bleeder screw just enough to allow a flow of fluid to leave the caliper. Watch for air bubbles to exit the submerged end of the tube. When the fluid flow slows after a couple of seconds, close the screw and have your assistant release the pedal.
12 Repeat Steps 10 and 11 until no more air is seen leaving the tube, then tighten the bleeder screw and proceed to the left rear wheel, the right front wheel and the left front wheel, in that order, and perform the same procedure. Be sure to check the fluid in the master cylinder reservoir frequently.
13 Never reuse old brake fluid. It contains moisture which will allow the fluid to boil and render the brakes inoperative.
14 Refill the master cylinder with fluid at the end of the operation.
15 Check the operation of the brakes. The pedal should feel solid when depressed, with no sponginess. If necessary, repeat the entire process. **Warning:** *Do not operate the vehicle if you are in doubt about the effectiveness of the brake system.*

10 First, remove the line you intend to replace, lay it on a clean workbench and measure it carefully. Obtain a new line of the same length and bend it to match the pattern of the old line. **Warning:** *Do not crimp or damage the line.* No bend should have a smaller radius than 9/16-inch. Make sure the protective coating on the new line is undamaged at the bends.
11 When installing the new line, make sure it's well supported by the brackets, the routing matches the original and there's plenty of clearance between moving or hot components.
12 After installation, check the master cylinder fluid level and add fluid as necessary. Bleed the brake system as outlined in Section 16 and test the brakes carefully before driving the vehicle. Be sure there are no leaks.

16 Brake hydraulic system - bleeding

Refer to illustration 16.9
Warning: *Wear eye protection when bleeding the brake system. If the fluid comes in contact with your eyes, immediately rinse them with water and seek medical attention.*

Chapter 10
Suspension and steering systems

Contents

Balljoints - check and replacement	7
Front control arm (3-Series) - inspection, bushing replacement and control arm removal and installation	3
Front control arm and thrust arm (5-Series) - inspection, removal and installation and bushing replacement	4
Front hub and wheel bearing assembly - removal and installation	8
Front stabilizer bar - removal and installation	2
Front strut assembly - removal and installation	5
General information	1
Power steering fluid level check	See Chapter 1
Power steering pump - removal and installation	22
Power steering system - bleeding	23
Rack-and-pinion steering gear (3-Series) - removal and installation	19
Rear coil springs (3-Series) - removal and installation	10
Rear shock absorbers (3-Series) - removal and installation	9
Rear stabilizer bar - removal and installation	12
Rear shock absorber/coil spring assembly (5-Series) - removal and installation	11
Rear trailing arms (3-Series) - removal and installation	13
Rear trailing arms (5-Series) - removal and installation	14
Rear wheel bearings - replacement	15
Steering and suspension check	See Chapter 1
Steering gearbox (5-Series) - removal and installation	21
Steering gear boots (3-Series) - replacement	18
Steering linkage (5-Series) - removal and installation	20
Steering system - general information	16
Steering wheel - removal and installation	24
Strut or shock absorber/coil spring - replacement	6
Suspension and steering checks	See Chapter 1
Tie-rod ends - removal and installation	17
Tire and tire pressure checks	See Chapter 1
Tire rotation	See Chapter 1
Wheel alignment - general information	26
Wheel bearing lubrication	See Chapter 1
Wheels and tires - general information	25

Specifications

General
Power steering fluid type .... See Chapter 1

Torque specifications

Front suspension | Ft-lbs
Strut damper rod nut	48
Strut cartridge threaded collar	96
Strut upper mounting nuts	16
Front control arm (3-Series)	
Control arm-to-steering knuckle balljoint nut	47
Control arm-to-subframe balljoint nut	61
Control arm bushing bracket bolts	30

Front suspension (continued)

	Ft-lbs
Lower control arm (5-Series)	
Control arm-to-steering arm balljoint stud nut	63
Control arm pivot bolt	57
Thrust arm (5-Series)	
Thrust arm-to-steering arm balljoint stud nut	63
Thrust arm through-bolt	96
Front hub (wheel bearing) nut	
3-Series	210
5-Series	214
Steering arm-to-strut bolts (5-Series)	48
Stabilizer bar (3-Series)	
Stabilizer bar-to-connecting link bolt	30
Stabilizer bar mounting brackets-to-subframe	16
Connecting link-to-bracket	16
Connecting link bracket-to-control arm	30
Stabilizer bar (5-Series)	
Stabilizer bar mounting brackets	16
Stabilizer bar link-to-strut housing locknut	
Yellow chrome	15
White chrome	24
White	43

Rear suspension

Rear shock absorber (3-Series)	
Shock absorber-to-upper mounting bracket	9 to 11
Shock absorber-to-trailing arm	52 to 63
Rear shock absorber (5-Series)	
Lower mounting bolt	92 to 105
Upper mounting nut	16 to 18
Upper spring mount-to-shock absorber locknut	16 to 18
Trailing arms (3-Series)	
Trailing arm-to-lower mount	52 to 63
Trailing arm-to-stabilizer bar	16 to 17
Trailing arms (5-Series)	
Trailing arm-to-rear axle carrier (rubber bushing through-bolt and nut)	49
Trailing arm-to-axle carrier connecting link (1983 on)	93
Rear wheel bearing drive flange axle nut (5-Series)	
M22	129 to 155
M27	173 to 192

Steering system

Steering wheel-to-steering shaft nut	58
Intermediate shaft-to-steering gear pinion shaft	
U-joint pinch bolt	16
Rack-and-pinion steering gear-to-subframe bolts (3-Series)	30
Steering gearbox-to-front suspension subframe bolts (5-Series)	31
Tie-rod end-to-steering arm nut	27
Tie-rod end clamping bolt	10
Pitman arm-to-steering gearbox (5-Series)	103
Steering linkage balljoints (all)	27

1 General information

Refer to illustrations 1.1a, 1.1b, 1.2 and 1.3

Warning: *Whenever any of the suspension or steering fasteners are loosened or removed, they must be inspected and if necessary, replaced with new ones of the same part number or of original equipment quality and design. Torque specifications must be followed for proper reassembly and component retention. Never attempt to heat, straighten or weld any suspension or steering component. Instead, replace any bent or damaged part with a new one.*

The front suspension **(see illustrations)** is a MacPherson strut design. The struts are secured at the upper ends to reinforced areas at the top of the wheel housings and at the lower ends to the steering arms/control arms. A stabilizer bar is attached to the control arms via a connecting link, and to the suspension subframe (3-Series models) or the underbody (5-Series models).

The independent rear suspension system on 3-Series models **(see illustration)** features coil springs and telescopic shock absorbers. The upper ends of the shocks are attached to the body; the lower ends are connected to trailing arms. A stabilizer bar is attached to the trailing arms via links and to the body with clamps.

The independent rear suspension system on 5-Series models **(see illustration)** uses coil-over shock absorber units instead of separate shocks and coil springs. The upper ends are attached to the body; the lower ends are connected to the trailing arms. The rear suspension of 5-Series models is otherwise similar to that of 3-Series models: Two trailing arms connected by a stabilizer bar.

The steering system consists of the steering wheel, a steering column, a universal joint shaft, the steering gear, the power steering pump and the steering linkage, which connects the steering gear to the steering arms. On 3-Series models, a rack-and-pinion steering gear is attached directly to the steering arms via the tie-rods and tie-rod ends. On 5-Series models, a recirculating-ball steering gearbox is connected to the steering arms via a Pitman arm, a center tie-rod, the outer tie-rods and the tie-rod ends.

Chapter 10 Suspension and steering systems

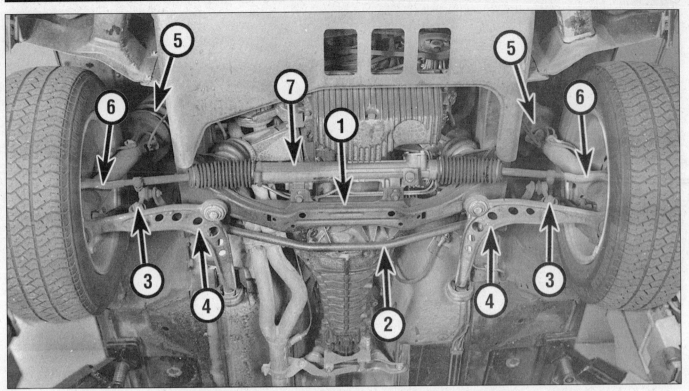

1.1a Front suspension and steering components (3-Series models)

1. Subframe
2. Stabilizer bar
3. Stabilizer bar link
4. Control arm
5. Strut
6. Tie-rod end
7. Steering gear

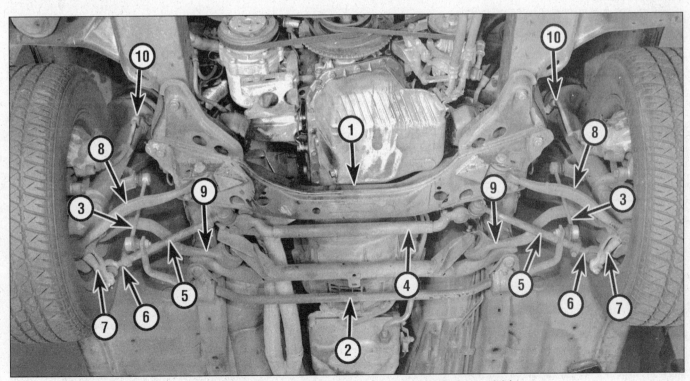

1.1b Front suspension and steering components (5-Series models)

1. Subframe
2. Stabilizer bar
3. Stabilizer bar link
4. Center tie-rod
5. Outer tie-rod
6. Tie-rod end
7. Steering arm
8. Control arm
9. Thrust arm
10. Strut

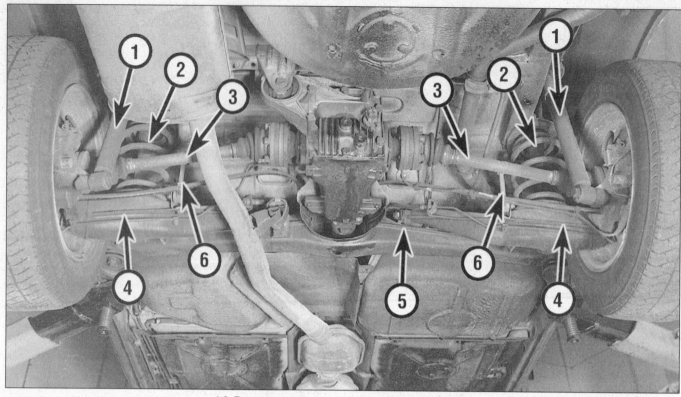

1.2 Rear suspension components (3-Series models)

1 Shock absorber
2 Coil spring
3 Driveaxle
4 Trailing arm
5 Rear axle carrier
6 Stabilizer bar link

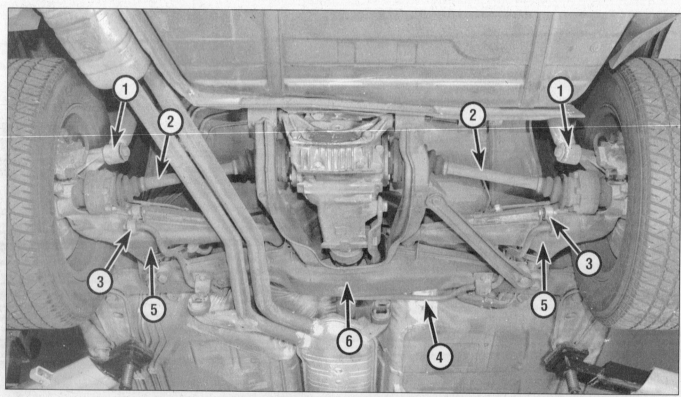

1.3 Rear suspension components (5-Series models)

1 Shock absorber/coil spring assembly
2 Driveaxle
3 Stabilizer bar link
4 Stabilizer bar
5 Trailing arm
6 Rear axle carrier

Chapter 10 Suspension and steering systems

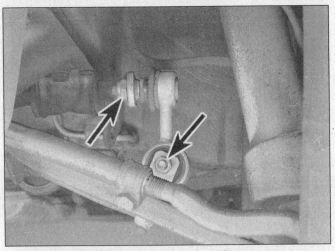

2.2a On 3-Series models, remove the nut (upper arrow) that attaches the stabilizer bar to the upper end of the connecting link (left side shown, right side similar) (if you're replacing the control arm, remove the lower nut (lower arrow) and disconnect the link assembly and bracket from the arm)

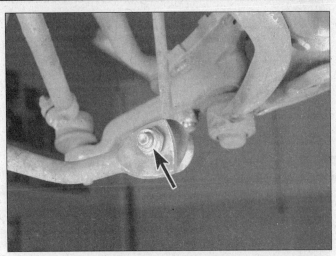

2.2b On 5-Series models, remove the nut (arrow) that fastens the stabilizer bar to the connecting link (left side shown, right side similar)

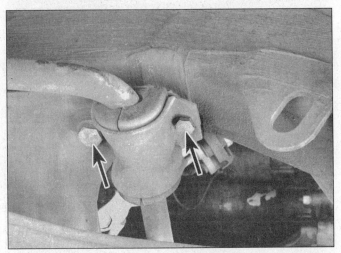

2.4 Remove the bolts (arrows) from the stabilizer bar brackets to detach the stabilizer bar from the subframe (3-Series model shown, 5-series similar)

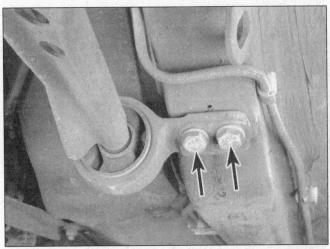

3.4 Remove the two bolts (arrows) which attach the bushing bracket to the underbody

2 Front stabilizer bar - removal and installation

Refer to illustrations 2.2a, 2.2b and 2.4

1 Raise the front of the vehicle and support it securely on jackstands.
2 If you're removing or replacing the stabilizer bar itself, or disconnecting the bar to replace the strut assembly on a 3-Series models, disconnect it from the stabilizer bar links connecting the bar to the control arms **(see illustrations)**. If you're replacing the strut assembly on a 5-Series model, disconnect the stabilizer bar link from the strut housing.
3 On 3-Series models, disconnect the left (driver's side) control arm rubber bushing from the underbody (see Section 3).
4 Remove the bolts from the stabilizer bar brackets which attach the stabilizer bar to the subframe **(see illustration)**.
5 Remove the stabilizer bar from the vehicle. On 1982 through 1986 models, separate the stabilizer bar from the strut bar bracket.
6 Installation is the reverse of the removal procedure. Be sure to tighten all fasteners to the torque listed in this Chapter's Specifications.

3 Front control arm (3-Series) - inspection, bushing replacement and control arm removal and installation

Inspection

1 Raise the front end of the vehicle and place it securely on jackstands.
2 Grip the top and bottom of each balljoint with a pair of large water pump pliers and squeeze to check for freeplay, or insert a prybar or large screwdriver between the control arm and the subframe or strut housing. If there's any freeplay, replace the control arm (the balljoints can't be replaced separately).
3 Inspect the rubber bushing. If it's cracked, dry, torn or otherwise deteriorated, replace it (see below).

Bushing replacement

Refer to illustration 3.4

Note: *Rubber bushings should always be replaced in pairs. Make sure both replacement bushings have the same markings (indicating they're manufactured by the same firm).*

4 Remove the two bolts **(see illustration)** which attach the bushing bracket to the underbody.

Chapter 10 Suspension and steering systems

3.12 Remove the self-locking nut from the balljoint stud protruding through the top of the subframe (not shown in this photo, but it's directly above the balljoint) and separate the balljoint from the subframe

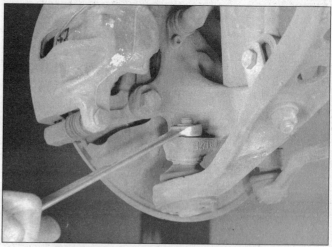

3.13a Remove the self-locking nut from the balljoint stud which attaches the outer end of the control arm to the steering knuckle . . .

5   Using a puller, remove the bracket and bushing from the end of the control arm. If the puller slips on the end of the control arm, centerpunch the control arm to give the puller bolt a place to seat.

6   Note the orientation of the old bushing. This is exactly how the new bushing should be oriented when it's installed. Press the old rubber bushing out of the bracket, or have it pressed out by an automotive machine shop.

7   Coat the end of the control arm with BMW's special lubricant (Part. No. 81 22 9 407 284) and press the new bushing and bracket onto the arm - or have it pressed on at an automotive machine shop - all the way to the stop. **Caution:** *Don't try to use some other type of lubricant; 30 minutes after it's applied, this lubricant loses its properties and the bushing is permanently located in its proper position.* Make sure the new bushing is pressed on so it's oriented exactly the same way as the old bushing.

8   Install the bracket bolts and tighten them to the torque listed in this Chapter's Specifications.

9   Lower the vehicle and leave it at rest for at least 30 minutes (this will give the special lubricant time to dry).

Control arm removal and installation

Refer to illustrations 3.12, 3.13a and 3.13b

Note: *If either balljoint is worn or damaged, the only way to replace it is to replace the control arm. If you're installing a new control arm, the bushing must also be replaced. The old bushing can't be removed from the old control arm and reused in the new control arm.*

10   Loosen but do not remove the lug bolts, raise the front of the vehicle and support it on jackstands. Remove the lug bolts and the front wheel.

11   Remove the two bolts which attach the rubber bushing bracket to the underbody **(see illustration 3.4).**

12   Remove the nut which secures the control arm balljoint to the subframe and remove the balljoint stud from the subframe. **Note:** *It may be necessary to use a pickfork-type balljoint separator to separate the balljoint from the subframe* **(see illustration),** *but the use of this tool will probably result in damage to the balljoint dust boot. To lessen the chances of damaging the boot, apply a coat of grease to the boot and the tool. If the boot does become damaged (and you're reinstalling the same control arm and balljoint), be sure to install a new boot.*

13   Loosen the nut which secures the outer control arm balljoint to the steering knuckle **(see illustration)** and detach the balljoint stud from the knuckle **(see illustration).** Again, if you're replacing the control arm, it's okay to use a picklefork to break the stud loose from the strut housing

14   Remove the control arm.

15   If you're replacing the control arm, you'll have to install a new

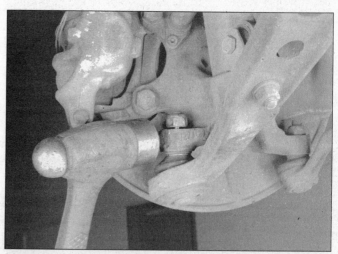

3.13b . . . give the steering knuckle a few sharp knocks with a hammer to pop the balljoint stud loose from the strut housing and remove the control arm (you can also use a pickle fork)

bushing (see above). The old bushing can't be removed, then reused in another control arm.

16   Installation is the reverse of removal. Be sure to use new self-locking nuts on the balljoint studs and tighten them, and the bushing bracket bolts, to the torque listed in this Chapter's Specifications.

17   When you're finished, have the front wheel alignment checked by an alignment shop or a dealer service department.

4   Front control arm and thrust arm (5-Series) - inspection, removal and installation and bushing replacement

Inspection

1   Inspect the thrust arm rubber bushing **(see illustration 4.6b).** If the bushing is cracked, torn or otherwise deteriorated, replace it. The control arm bushing can't be inspected until the control arm is removed.

2   Raise the vehicle and place it securely on jackstands.

3   To inspect the control arm and thrust arm balljoints for wear, grip the top and bottom of each balljoint with a pair of large water pump pliers and try to squeeze them, or use a prybar or large screwdriver to lever them up and down. If there's any freeplay, replace the control arm or thrust arm. The balljoints can't be replaced separately.

Chapter 10 Suspension and steering systems

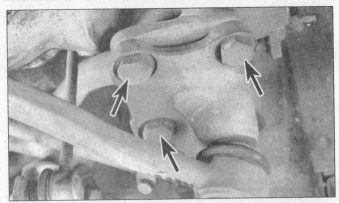

4.5 If you're removing the control arm, remove the three bolts (arrows) from the steering arm and separate the strut assembly from the arm

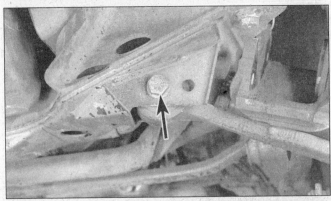

4.6a If you're removing the control arm, remove the self-locking nut and the through-bolt (arrow) that attach the inner end of the arm to the vehicle

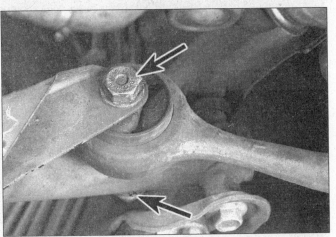

4.6b If you're removing the thrust arm, remove the nut and bolt (arrows) that secure the rear end of the arm

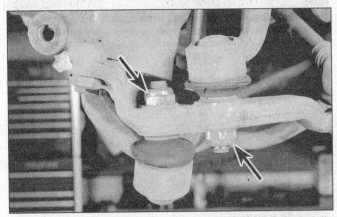

4.7a Remove the self-locking nut (control arm nut, left arrow; thrust arm nut, right arrow) from the balljoint, then support the steering arm and press or knock the balljoint out of the steering arm

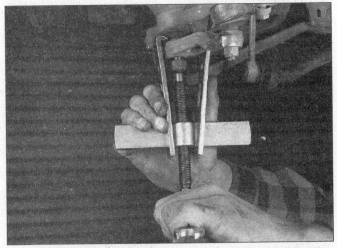

4.7b You can separate the thrust arm balljoint from the steering arm with a puller . . .

4.7c . . . but you'll have to use a hammer (or picklefork) to knock the control arm balljoint stud loose from the steering arm because there's no room to use a puller

Removal

Refer to illustrations 4.5, 4.6a, 4.6b, 4.7a, 4.7b and 4.7c

Note: *If a balljoint is worn or damaged, the only way to replace it is to replace the control arm or thrust arm. If you're installing a new control arm or thrust arm, the bushing must also be replaced. The old bushing can't be removed from the old control arm or thrust arm and reused in the new arm.*

4　Loosen the wheel lug bolts, raise the vehicle and support it securely on jackstands. Remove the wheel.
5　If you're removing the control arm, remove the three bolts from the steering arm **(see illustration)** and separate the strut assembly from the arm.
6　Remove the nut and the through-bolt that attach the rear mount of the control arm or the thrust arm **(see illustrations)**.
7　Remove the nut from the balljoint **(see illustration)**. Support the steering arm and press or knock the balljoint out of the steering arm **(see illustrations)**.

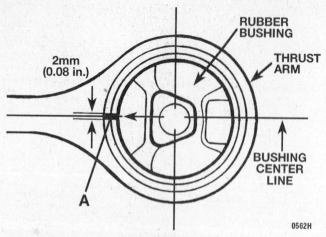

4.8 Correct orientation of the bushing for the 5-Series thrust arm: The arrow on the rubber bushing is aligned with the mark on the arm and the center of the bushing is concentric with the bore

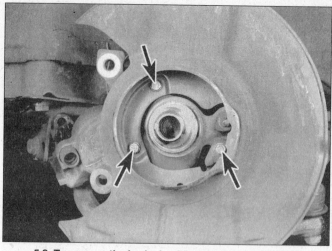

5.8 To remove the brake backing plate, remove these three bolts (arrows)

11   Have the front end alignment checked at a dealer service department or alignment shop.

5 Front strut assembly - removal and installation

Removal

Refer to illustrations 5.8 and 5.13

Note: *Although strut assemblies don't always fail or wear out simultaneously, replace both left and right struts at the same time to prevent handling peculiarities and abnormal ride quality.*

1   Loosen but do not remove the front wheel lug bolts.
2   Raise the front of the vehicle and support it on jackstands.
3   Remove the front wheel.
4   Detach all brake hoses and electrical wires attached to the strut housing.
5   Disconnect the electrical connections for the ABS system, if equipped.
6   If you're removing the left strut, disconnect the electrical connector for the brake pad wear sensor.
7   Remove the ABS sensor (see Chapter 9), if equipped. Remove the brake disc (see Chapter 9).
8   Remove the brake backing plate **(see illustration)**.
9   On 3-Series models, disconnect the stabilizer bar from its connecting link (see Section 2). On 5-Series models, disconnect the stabilizer bar link from the strut housing (see Section 2).
10 On 3-Series models, disconnect the control arm balljoint from the steering knuckle (see Section 3) and the tie-rod end from the steering arm (see Section 17).
11 On 5-Series models, disconnect the bolts that attach the steering arm to the strut housing **(see illustration 4.5)**.
12 Pull out the lower end of the strut housing far enough to clear the end of the control arm (3-Series) or the steering arm (5-Series).
13 Support the weight of the strut and remove the three mounting nuts at the top of the strut, located inside the engine compartment **(see illustration)** and remove the strut.
14 Remove the strut assembly. If you're replacing the strut cartridge, see Section 6.

5.13 Support the weight of the strut and remove the three mounting nuts (arrows) at the top of the strut (5-Series shown, 3-Series similar)

Bushing inspection and replacement

Refer to illustration 4.8

8   If the bushing is cracked, torn or otherwise deteriorated, take the arm to a BMW dealer service department or an automotive machine shop and have it pressed out and a new bushing pressed in. Bushings should always be replaced in pairs (a new bushing should be installed in each arm, and both bushings should have the same manufacturer markings). If you're installing a new thrust arm bushing, make sure it's correctly oriented **(see illustration)**.

Installation

9   Installation is the reverse of removal. Be sure to use new self-locking nuts on the balljoint stud nut and the through-bolt. Don't forget to install the washers on both sides of the through-bolt. If you're installing the control arm, be sure to install the steering arm mounting bolts with a bolt locking compound like Loctite 270 (or its equivalent). Don't tighten the through-bolt to the final torque yet. **Note:** *Thrust arms are marked with an L for the left (driver's) side and an R for the right (passenger's) side. Be sure to check the marking before installing a replacement arm.*
10 Support the control arm with a floor jack and raise it to simulate normal ride height, then tighten the through-bolt to the torque listed in this Chapter's Specifications. Install the wheel and tighten the lug bolts to the torque listed in the Chapter 1 Specifications.

Installation

Refer to illustration 5.15

15 Installation is the reverse of removal. On 3-Series models, be sure to use new self-locking nuts on the control arm balljoint, the tie-rod end balljoint and the upper strut mount. On 5-Series models, make sure the tang in the steering arm is mated with the notch in the strut housing **(see illustration)**. BMW recommends using a thread locking

Chapter 10 Suspension and steering systems

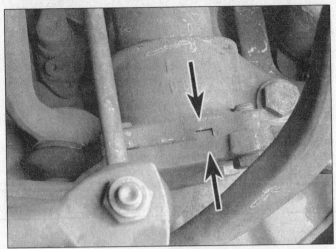

5.15 On 5-Series models, make sure the tang in the steering arm is mated with the notch in the strut housing (arrows)

6.3 Following the tool manufacturer's instructions, install the spring compressor on the spring and compress it sufficiently to relieve all pressure from the suspension support

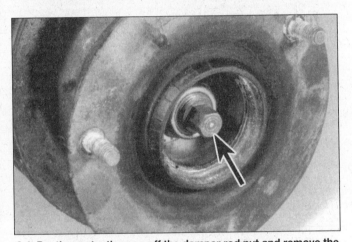

6.4 Pry the protective cap off the damper rod nut and remove the large nut (arrow) - to prevent the damper rod from turning, place a wrench on the hex-shaped end of the shaft

6.9 Loosen and remove the threaded collar (arrow) and pull the old strut cartridge from the strut housing - on all struts except gas-charged units, pour the old oil from the strut housing

compound such as Loctite 270 (or its equivalent) on the steering arm mounting bolts. On all models, tighten the fasteners to the torque listed in this Chapter's Specifications.

16 When you're done, drive the vehicle to a dealer service department or an alignment shop and have the alignment checked, and if necessary, adjusted.

6 Strut or shock absorber/coil spring - replacement

Refer to illustrations 6.3, 6.4, 6.9 and 6.11

Note: *This section applies to all front strut assemblies and, on 5-Series models, the rear coil-over shock absorber assemblies.*

1 If the struts, shock absorbers or coil springs exhibit the telltale signs of wear (leaking fluid, loss of damping capability, chipped, sagging or cracked coil springs) explore all options before beginning any work. Strut or shock absorber assemblies complete with springs may be available on an exchange basis, which eliminates much time and work. Whichever route you choose to take, check on the cost and availability of parts before disassembling the vehicle. **Warning:** *Disassembling a strut or coil-over shock absorber assembly is a potentially dangerous undertaking and utmost attention must be directed to the job, or serious injury may result. Use only a high quality spring compressor and carefully follow the manufacturer's instructions furnished with the tool. After removing the coil spring from the strut assembly, set it aside in a safe, isolated area.*

2 Remove the strut or shock absorber assembly (see Section 5 or 11). Mount the assembly in a vise. Line the vise jaws with wood or rags to prevent damage to the unit and don't tighten the vise excessively.

3 Following the tool manufacturer's instructions, install the spring compressor (which can be obtained at most auto parts stores or equipment yards on a daily rental basis) on the spring and compress it sufficiently to relieve all pressure from the suspension support **(see illustration)**. This can be verified by wiggling the spring.

4 Pry the protective cap off the damper rod self-locking nut. Loosen the nut **(see illustration)** with an offset box end wrench while holding the damper rod stationary with another wrench.

5 Remove the nut, the strut bearing, the insulator and the large washer. Inspect the bearing for smooth operation. If it doesn't turn smoothly, replace it. Check the rubber insulator for cracking and general deterioration. If there is any separation of the rubber, replace the insulator.

6 Lift off the upper spring retainer and the rubber ring at the top of the spring. Check the rubber ring for cracking and hardness. Replace it if necessary.

7 Carefully lift the compressed spring from the assembly and set it in a safe place, such as a steel cabinet. **Warning:** *Never place your head near the end of the spring!*

8 Slide the protective tube and rubber bumper off the damper rod. If either of them is damaged or worn, replace it.

9 If you're working on a front strut loosen and remove the threaded collar **(see illustration)** and pull the old strut cartridge from the strut housing. Pour the old oil from the strut housing.

Chapter 10 Suspension and steering systems

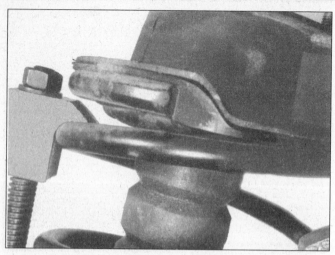

6.11 Make sure you align the end of the coil spring with the shoulder of the rubber ring and with the spring retainer

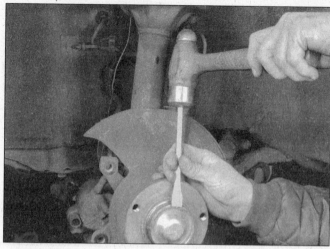

8.2 Using a hammer and chisel, knock out the dust cap in the center of the hub

8.3 Using a chisel, knock out the staked portion of the hub nut

10 On all struts except gas-charged units, fill the strut housing with 3/4-ounce (3-Series) or one-ounce (5-Series) of engine oil (the oil helps cool the shock absorber by transferring heat to the strut housing).
Note: *It doesn't matter what viscosity or grade of engine oil is used.*
11 Installation is otherwise the reverse of removal. Tighten the threaded collar to the torque listed in this Chapter's Specifications. Make sure you align the end of the coil spring with the shoulder of the rubber ring and with the spring retainer **(see illustration)**. Tighten the damper rod nut to the torque listed in this Chapter's Specifications.
12 Install the strut or shock absorber assembly (see Section 5 or 11).

7 Balljoints - check and replacement

Check

Note: *On 3-Series models, there are two balljoints on each control arm - one between the middle of the arm and the subframe and the other between the outer end of the arm and the steering knuckle. On 5-Series models, there are balljoints on the outer ends of the control arm and the thrust arm.*
1 Raise the vehicle and support it securely on jackstands.
2 Visually inspect the rubber boot between the balljoint and the subframe or steering knuckle, etc. for cuts, tears or leaking grease. If you note any of these conditions, replace the control arm or thrust arm.
3 Place a large prybar under the balljoint and attempt to push the balljoint up. Next, position the prybar between the arm and the subframe or between the arm and steering knuckle. If you can see or feel any movement during either check, a worn out balljoint is indicated.
4 Have an assistant grasp the tire at the top and bottom and shake the top of the tire with an in-and-out motion. Touch the balljoint stud nut. If any looseness is felt, suspect a worn out balljoint stud or a widened hole in the subframe or steering knuckle. If the latter problem exists, the subframe or steering arm (5-Series) or steering knuckle (3-Series), which is integral with the strut housing should be replaced as well as the balljoint.

Replacement

Note: *None of these balljoints can be serviced or replaced individually. If one of them is worn, the entire arm must be replaced.*

8 Front hub and wheel bearing assembly - removal and installation

Removal

Refer to illustrations 8.2, 8.3, 8.8a and 8.8b
1 Loosen the wheel lug bolts, raise the front of the vehicle, place it securely on jackstands, remove the lug bolts and remove the wheel.
2 Using a hammer and chisel, remove the dust cap from the center of the wheel hub **(see illustration)**.
3 Unstake the hub nut **(see illustration)**.
4 Remove the hub cap from the center of the wheel. Remount the wheel and lower the vehicle to the ground. Loosen, but do not remove, the hub nut. **Warning:** *Always loosen and tighten the hub nut with the vehicle on the ground. The leverage needed to loosen the nut could topple the vehicle off a lift or a jackstand.*
5 Raise the front of the vehicle, support it securely on jackstands and remove the front wheel.
6 Remove the front brake caliper and mounting bracket (see Chapter 9). Do NOT disconnect the brake hose. Hang the caliper out of the way with a piece of wire.
7 Remove the brake disc (see Chapter 9).
8 Remove the hub nut and pull the hub and bearing assembly off the stub axle. You may have to tap it off if it's stuck **(see illustration)**. If the inner race of the bearing remains on the stub axle (which it probably will), remove the dust shield (rubber boot) behind the bearing and use a puller to remove the inner race **(see illustration)**.

Installation

Refer to illustration 8.10
9 Install a new rubber dust boot.
10 Push the new hub and bearing onto the stub axle. If it's necessary

Chapter 10 Suspension and steering systems

8.8a If the hub sticks, knock it loose with a hammer

8.8b If the inner race of the bearing sticks to the stub axle, use a puller to get it off

8.10 Use a large socket or a suitable piece of pipe to drive against the inner race of the new bearing

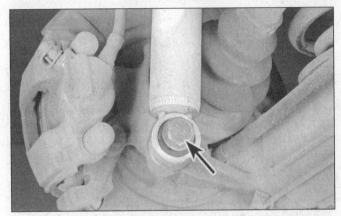

9.3 Remove the shock absorber lower mounting bolt (arrow)

to use force, press or drive only against the bearing inner race **(see illustration)**.

11 Install a new hub nut and tighten it slightly but not completely yet.
12 Install the brake disc, its countersunk retaining screw and the brake caliper (see Chapter 9).
13 Install the wheel and lower the vehicle to the ground.
14 Tighten the hub nut to the torque listed in this Chapter's Specifications. Again, make sure you do this with the vehicle on the ground, not up on jackstands.
15 Raise the front of the vehicle and place it securely on jackstands. Remove the wheel.
16 Stake the collar of the nut into the groove of the spindle.
17 Apply Loctite 638 sealant (or its equivalent) to a NEW grease cap and install the cap by driving it into place with a soft-face mallet.
18 Install the wheel and lug bolts. Lower the vehicle to the ground and tighten the lug bolts to the torque listed in the Chapter 1 Specifications.

9 Rear shock absorbers (3-Series) - removal and installation

Refer to illustrations 9.3 and 9.4

Note: *Although shock absorbers don't always wear out simultaneously, replace both left and right shocks at the same time to prevent handling peculiarities and abnormal ride quality.*

1 If a shock absorber is to be replaced with a new one, it is recommended that both shocks on the rear of the vehicle be replaced at the same time.
2 Raise the rear of the vehicle and support it securely on jack-

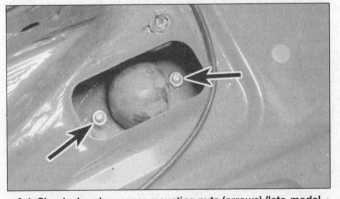

9.4 Shock absorber upper mounting nuts (arrows) (late-model convertible shown; on earlier model coupes and sedans, upper nuts are accessible from trunk; on later models, they're behind the back of the rear seat, up under the package tray

stands. Support the trailing arm with a floor jack. Place a block of wood on the jack head to serve as a cushion.
3 Remove the shock absorber lower mounting bolt **(see illustration)**.
4 On some models, working inside the trunk, you can remove the trim to access the upper mounting nuts; on later models, you'll have to remove the rear seatback to get at the upper mounting nuts. On convertibles, simply remove the top from its recessed well behind the passenger compartment and remove the small rubber access cover. As you remove the mounting nuts **(see illustration)**, have an assistant support the shock from below so it doesn't fall out.

Chapter 10 Suspension and steering systems

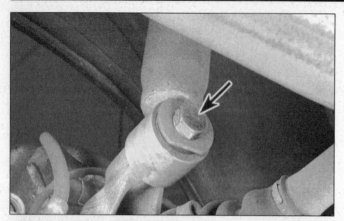

11.2 Remove the shock absorber lower mounting bolt (arrow)

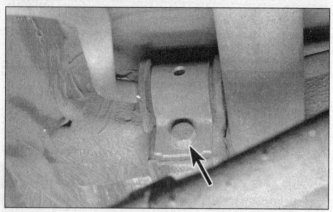

12.2 Rear stabilizer bar bracket bolts (arrows) (3-Series)

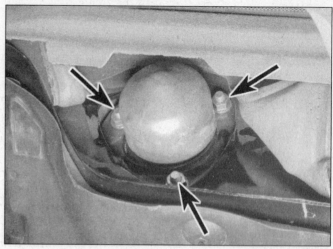

11.3 Shock absorber upper mounting nuts (arrows) on a 1989 or later 5-Series model

bar from its connecting links, or disconnect the links from the trailing arms (see Section 12).
6   Loop a chain through the coil spring and bolt the chain together to prevent the coil spring from popping out when the trailing arm is lowered. Be sure to leave enough slack in the chain to allow the spring to extend completely.
7   Disconnect the shock absorber lower mounting bolt (see Section 9), carefully lower the trailing arm and remove the coil spring.
8   Installation is the reverse of removal. As the trailing arm is raised back up, make sure the spring seats properly.

11 Rear shock absorber/coil spring assembly (5-Series) - removal and installation

Refer to illustrations 11.2 and 11.3
Note: *Although shock absorbers don't always wear out simultaneously, replace both left and right shocks at the same time to prevent handling peculiarities and abnormal ride quality.*
1   Loosen the wheel lug bolts, raise the vehicle and support it securely on jackstands. Remove the wheels.
2   Remove the shock lower mounting bolt **(see illustration)**.
3   On pre-1989 models, peel back the trim inside the trunk far enough to access the upper mounting nuts. To get at the upper mounting nuts on 1989 and later models, remove the lower part of the rear seat (see Chapter 11), remove the two bolts holding the upper part of the rear seat and remove the upper part of the rear seat. Support the trailing arm with a jack and remove the upper mounting nuts **(see illustration)**. Lower the jack and remove the shock and the gasket.
4   Installation is the reverse of removal. Don't forget to install the gasket between the upper end of the shock and the body. Tighten the upper nuts to the torque listed in this Chapter's Specifications. Don't tighten the lower bolt until the vehicle is lowered.
5   Lower the vehicle, allowing it to sit at normal ride height, and tighten the lower bolt to the torque listed in this Chapter's Specifications.

12 Rear stabilizer bar - removal and installation

Refer to illustrations 12.2, 12.3a and 12.3b
Note: *The rear stabilizer bar is mounted basically the same way on all models. Follow these general removal and installation procedures keeping in mind any variations.*
1   Raise the rear of the vehicle and support it securely on jackstands. Block the front wheels to keep the vehicle from rolling.
2   Remove the stabilizer bar bracket bolts or nuts **(see illustrations)**.

---

5   Look for oil leaking past the seal in the top of the shock body. Inspect the rubber bushings in the shock eye. If they're cracked, dried or torn, replace them. To test the shock, grasp the shock body firmly with one hand and push the damper rod in and out with the other. The strokes should be smooth and firm. If the rod goes in and out easily, or unevenly, the shock is defective and must be replaced.
6   Install the shocks in the reverse order of removal, but don't tighten the mounting bolts and nuts yet.
7   Bounce the rear of the vehicle a couple of times to settle the bushings, then tighten the nuts and bolts to the torque values listed in this Chapter's Specifications.

10 Rear coil springs (3-Series) - removal and installation

Note: *Although coil springs don't always wear out simultaneously, replace both left and right springs at the same time to prevent handling peculiarities and abnormal ride quality.*
1   Loosen the wheel lug bolts. Raise the rear of the vehicle and support it securely on jackstands. Make sure the stands don't interfere with the rear suspension when it's lowered and raised during this procedure. Remove the wheels.
2   Disconnect the hangers and brackets which support the rear portion of the exhaust system and temporarily lower the exhaust system (see Chapter 4). Lower the exhaust system only enough to lower the suspension and remove the springs. Suspend it with a piece of wire.
3   Support the differential with a floor jack, remove the differential rear mounting bolt, push the differential down and wedge it into this lowered position with a block of wood (see Chapter 8). This reduces the drive angle, preventing damage to the CV joints when the trailing arms are lowered to remove the springs.
4   Place a floor jack under the trailing arm.
5   If the vehicle is equipped with a rear stabilizer bar, disconnect the

Chapter 10 Suspension and steering systems

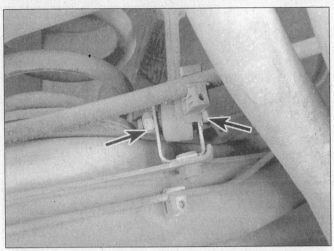

12.3a A nut and bolt (arrows) connect each rear stabilizer bar link to the rear trailing arms (3-Series)

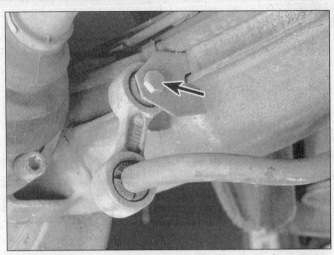

12.3b Bolt (arrow) connecting rear stabilizer bar link to trailing arm (5-Series)

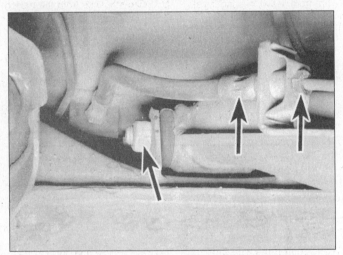

13.3 Disconnect the rear brake hose (arrow) from the metal brake line fitting (arrow) at this bracket on the trailing arm, then plug the line and hose immediately to prevent brake fluid leaks; the other arrow points to the nut for the inner pivot bolt

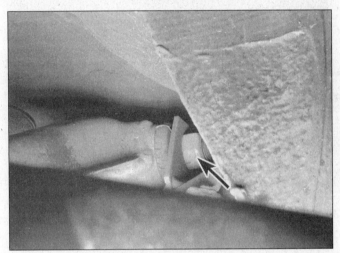

13.6 Nut (arrow) for the outer pivot bolt

3 Disconnect the stabilizer bar from the link at each end of the bar **(see illustrations)** and detach the stabilizer bar.
4 Inspect and, if necessary, replace any worn or defective bolts, washers, bushings or links.
5 Installation is the reverse of removal. Tighten all fasteners securely.

13 Rear trailing arms (3-Series) - removal and installation

Refer to illustrations 13.3 and 13.6

1 Loosen the wheel lug bolts, raise the rear of the vehicle and support it securely on jackstands. Remove the wheel(s).
2 Remove the driveaxle (see Chapter 8).
3 Disconnect the rear brake hose from the metal brake line at the bracket on the trailing arm **(see illustration)**. **Note:** *For information on disconnecting brake hose-to-metal line connections, see Chapter 9. Plug the line and hose to prevent brake fluid from leaking out.*
4 Disconnect the parking brake cable (see Chapter 9).
5 Disconnect the lower end of the shock absorber from the trailing arm (see Section 9) and lower the trailing arm.
6 Remove the trailing arm pivot bolts **(see illustration)** and remove the trailing arm.

7 Inspect the pivot bolt bushings. If they're cracked, dried out or torn, take the trailing arm to an automotive machine shop and have them replaced. Each bushing has a larger diameter shoulder on one end. Make sure this larger diameter shoulder on each bushing faces away from the trailing arm, i.e. the inner bushing shoulder faces the center of the vehicle and the outer bushing shoulder faces away from the vehicle.
8 Installation is the reverse of removal. Support the trailing arm with a floor jack and raise it to simulate normal ride height, then tighten the fasteners to the torque listed in this Chapter's Specifications. Be sure to bleed the brakes as described in Chapter 9.

14 Rear trailing arm (5-Series) - removal and installation

Refer to illustrations 14.3 and 14.8

1 Loosen the wheel lug bolts, raise the rear of the vehicle and support it securely on jackstands. Remove the wheel(s).
2 Remove the driveaxle (see Chapter 8).
3 Disconnect the rear brake hose from the metal brake line at the bracket on the trailing arm **(see illustration)**. **Note:** *For information on disconnecting brake hose-to-metal line connections, see Chapter 9. Plug the line and hose to prevent brake fluid from leaking out.*
4 Disconnect the parking brake cable from the parking brake actuator and unclip the parking brake cable from the trailing arm (see Chapter 9).

Chapter 10 Suspension and steering systems

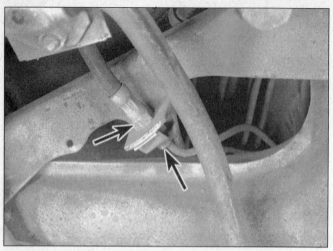

14.3 Disconnect the rubber brake hose (arrow) from the fitting on the metal brake line (arrow) at this bracket

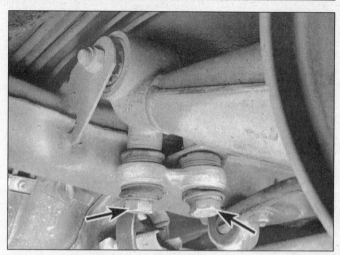

14.8 On 1983 and later models, remove one of these trailing arm-to-axle carrier bolts (it doesn't make any difference which one you remove, one attaches the link to the trailing arm and the other attaches the link to the axle carrier)

5 Remove the ABS wheel speed sensor, if equipped, from the trailing arm, and unclip the sensor wire harness from the arm. Position the sensor aside so it won't be damaged during removal of the trailing arm.
6 If you're removing the right trailing arm, unplug the connector for the brake pad wear sensor.
7 Disconnect the rear stabilizer bar from the trailing arm (see Section 12).
8 On 1983 and later models, remove one of the rear axle carrier bolts **(see illustration)**.
9 Disconnect the shock absorber lower mounting bolt (see Section 11).
10 Remove the two trailing arm pivot bolts and nuts and remove the trailing arm from the vehicle.
11 Inspect the pivot bolt bushings. If they're cracked, dried out or torn, take the trailing arm to an automotive machine shop and have them replaced. The bushing inner sleeve is longer on side. Make sure the bushings are installed with the longer side of the bushing sleeve facing toward the center of the vehicle.
12 Installation is the reverse of removal. Install the inner pivot bolt first. Don't tighten the nuts on the pivot bolts or the shock absorber yet.
13 Bleed the brakes as described in Chapter 9.
14 Support the trailing arm with a floor jack and raise it to simulate normal ride height, then tighten the bolts and nuts to the torque listed in this Chapter's Specifications.

15 Rear wheel bearings - replacement

3-Series

Refer to illustration 15.4

1 Loosen the driveaxle nut and the rear wheel lug bolts, raise the rear of the vehicle and place it securely on jackstands. Remove the rear wheel. **Note:** *Depending on the type of rear wheel, it may be necessary to remove the wheel first, remove the hubcap, then reinstall the wheel and loosen the driveaxle nut.*
2 Remove the drive axle (see Chapter 8).
3 On models with rear disc brakes, remove the brake caliper and mounting bracket. Don't disconnect the hose. Hang the caliper out of the way with a piece of wire. Remove the brake disc (see Chapter 9). Working from behind, drive the wheel hub out of the wheel bearing with a large socket or a piece of pipe. If the bearing inner race sticks to the hub (which it probably will), use a puller to remove the race from the hub.
4 Remove the large snap-ring **(see illustration)** that holds the wheel bearing in the wheel bearing housing, then drive out the bearing

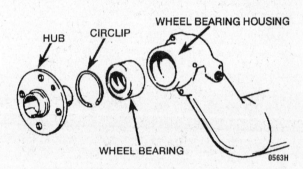

15.4 An exploded view of the 3-Series rear wheel bearing assembly

with a large socket or piece of pipe.
5 Installation is basically the reverse of removal, with the following exceptions.
 a) Be extremely careful where you place the socket or piece of pipe when you drive the new bearing into the housing. It should be butted up against the inner race of the bearing. Driving the new bearing into place by pounding on the outer race will ruin the bearing.
 b) Install the wheel and lower the vehicle to the ground before attempting to tighten the driveaxle nut to the torque listed in the Chapter 8 Specifications.

5-Series

Refer to illustrations 15.7, 15.8, 15.11, 15.12, 15.13a, 15.13b, 15.14a, 15.14b, 15.14c and 15.14d

6 Raise the rear of the vehicle and support it securely on jackstands. Disconnect the outer CV joint from the drive flange (see Chapter 8). Support the outer end of the driveaxle with a piece of wire - don't let it hang, as this could damage the inner CV joint.
7 Pry out the lockplate that secures the drive flange nut **(see illustration)**. Once you've pried out an edge of the lockplate, pull it out with a pair of needle-nose pliers.
8 Lower the vehicle and break loose the drive flange nut, but don't remove it. You'll need a breaker bar **(see illustration)**. **Warning**: *Don't attempt to loosen this nut with the vehicle on jackstands. The force required to loosen the nut could topple the vehicle from the stands.*
9 Loosen the rear wheel lug bolts, raise the rear of the vehicle again, place it securely on jackstands and remove the wheel.

Chapter 10 Suspension and steering systems

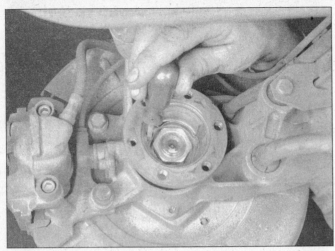

15.7 Pry out the lockplate that secures the drive flange nut - once you've pried out an edge of the lockplate, pull it out with a pair of needle-nose pliers

15.8 Lower the vehicle and break loose the drive flange nut with a breaker bar

15.11 Remove the drive flange with a puller

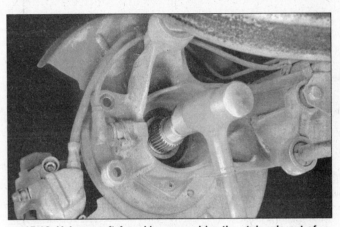

15.12 Using a soft-faced hammer, drive the stub axle out of the bearing

15.13a Remove the large snap-ring that retains the bearing in the housing . . .

15.13b . . . then drive the bearing out of the housing with a large socket or a suitable piece of pipe

10 Remove the brake caliper and the brake disc (see Chapter 9). Hang the caliper out of the way with a piece of wire.
11 Remove the drive flange nut. Using a suitable puller, remove the drive flange **(see illustration)**.
12 Using a soft-faced hammer, drive the stub axle out of the bearing

(see illustration). If the bearing inner race comes off with the stub axle (which it probably will), use a puller to remove the race from the stub axle. If you can't get the race off with a puller, take the stub axle to an automotive machine shop and have it pressed off.
13 Remove the large snap-ring that retains the bearing in the bearing housing **(see illustration)**, then drive the bearing out of the bearing housing with a large socket or a suitable piece of pipe **(see illustration)**.

Chapter 10 Suspension and steering systems

15.14a To install the new bearing, use a large socket or a piece of pipe with an outside diameter the same diameter as the outer race of the bearing - don't apply force to the inner race - and make sure the bearing is fully seated against the back of its bore

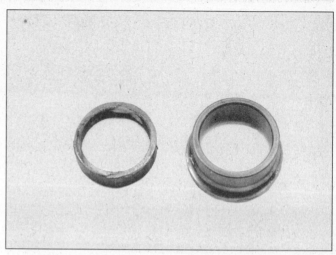

15.14b You can fabricate your own spacer tool from a piece of 1/2-inch long, 1-1/2 inch inside diameter pipe (left); you'll also need to use the old inner race (right)

15.14c Hold the stub axle flange with a large prybar while tightening the nut

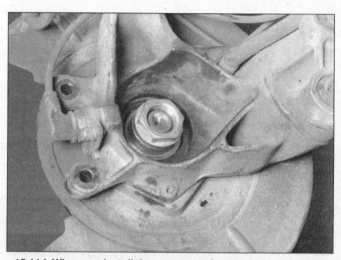

15.14d When you install the spacer, make sure it's centered on the inner race of the wheel bearing before tightening the nut

14 Installation is basically the reverse of removal, with the following recommendations:
  a) Use a large socket or a suitable piece of pipe to install the new bearing with an outside diameter the same diameter as the outer race **(see illustration)**. Don't apply force to the inner race. Make sure the bearing is fully seated against the back of the bore. Install the snap-ring, making sure it's fully seated into its groove.
  b) BMW shops use special tools (Part Nos. 23 1 1300, 33 4 080 and 33 4 020) to pull the stub axle into the bearing because the smooth portion between the splined portion of the stub axle and the flange is a press-fit, and knocks out the wheel bearing inner race during installation of the stub axle. However, you can install the stub axle without these tools, using the old inner race and a piece of pipe 1/2-inch long by 1-1/2 inches inside diameter **(see illustration)**. First, insert the stub axle through the new bearing until the threaded portion protrudes from the inner race. Install the nut and tighten it until the splined portion of the stub axle shaft bottoms against the nut. You'll need to hold the stub axle flange with a prybar or a large screwdriver while tightening the nut **(see illustration)**. Remove the nut, install your piece of 1/2 x 1-1/2 inch pipe, centered on the inner race and reinstall the nut **(see illustration)**. Tighten the nut again until it bottoms against the splines. Remove the nut, install the old inner race, install the nut and tighten it once more until it bottoms against the splines. Remove the nut, remove the old inner race, install your piece of pipe, install the old inner race, install the nut and tighten it until it bottoms against the splines. Remove the nut, the old race and the pipe. Install the drive flange, install the nut and tighten it securely, but don't attempt to tighten it to the final torque until the vehicle is lowered to the ground.
  c) Install the wheel and lower the vehicle to the ground before tightening the stub axle nut to the torque listed in this Chapter's Specifications.
15 The remainder of installation is the reverse of removal.

16 Steering system - general information

On 3-Series models, the steering wheel and steering column are connected to a power-assisted rack-and-pinion steering gear via a short universal joint shaft. When the steering wheel is turned, the steering column and U-joint turn a pinion gear shaft on top of the rack. The pinion gear teeth mesh with the gear teeth of the rack, so the rack moves right or left in the housing when the pinion is turned. The movement of the rack is transmitted through the tie-rods and tie-rod ends to the steering arms, which are an integral part of the strut housings.

Chapter 10 Suspension and steering systems

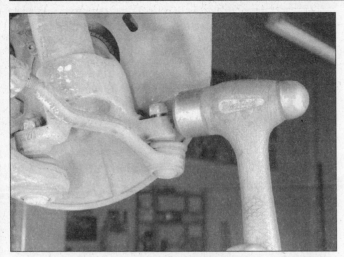

17.2 Loosen the nut on the tie-rod balljoint stud and give the steering arm a few whacks with a hammer to pop the balljoint stud loose, remove the nut and separate the balljoint stud from the steering arm

17.3 Loosen the clamp bolt (arrow) that locks the tie-rod end to the inner tie-rod, paint an alignment mark on the threads to ensure the tie-rod end is installed in the same position and unscrew the tie-rod end from the inner tie-rod

On 5-Series models, the upper part of the steering system is identical to a 3-Series. Instead of a rack-and-pinion setup, these models use a power-assisted recirculating ball steering gearbox which steers the front wheels via steering linkage consisting of a Pitman arm, an idler arm, a center tie-rod, a pair of inner tie-rods and two tie-rod ends.

Both steering systems are power-assisted. Hydraulic pressure - provided by an engine-driven pump which delivers power steering fluid to the rack-and-pinion steering gear or the recirculating-ball steering gearbox - enhances steering response and reduces steering effort.

Aside from maintaining the proper level of power steering fluid in the system and checking the tension of the drivebelt (see Chapter 1), the steering system requires no maintenance. However, on high-mileage vehicles, the tie-rod end balljoints, the U-joints on either end of the U-joint shaft and the rubber coupling between the steering column and the U-joint shaft will wear, develop excessive play and cause the steering to feel somewhat loose. At this point, you'll have to replace these items; they can't be serviced.

Before you conclude that the steering system needs work, however, always check the tires (see Section 25) and tire pressures (see Chapter 1). Also inspect the bearings in the strut upper mounts (see Section 5), the front hub bearings (see Section 8) and other suspension parts, which may also be contributing to an imprecise steering feel.

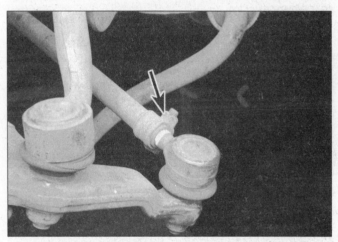

17.6 Measure the length of the tie-rod and record your measurement, or paint an alignment mark on the threads to ensure the tie-rod end is installed in the same position, then loosen the clamp bolt (arrow)

17 Tie-rod ends - removal and installation

1 Loosen but do not remove the wheel lug bolts, raise the front of the vehicle and secure it on jackstands. Remove the front wheel.

3-Series models

Refer to illustrations 17.2 and 17.3

2 Loosen the nut on the tie-rod balljoint stud and give the steering arm a few whacks with a hammer **(see illustration)** to pop the balljoint stud loose. If this method doesn't work, a picklefork-type balljoint separator can be used, but this will probably destroy the boot on the tie-rod end. Remove the nut and separate the balljoint stud from the steering arm.

3 Loosen the clamp bolt that locks the tie-rod end to the inner tie-rod, then measure the length of the tie-rod end or paint an alignment mark on the threads to ensure the tie-rod end is installed in the same position **(see illustration)**. Unscrew the tie-rod end from the inner tie-rod.

4 Installation is the reverse of removal. Make sure the mark you made on the threads of the tie-rod end is aligned with the clamp on the end of the inner tie-rod.

5 Have the toe-in checked and, if necessary, adjusted at a dealer service department or alignment shop.

5-Series models

Refer to illustrations 17.6 and 17.7

6 Measure the length of the tie-rod and record your measurement, or paint an alignment mark on the threads to ensure the tie-rod end is installed in the same position **(see illustration)**. Loosen the clamp bolt.

7 Use a puller to separate the tie-rod end from the steering arm **(see illustration)**.

8 Unscrew the tie-rod end.

9 Installation is the reverse of removal. Make sure you align the paint mark on the threads of the tie-rod end with the clamp on the outer tie-rod.

18 Steering gear boots (3-Series) - replacement

1 Remove the tie-rod ends (see Section 17).

2 Cut the boot clamps at both ends of the old boots and slide off the boots.

3 While the boots are removed, inspect the seals in the end of the steering gear. If they're leaking, replace the steering gear (see Section 19).

4 Slide the new boots into place and install new boot clamps.

5 Install the tie-rod ends (see Section 17).

17.7 Use a puller to separate the tie-rod end from the steering arm

19.6 Rack-and-pinion steering gear mounting bolts (3-Series models) (self-locking nuts not visible in this photo)

19 Rack-and-pinion steering gear (3-Series) - removal and installation

Refer to illustration 19.6

1 Loosen but do not remove the wheel lug bolts, raise the vehicle and support it securely on jackstands. Remove the front wheels.
2 Mark the lower universal joint on the steering shaft and the pinion shaft to ensure proper alignment when they're reassembled. Remove the nut and bolt that attach the lower end of the U-joint shaft to the steering gear pinion shaft. Loosen the bolt and nut at the upper end of the U-joint shaft. Slide the U-joint shaft up a little, disengage it from the pinion shaft and remove it. Inspect the U-joints and the rubber coupling for wear. If any of them are worn or defective, replace the U-joint shaft.
3 Using a large syringe or hand pump, empty the power steering fluid reservoir.
4 Remove the banjo bolts and disconnect the power steering pressure and return lines from the steering gear. Place a container under the lines to catch spilled fluid. Plug the lines to prevent excessive fluid loss and contamination. Discard the sealing washers (new ones should be used when reassembling).
5 Disconnect the tie-rod ends from the steering knuckle arms (see Section 17).
6 Remove the nuts and bolts from the steering gear mounting brackets **(see illustration)**. Discard the old nuts.
7 Withdraw the assembly from beneath the vehicle. Take care not to damage the steering gear boots.
8 Installation is the reverse of removal. Make sure the marks you made on the lower U-joint and the pinion shaft are aligned before you tighten the clamping bolts for the upper and lower U-joints. Use new self-locking nuts on the steering rack mounting bolts Nd new sealing washers on the hydraulic line fittings. Tighten the mounting bolts, the tie-rod end nuts and the U-joint shaft clamping bolts to the torque values listed in this Chapter's Specifications.
9 After lowering the vehicle, fill the reservoir with the recommended fluid (see Chapter 1).
10 Bleed the power steering system (see Section 23).
11 It's a good idea to have the front wheels aligned by a dealer service department or alignment shop after reassembly.

20 Steering linkage (5-Series) - removal and installation

Refer to illustrations 20.4 and 20.7

1 Raise the vehicle and place it securely on jackstands.
2 Firmly grasp each front tire at the top and bottom, then at the front and rear, and check for play in the steering linkage by rocking the

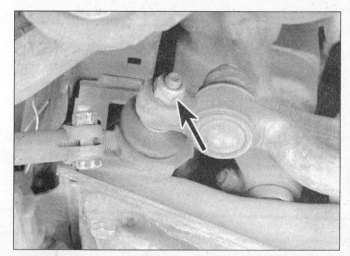

20.4 To remove an outer tie-rod, remove this nut (arrow) from the end of the center tie-rod and use a balljoint separator to separate the balljoint stud from the center tie-rod (if you're replacing the inner tie-rod end, mark the threads with paint before loosening the clamp bolt and nut

tire back and forth. There should be little or no play in any of the linkage balljoints. Inspect the Pitman arm, the idler arm, the center tie-rod, the inner tie-rods, the tie-rod ends and the steering arms for any obvious damage. Try forcing the linkage parts in opposite directions from one another. There should be no play between any of them. If any of the parts are bent or damaged in any way, or if any of the balljoints are worn, replace the part.
3 Before dismantling the steering linkage, obtain a suitable balljoint separator. A two-jaw puller or a wedge-type tool, or "pickle fork," will work (although the wedge-type tends to tear the balljoint boots). Sometimes, you can also jar a balljoint taper pin free from its eye by striking opposite sides of the eye simultaneously with two large hammers, but the space available to do this is limited and the balljoint stud sometimes sticks to the eye because of rust and dirt.
4 To remove the outer tie-rods, disconnect the tie-rod ends from the steering arms (see Section 17). Remove the nut that attaches the balljoint on the inner end of each outer tie rod to the center tie-rod **(see illustration)**. Using a balljoint separator, disconnect the outer tie-rods from the center tie-rod. If you're replacing the balljoint at either end of the outer tie-rods, paint or scribe alignment marks on the threads to mark their respective positions as a guide to adjustment during reassembly **(see illustration 17.3)**.
5 To remove the center tie-rod, remove the nuts that attach the

Chapter 10 Suspension and steering systems

20.7 To unbolt the idler arm from the subframe crossmember, remove this nut (arrow)

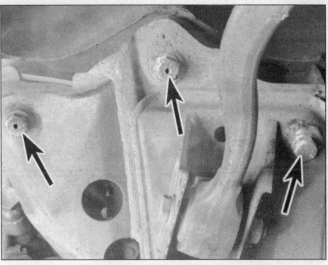

21.7 Subframe crossmember bolts (arrows)

center tie-rod balljoints to the Pitman arm and the idler arm and use your balljoint separator to disconnect the balljoints from the two arms.

6 To remove the Pitman arm, you'll have to remove the steering gearbox first (see Section 21). Look for match marks between the sector shaft and arm. If there aren't any, scribe a mark across the bottom face of both parts. Remove the Pitman arm pinch bolt and nut, then remove the arm with a puller.

7 To unbolt the idler arm, remove the small cover bolted to the top of the subframe crossmember, put a wrench on the bolt and remove the nut recessed into the underside of the subframe crossmember **(see illustration)**. Check the idler arm rubber bushing for wear. If it's damaged or worn, replace it.

8 Check each balljoint for excessive play or stiffness and for split or deteriorated rubber dust boots. Replace all worn or damaged balljoints. The inner and outer tie-rod ends on the outer tie-rods can be replaced individually; if either balljoint on the center tie-rod is damaged or worn, you must replace the center tie-rod.

9 Reassembly is the reverse of the disassembly procedure, but observe the following points:
 a) Realign the match marks on the Pitman arm and the steering gearbox sector shaft when reassembling them.
 b) If you're installing new inner or outer tie-rod ends on the outer tie-rods, position them so that the match marks made during disassembly are aligned and make sure they are equally spaced on each side.
 c) Position the tie-rod end balljoint studs on the outer tie-rods at an angle of 90-degrees to each other.
 d) Make sure the left and right outer tie-rods are equal in length when they are installed.
 e) Tighten all retaining bolts to the torque values listed in this Chapter's Specifications.
 f) When reassembly of the linkage is complete, have the front wheel alignment checked, and if necessary, adjusted.

21 Steering gearbox (5-Series) - removal and installation

Refer to illustrations 21.7, 21.9, 21.10a and 21.10b

Note: *If you find that the steering gearbox is defective, it is not recommended that you overhaul it. Because of the special tools needed to do the job it is best to let your dealer service department overhaul it for you (or replace it with a factory rebuilt unit). However, you can remove and install it yourself by following the procedure outlined here.*

1 Depress the brake pedal about 20 times to depressurize the hydraulic system.

2 Using a large syringe or hand pump, empty the power steering fluid reservoir (see Chapter 1).

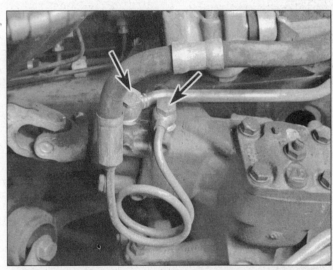

21.9 Disconnect the power steering pressure line fitting (right arrow) and the return line fitting (left arrow) (note that the return line fitting banjo bolt is larger than the bolt for the pressure line fitting)

3 Raise the front of the vehicle and support it securely on jackstands.

4 Support the front of the engine with a floor jack. Place a block of wood between the jack head and the oil pan to protect the oil pan from damage.

5 Remove the pivot bolts from the inner ends of the front control arms (see Section 4).

6 Remove the nuts from the left and right engine mounts (see Chapter 2).

7 Remove the mounting bolts (two on each side on earlier models, three on each side on later models) from the subframe crossmember **(see illustration)** and remove the subframe.

8 Remove the nuts and bolts that secure the universal joint shaft to the steering gearbox worm shaft. Slide the U-joint shaft up and off the worm shaft. Inspect the U-joint shaft for wear. If it's stiff or worn, replace it.

9 Remove the banjo bolts and disconnect the hydraulic pressure line and the return line from the gearbox **(see illustration)**. Plug the ends of the lines to prevent fluid loss and contamination. Discard the sealing washers - new ones should be used when reassembling.

21.10a This bolt (arrow) attaches the steering gearbox to the subframe crossmember (the nut, not visible in this photo, is accessed through a hole in the crossmember) (engine removed for clarity)

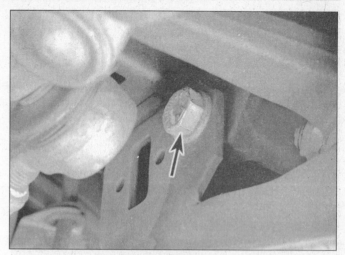

21.10b This bolt (arrow) attaches the steering gearbox to the frame (the nut, not visible in this photo, is on the front side of the gearbox)

22.6a Typical 3-Series power steering pump adjusting bolt (arrow) . . .

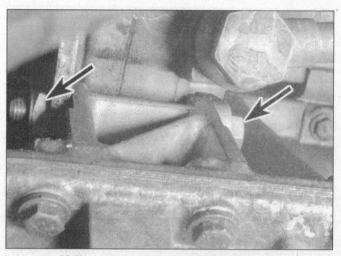

22.6b . . . and mounting nut and bolt (arrows)

10 Remove the steering gearbox retaining bolts **(see illustrations)** and remove the steering gearbox.
11 If it's necessary to detach the Pitman arm from the gearbox sector shaft (to have the box serviced or to switch the arm to a new or rebuilt unit), make a match mark across the two for correct reassembly. Remove the Pitman arm retaining nut and washer. Use a puller to withdraw the arm if necessary.
12 Install the Pitman arm by aligning the match marks made during removal, then tighten the nut to the torque listed in this Chapter's Specifications.
13 When installed, the Pitman arm must not have any measurable endplay within 100-degrees from the neutral position. If play exists, have the following parts checked:
   a) Sector shaft and bearings (for wear)
   b) Thrust washer and adjuster bolt head (for wear)
   c) Ball nut and worm shaft (for wear)
14 Install the steering gearbox. Align the mark on the pinion gear shaft with the mark on the U-joint shaft and tighten the steering gearbox bolts to the torque listed in this Chapter's Specifications.
15 The remainder of installation is the reverse of removal. Be sure to use new self-locking nuts on the U-joint shaft, the center tie-rod, the steering gearbox and the crossmember. Also, use new sealing washers on the hydraulic line fittings.

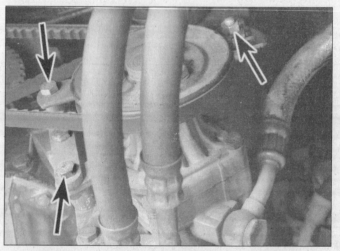

22.6c Typical 5-Series power steering pump mounting bolts (arrows)

16 Refer to Chapter 1 and fill the power steering reservoir with the recommended fluid, then bleed the system as described in Section 23. Check for leakage from the lines and connections.

Chapter 10 Suspension and steering systems

24.3 After removing the steering wheel nut, mark the relationship of the steering wheel to the steering shaft (arrows) to ensure proper alignment during reassembly

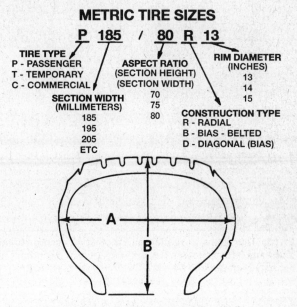

25.1 Metric tire size code

A = Section width   B = Section height

22 Power steering pump - removal and installation

Refer to illustrations 22.6a, 22.6b and 22.6c

1 Raise the vehicle and support it securely on jackstands. Remove the engine under-cover.
2 On 533i and 535i models, discharge the hydraulic system by depressing the brake pedal about 20 times before loosening the hydraulic line fittings.
3 Disconnect the fluid return hose and drain the power steering fluid from the reservoir into a clean container. Disconnect the pressure line from the pump.
4 Push on the power steering pump drivebelt by hand to increase the tension, then unscrew the pulley nut.
5 Loosen the power steering pump drivebelt tensioner bolt and remove the drivebelt (see Chapter 1).
6 Remove the mounting bolts **(see illustrations)** and detach the power steering pump.
7 Installation is the reverse of removal. Tighten the fasteners securely. Adjust the drivebelt tension (see Chapter 1).
8 Top up the fluid level in the reservoir (see Chapter 1) and bleed the system (see Section 23).

23 Power steering system - bleeding

1 To bleed the power steering system, begin by checking the power steering fluid level and adding fluid if necessary (see Chapter 1).
2 Raise and support the front of the vehicle on jackstands.
3 Turn the steering wheel from lock-to-lock several times and recheck the fluid level.
4 Start the engine and run it at 1000 rpm or less. Turn the steering wheel from lock-to-lock again (three or four times) and recheck the fluid level one more time. **Note:** *On 533i and 535i models, pump the brake pedal five or six times before turning the steering wheel. Once the fluid level remains constant, continue turning the wheel back and forth until no more bubbles appear in the fluid in the reservoir.*
5 Lower the front of the vehicle to the ground. Run the engine and again turn the wheels from lock-to-lock several more times. Recheck the fluid level. Position the wheels straight ahead.

24 Steering wheel - removal and installation

Refer to illustration 24.3

Warning: *If the vehicle is equipped with an airbag, do not attempt this procedure. Have it performed by a dealer service department or other qualified repair shop.*

1 Disconnect the negative battery cable. **Caution:** *If the radio in your vehicle is equipped with an anti-theft system, make sure you have the correct activation code before disconnecting the battery.* **Note:** *If, after connecting the battery, the wrong language appears on the instrument panel display, refer to page 0-7 for the language resetting procedure.*
2 Using a small screwdriver, pry off the BMW emblem in the center of the steering wheel.
3 Remove the steering wheel nut and mark the relationship of the steering wheel hub to the shaft **(see illustration)**.
4 On all 3-Series models and on 1986 and later 5-Series models, turn the ignition key to the first position to unlock the ignition lock.
5 Remove the steering wheel from the steering shaft. If the wheel is difficult to remove from the shaft, use a steering wheel puller to remove it - don't beat on the shaft.
6 Installation is the reverse of removal. Be sure to align the match marks you made on the steering wheel and the shaft. Tighten the steering wheel nut to the torque listed in this Chapter's Specifications.

25 Wheels and tires - general information

Refer to illustration 25.1

1 All vehicles covered by this manual are equipped with metric-sized steel belted radial tires **(see illustration)**. Use of other size or type of tires may affect the ride and handling of the vehicle. Don't mix different types of tires, such as radials and bias belted, on the same vehicle as handling may be seriously affected. It's recommended that tires be replaced in pairs on the same axle, but if only one tire is being replaced, be sure it's the same size, structure and tread design as the other.
2 Because tire pressure has a substantial effect on handling and wear, the pressure on all tires should be checked at least once a month or before any extended trips (see Chapter 1).
3 Wheels must be replaced if they are bent, dented, leak air, have elongated bolt holes, are heavily rusted, out of vertical symmetry or if the lug bolts won't stay tight. Wheel repairs that use welding or peening are not recommended.
4 Tire and wheel balance is important in the overall handling, braking and performance of the vehicle. Unbalanced wheels can adversely

affect handling and ride characteristics as well as tire life. Whenever a tire is installed on a wheel, the tire and wheel should be balanced by a shop with the proper equipment.

26 Wheel alignment - general information

Refer to illustration 26.1

A wheel alignment refers to the adjustments made to the wheels so they are in proper angular relationship to the suspension and the ground. Wheels that are out of proper alignment not only affect vehicle control, but also increase tire wear. The front end angles normally measured are camber, caster and toe-in **(see illustration)**. Front-wheel toe-in is adjustable on all models; caster is not adjustable. Camber is only adjustable by replacing the strut upper mount with a special eccentric version. Toe-in is adjustable on the rear wheels, but only by replacing the trailing arm outer bushings with special eccentric bushings.

Getting the proper wheel alignment is a very exacting process, one in which complicated and expensive machines are necessary to perform the job properly. Because of this, you should have a technician with the proper equipment perform these tasks. We will, however, use this space to give you a basic idea of what is involved with a wheel alignment so you can better understand the process and deal intelligently with the shop that does the work.

Toe-in is the turning in of the wheels. The purpose of a toe specification is to ensure parallel rolling of the wheels. In a vehicle with zero toe-in, the distance between the front edges of the wheels will be the same as the distance between the rear edges of the wheels. The actual amount of toe-in is normally only a fraction of an inch. On the front end, toe-in is controlled by the tie-rod end position on the tie-rod. On the rear end, toe-in can only be adjusted by installing special eccentric bushings in the trailing arm outer mount. Incorrect toe-in will cause the tires to wear improperly by making them scrub against the road surface.

Camber is the tilting of the wheels from vertical when viewed from one end of the vehicle. When the wheels tilt out at the top, the camber is said to be positive (+). When the wheels tilt in at the top the camber is negative (-). The amount of tilt is measured in degrees from vertical and this measurement is called the camber angle. This angle affects the amount of tire tread which contacts the road and compensates for changes in the suspension geometry when the vehicle is cornering or traveling over an undulating surface.

Caster is the tilting of the front steering axis from the vertical. A tilt toward the rear is positive caster and a tilt toward the front is negative caster. Caster is not adjustable on the vehicles covered by this manual.

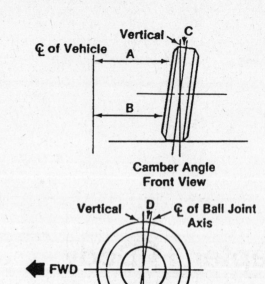

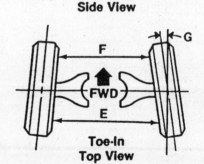

26.1 Front end alignment details

1   A minus B = C (degrees camber)
2   E minus F = toe-in (measured in inches)
3   G = toe-in (expressed in degrees)

Chapter 11 Body

Contents

Body - maintenance 2	Hood - removal, installation and adjustment 10
Body repair - major damage 6	Latch, lock cylinder and handles - removal, installation
Body repair - minor damage 5	and adjustment 15
Bumpers - removal and installation 11	Radiator grille - removal and installation 9
Center console removal and installation 22	Seat belt check 21
Door - removal, installation and adjustment 13	Seats - removal and installation 20
Door trim panel - removal and installation 12	Steering column cover - removal and installation 19
Door window glass - removal and installation 16	Upholstery and carpets - maintenance 4
Exterior mirror - removal and installation 18	Trunk lid - removal, installation and adjustment 14
Fixed glass - replacement 8	Vinyl trim - maintenance 3
General information 1	Window regulator - removal and installation 17
Hinges and locks - maintenance 7	

1 General information

These models feature a "unibody" layout, where the floorpan and body components are welded together and attached to separate front and rear subframe assemblies. Certain components are particularly vulnerable to accident damage and can be unbolted and repaired or replaced. Among these parts are the body moldings, bumpers, hood, doors, liftgate, liftgate and all glass.

Only general body maintenance procedures and body panel repair procedures within the scope of the do-it-yourselfer are included in this Chapter.

2 Body - maintenance

1   The condition of your vehicle's body is very important, because the resale value depends a great deal on it. It's much more difficult to repair a damaged body than it is to repair mechanical components. The hidden areas of the body, such as the fenderwells, the frame and the engine compartment, are equally important, although they don't require as frequent attention as the rest of the body.
2   Once a year, or every 12,000 miles, it's a good idea to have the underside of the body steam cleaned. All traces of dirt and oil will be removed and the area can then be inspected carefully for rust, damaged brake lines, frayed electrical wires, damaged cables and other problems. The front suspension components should be greased after completion of this job.
3   At the same time, clean the engine and the engine compartment with a steam cleaner or water soluble degreaser.
4   The fenderwells should be given close attention, since undercoating can peel away and stones and dirt thrown up by the tires can cause the paint to chip and flake, allowing rust to set in. If rust is found, clean down to the bare metal and apply an anti-rust paint.
5   The body should be washed about once a week (or when dirty). Wet the vehicle thoroughly to soften the dirt, then wash it down with a soft sponge and plenty of clean soapy water. If the surplus dirt is not washed off very carefully, it can wear down the paint.
6   Spots of tar or asphalt thrown up from the road should be removed with a cloth soaked in solvent.
7   Once every six months, wax the body and chrome trim. If a chrome cleaner is used to remove rust from any of the vehicle's plated parts, remember that the cleaner also removes part of the chrome, so use it sparingly.

3 Vinyl trim - maintenance

Don't clean vinyl trim with detergents, caustic soap or petroleum-based cleaners. Plain soap and water works just fine, with a soft brush to clean dirt that may be ingrained. Wash the vinyl as frequently as the rest of the vehicle.

After cleaning, application of a high quality rubber and vinyl protectant will help prevent oxidation and cracks. The protectant can also be applied to weatherstripping, vacuum lines and rubber hoses, which often fail as a result of chemical degradation, and to the tires.

4 Upholstery and carpets - maintenance

1   Every three months remove the carpets or mats and clean the interior of the vehicle (more frequently if necessary). Vacuum the upholstery and carpets to remove loose dirt and dust.

Chapter 11 Body

2 Leather upholstery requires special care. Stains should be removed with warm water and a very mild soap solution. Use a clean, damp cloth to remove the soap, then wipe again with a dry cloth. Never use alcohol, gasoline, nail polish remover or thinner to clean leather upholstery.
3 After cleaning, regularly treat leather upholstery with a leather wax. Never use car wax on leather upholstery.
4 In areas where the interior of the vehicle is subject to bright sunlight, cover leather seats with a sheet if the vehicle is to be left out for any length of time.

5 Body repair - minor damage

See photo sequence

Repair of minor scratches

1 If the scratch is superficial and does not penetrate to the metal of the body, repair is very simple. Lightly rub the scratched area with a fine rubbing compound to remove loose paint and built up wax. Rinse the area with clean water.
2 Apply touch-up paint to the scratch, using a small brush. Continue to apply thin layers of paint until the surface of the paint in the scratch is level with the surrounding paint. Allow the new paint at least two weeks to harden, then blend it into the surrounding paint by rubbing with a very fine rubbing compound. Finally, apply a coat of wax to the scratch area.
3 If the scratch has penetrated the paint and exposed the metal of the body, causing the metal to rust, a different repair technique is required. Remove all loose rust from the bottom of the scratch with a pocket knife, then apply rust inhibiting paint to prevent the formation of rust in the future. Using a rubber or nylon applicator, coat the scratched area with glaze-type filler. If required, the filler can be mixed with thinner to provide a very thin paste, which is ideal for filling narrow scratches. Before the glaze filler in the scratch hardens, wrap a piece of smooth cotton cloth around the tip of a finger. Dip the cloth in thinner and then quickly wipe it along the surface of the scratch. This will ensure that the surface of the filler is slightly hollow. The scratch can now be painted over as described earlier in this Section.

Repair of dents

4 When repairing dents, the first job is to pull the dent out until the affected area is as close as possible to its original shape. There is no point in trying to restore the original shape completely as the metal in the damaged area will have stretched on impact and cannot be restored to its original contours. It is better to bring the level of the dent up to a point about 1/8-inch below the level of the surrounding metal. In cases where the dent is very shallow, it is not worth trying to pull it back out at all.
5 If the back side of the dent is accessible, it can be hammered out gently from behind using a soft-face hammer. While doing this, hold a block of wood firmly against the opposite side of the metal to absorb the hammer blows and prevent the metal from being stretched.
6 If the dent is in a section of the body which has double layers, or some other factor makes it inaccessible from behind, a different technique is required. Drill several small holes through the metal inside the damaged area, particularly in the deeper sections. Screw long, self tapping screws into the holes just enough for them to get a good grip in the metal. Now the dent can be pulled out by pulling on the protruding heads of the screws with locking pliers.
7 The next stage of repair is the removal of the paint from the damaged area and from an inch or so of the surrounding metal. This is easily done with a wire brush or sanding disk in a drill motor, although it can be done just as effectively by hand with sandpaper. To complete the preparation for filling, score the surface of the bare metal with a screwdriver or the tang of a file or drill small holes in the affected area. This will provide a good grip for the filler material. To complete the repair, see the Section on *filling and painting*.

Repair of rust holes or gashes

8 Remove all paint from the affected area and from an inch or so of the surrounding metal using a sanding disk or wire brush mounted in a drill motor. If these are not available, a few sheets of sandpaper will do the job just as effectively.
9 With the paint removed, you will be able to determine the severity of the corrosion and decide whether to replace the whole panel, if possible, or repair the affected area. New body panels are not as expensive as most people think and it is often quicker to install a new panel than to repair large areas of rust.
10 Remove all trim pieces from the affected area except those which will act as a guide to the original shape of the damaged body, such as headlight shells, etc. Using metal snips or a hacksaw blade, remove all loose metal and any other metal that is badly affected by rust. Hammer the edges of the hole inward to create a slight depression for the filler material.
11 Wire brush the affected area to remove the powdery rust from the surface of the metal. If the back of the rusted area is accessible, treat it with rust inhibiting paint.
12 Before filling is done, block the hole in some way. This can be done with sheet metal riveted or screwed into place, or by stuffing the hole with wire mesh.
13 Once the hole is blocked off, the affected area can be filled and painted. See the following subsection on *filling and painting*.

Filling and painting

14 Many types of body fillers are available, but generally speaking, body repair kits which contain filler paste and a tube of resin hardener are best for this type of repair work. A wide, flexible plastic or nylon applicator will be necessary for imparting a smooth and contoured finish to the surface of the filler material. Mix up a small amount of filler on a clean piece of wood or cardboard (use the hardener sparingly). Follow the manufacturer's instructions on the package, otherwise the filler will set incorrectly.
15 Using the applicator, apply the filler paste to the prepared area. Draw the applicator across the surface of the filler to achieve the desired contour and to level the filler surface. As soon as a contour that approximates the original one is achieved, stop working the paste. If you continue, the paste will begin to stick to the applicator. Continue to add thin layers of paste at 20-minute intervals until the level of the filler is just above the surrounding metal.
16 Once the filler has hardened, the excess can be removed with a body file. From then on, progressively finer grades of sandpaper should be used, starting with a 180-grit paper and finishing with 600-grit wet-or-dry paper. Always wrap the sandpaper around a flat rubber or wooden block, otherwise the surface of the filler will not be completely flat. During the sanding of the filler surface, the wet-or-dry paper should be periodically rinsed in water. This will ensure that a very smooth finish is produced in the final stage.
17 At this point, the repair area should be surrounded by a ring of bare metal, which in turn should be encircled by the finely feathered edge of good paint. Rinse the repair area with clean water until all of the dust produced by the sanding operation is gone.
18 Spray the entire area with a light coat of primer. This will reveal any imperfections in the surface of the filler. Repair the imperfections with fresh filler paste or glaze filler and once more smooth the surface with sandpaper. Repeat this spray-and-repair procedure until you are satisfied that the surface of the filler and the feathered edge of the paint are perfect. Rinse the area with clean water and allow it to dry completely.
19 The repair area is now ready for painting. Spray painting must be carried out in a warm, dry, windless and dust free atmosphere. These conditions can be created if you have access to a large indoor work area, but if you are forced to work in the open, you will have to pick the day very carefully. If you are working indoors, dousing the floor in the work area with water will help settle the dust which would otherwise be in the air. If the repair area is confined to one body panel, mask off the surrounding panels. This will help minimize the effects of a slight mismatch in paint color. Trim pieces such as chrome strips, door handles, etc. will also need to be masked off or removed. Use masking tape and several thicknesses of newspaper for the masking operations.
20 Before spraying, shake the paint can thoroughly, then spray a

Chapter 11 Body

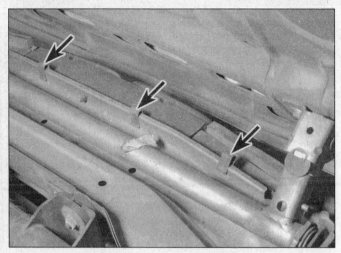

9.1 Open the hood and remove the grill retaining clips (arrows)

9.2a The center grille is held in place by two screws (arrows)

test area until the spray painting technique is mastered. Cover the repair area with a thick coat of primer. The thickness should be built up using several thin layers of primer rather than one thick one. Using 600-grit wet-or-dry sandpaper, rub down the surface of the primer until it is very smooth. While doing this, the work area should be thoroughly rinsed with water and the wet-or-dry sandpaper periodically rinsed as well. Allow the primer to dry before spraying additional coats.

21 Spray on the top coat, again building up the thickness by using several layers of paint. Begin spraying in the center of the repair area and then, using a circular motion, work outward until the whole repair area and about two inches of the surrounding original paint is covered. Remove all masking material 10 to 15 minutes after spraying on the final coat of paint. Allow the new paint at least two weeks to harden, then use a very fine rubbing compound to blend the edges of the new paint into the existing paint, Finally, apply a coat of wax.

6 Body repair — major damage

1 Major damage must be repaired by an auto body shop. These shops have the specialized equipment required to do the job properly.
2 If the damage is extensive, the frame must be checked for proper alignment or the vehicle's handling characteristics may be adversely affected and other components may wear at an accelerated rate.
3 Due to the fact that all of the major body components (hood, fenders, etc.) are separate and replaceable units, any seriously damaged components should be replaced rather than repaired. Sometimes the components can be found in a wrecking yard that specializes in used vehicle components, often at a considerable savings over the cost of new parts.

7 Hinges and locks - maintenance

Once every 3,000 miles, or every three months, the hinges and latch assemblies on the doors, hood and the tailgate or the liftgate should be given a few drops of light oil or lock lubricant. The door latch strikers should also be lubricated with a thin coat of grease to reduce wear and ensure free movement. Lubricate the door and the liftgate locks with spray-on graphite lubricant.

8 Fixed glass - replacement

Replacement of the windshield and fixed glass requires the use of special fast-setting adhesive/caulk materials and some specialized tools and techniques. These operations should be left to a dealer service department or a shop specializing in glass work.

9.2b Side grille screw locations (arrows)

9 Radiator grille - removal and installation

3-Series

Refer to illustrations 9.1, 9.2a and 9.2b
1 Detach the clips along the top of the grille **(see illustration)**.
2 Remove the screws and lift the center and side grilles out **(see illustrations)**.
3 Installation is the reverse of removal.

5-Series

1988 and earlier models
6 Remove the screws and detach the center and side grille pieces.
7 Installation is the reverse of removal.

1989 and later models

Center grille
Refer to illustration 9.10
8 Remove the screws and detach the headlight covers in the engine compartment for access.
9 Remove the screw and lift out the plastic cover behind the center grille for access to the clips.

These photos illustrate a method of repairing simple dents. They are intended to supplement *Body repair - minor damage* in this Chapter and should not be used as the sole instructions for body repair on these vehicles.

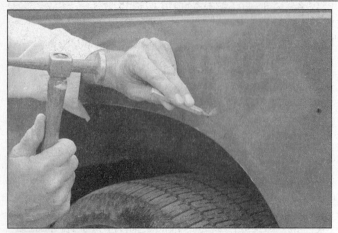

1 If you can't access the backside of the body panel to hammer out the dent, pull it out with a slide-hammer-type dent puller. In the deepest portion of the dent or along the crease line, drill or punch hole(s) at least one inch apart . . .

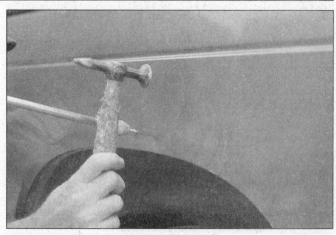

2 . . . then screw the slide-hammer into the hole and operate it. Tap with a hammer near the edge of the dent to help 'pop' the metal back to its original shape. When you're finished, the dent area should be close to its original contour and about 1/8-inch below the surface of the surrounding metal

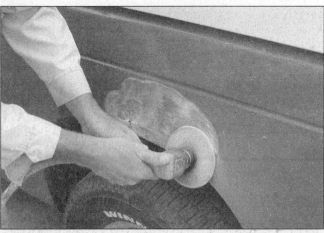

3 Using coarse-grit sandpaper, remove the paint down to the bare metal. Hand sanding works fine, but the disc sander shown here makes the job faster. Use finer (about 320-grit) sandpaper to feather-edge the paint at least one inch around the dent area

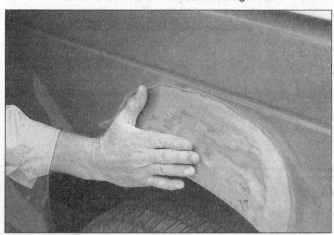

4 When the paint is removed, touch will probably be more helpful than sight for telling if the metal is straight. Hammer down the high spots or raise the low spots as necessary. Clean the repair area with wax/silicone remover

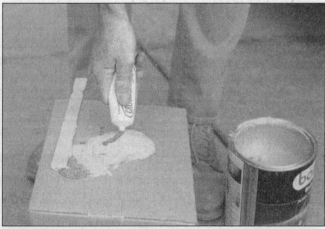

5 Following label instructions, mix up a batch of plastic filler and hardener. The ratio of filler to hardener is critical, and, if you mix it incorrectly, it will either not cure properly or cure too quickly (you won't have time to file and sand it into shape)

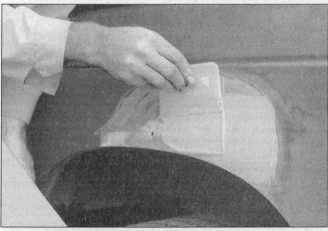

6 Working quickly so the filler doesn't harden, use a plastic applicator to press the body filler firmly into the metal, assuring it bonds completely. Work the filler until it matches the original contour and is slightly above the surrounding metal

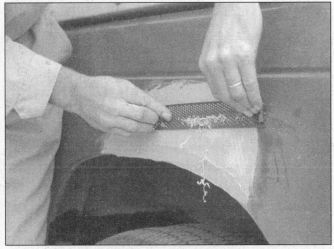
7 Let the filler harden until you can just dent it with your fingernail. Use a body file or Surform tool (shown here) to rough-shape the filler

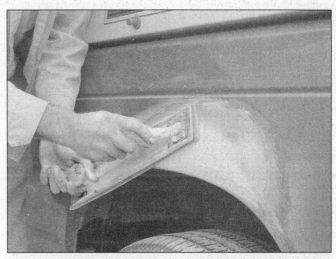
8 Use coarse-grit sandpaper and a sanding board or block to work the filler down until it's smooth and even. Work down to finer grits of sandpaper - always using a board or block - ending up with 360 or 400 grit

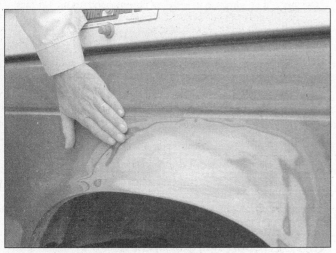
9 You shouldn't be able to feel any ridge at the transition from the filler to the bare metal or from the bare metal to the old paint. As soon as the repair is flat and uniform, remove the dust and mask off the adjacent panels or trim pieces

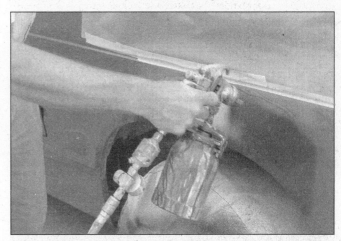
10 Apply several layers of primer to the area. Don't spray the primer on too heavy, so it sags or runs, and make sure each coat is dry before you spray on the next one. A professional-type spray gun is being used here, but aerosol spray primer is available inexpensively from auto parts stores

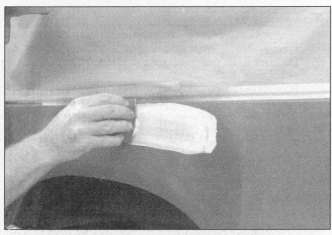
11 The primer will help reveal imperfections or scratches. Fill these with glazing compound. Follow the label instructions and sand it with 360 or 400-grit sandpaper until it's smooth. Repeat the glazing, sanding and respraying until the primer reveals a perfectly smooth surface

12 Finish sand the primer with very fine sandpaper (400 or 600-grit) to remove the primer overspray. Clean the area with water and allow it to dry. Use a tack rag to remove any dust, then apply the finish coat. Don't attempt to rub out or wax the repair area until the paint has dried completely (at least two weeks)

Chapter 11 Body

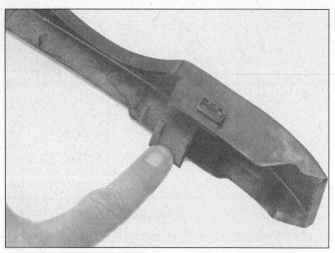

9.10 Detach the center grille valance by reaching under each headlight and pressing on the release lever

9.14 Remove the screws and pull the side grille assembly straight out

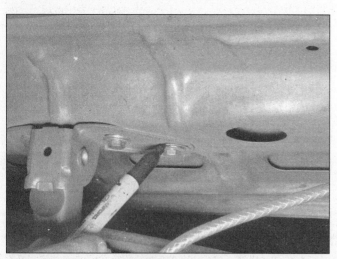

10.1 Use paint or a marking pen to mark on the hinge plate around the bolt heads - mark around the entire hinge plate before adjusting the hood

10.3a Use needle-nose pliers to pull off the hinge pin clip . . .

10 From the engine compartment, reach under the headlight housings and detach the clips retaining center grille valances, then push the grille forward **(see illustration)**.
11 Use a screwdriver to depress the clips, detach the grille assembly and remove it by pulling it straight out.
12 Install the center grille by placing it in position and pushing it straight back until it clips into places

Side grille

Refer to illustration 9.14
13 Remove the center grille.
14 Remove the screws and lift the side grille assembly out **(see illustration)**.
15 Installation is the reverse of removal.

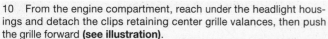

10 Hood - removal, installation and adjustment

Note: *The hood is heavy and somewhat awkward to remove and install - at least two people should perform this procedure.*

Removal and installation

3-Series

Refer to illustrations 10.1, 10.3a and 10.3b
1 Scribe or draw alignment marks around the bolt heads to ensure

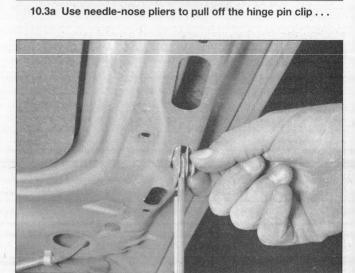

10.3b . . . and pull the hinge pin out while supporting the hood

proper alignment on reinstallation **(see illustration)**.
2 Disconnect the ground cable and windshield washer hose from the hood.
3 Detach the hood hinge rod clip and remove the pin **(see illustrations)**. Be sure to support the hood while doing this.

Chapter 11 Body   11-7

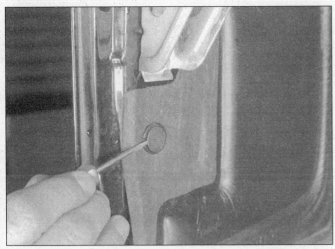
10.8a Use a small screwdriver to pry out the clip pin, then . . .

10.8b . . . pry the retainer out

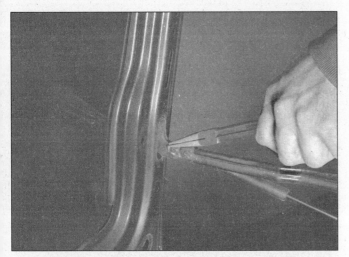
10.9 Pull off the hinge pin clip with needle nose pliers

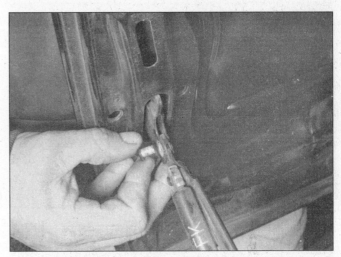
10.10 Support the hood and withdraw the hinge pin

10.12 Unscrew the hinge through-bolts (arrow)

6 Lift the hood off.
7 Installation is the reverse of the removal.

5-Series

Refer to illustrations 10.8a, 10.8b, 10.9, 10.10 and 10.12

8 Open the hood. On some later models it may be necessary to pry out the plastic clips and remove the insulation pad for access to the hood light connector and windshield washer hoses **(see illustrations)**.
9 Detach the clips and disconnect the hood props **(see illustration)**.
10 Remove the pins from the hinges **(see illustration)**.
11 Have an assistant hold onto the hood on one side while you hold the other side.
12 Remove the hood-to-hinge through-bolt on your side of the hood, then hold your side of the hood while your assistant removes the through-bolt on the other side **(see illustration)**.

Adjustment

Refer to illustrations 10.15, 10.16 and 10.17

13 The hood can be adjusted to obtain a flush fit between the hood and fenders after loosening the hood hinge bolts. On some 5-Series models, it will be necessary to remove the side grille sections for access to the hinge bolts.
14 Move the hood from side-to-side or front-to-rear until the hood is properly aligned with the fenders at the front. Tighten the bolts securely.

4 Have an assistant hold onto the hood on one side while you hold the other side.
5 Remove the hood-to-hinge assembly bolts on your side of the hood, then hold your side of the hood while your assistant removes the hood-to-hinge bolts on the other side.

Chapter 11 Body

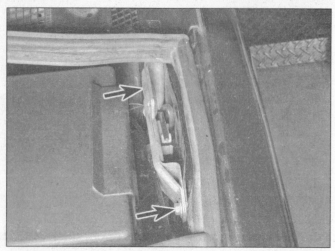

10.15 Loosen the bolts (arrows) and raise or lower the catch to adjust the hood height

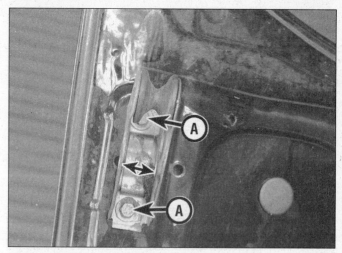

10.16 Loosen the bolts (A) and adjust the roller guide side-to-side until it engages securely in the catch

10.17 Screw the hood stops in-or-out to adjust the hood vertically

12.3 On manual regulator models, pry off the window crank trim piece for access to the retaining screw

15   The rear height of the hood can be adjusted by loosening the bolts and raising or lowering the catch **(see illustration)**. After adjustment, tighten the bolts securely.
16   Side-to-side adjustment of the hood can be made by loosening the roller guide bolt nuts and moving the guide position until it slides into the catch properly **(see illustration)**.
17   After adjustment, screw the stop pads in or out to support the hood in its new position **(see illustration)**.
18   The hood mechanism should be lubricated periodically with white lithium base grease to prevent sticking or jamming.

11 Bumpers - removal and installation

1   Detach the bumper cover (if equipped).
2   Disconnect any wiring or other components that would interfere with bumper removal.
3   Support the bumper with a jack or jackstand. Alternatively, have an assistant support the bumper as the bolts are removed.
4   Remove the retaining bolts and detach the bumper.
5   Installation is the reverse of removal.
6   Tighten the retaining bolts securely.
7   Install the bumper cover and any other components that were removed.

12 Door trim panel - removal and installation

Refer to illustration 12.3

1   Disconnect the negative cable from the battery. **Caution:** *If the radio in your vehicle is equipped with an anti-theft system, make sure you have the correct activation code before disconnecting the battery. Refer to the information on page 0-7 at the front of this manual before detaching the cable.* **Note**: *If, after connecting the battery, the wrong language appears on the instrument panel display, refer to page 0-7 for the language resetting procedure.*
2   Remove all door trim panel retaining screws and door pull/armrest assemblies.
3   On models with manual window regulators, remove the window regulator crank **(see illustration)**. On power regulator models, pry off the control switch assembly and unplug it.
4   Disengage the trim panel-to-door retaining clips. Work around the outer edge until the panel is free.
5   Once all of the clips are disengaged, detach the trim panel, unplug any electrical connectors and remove the trim panel from the vehicle.
6   For access to the inner door, carefully peel back the plastic water shield.
7   Prior to installation of the door panel, be sure to reinstall any clips

Chapter 11 Body

13.4 Detach the circlip (arrow) from the tapered end of the pin

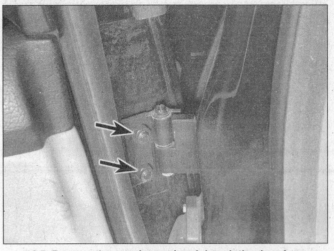

13.5 Remove the nuts (arrows) and detach the door from the hinges

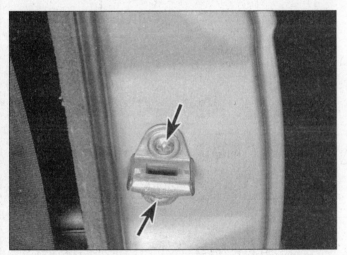

13.7 The door lock striker position can be adjusted after loosening the screws (arrows)

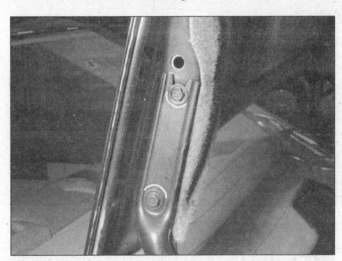

14.3 Mark around the hinge bolts so you can reinstall the trunk lid in its original location - unscrew or loosen the trunk lid-to-hinge bolts to remove or adjust it

in the panel which may have come out during the removal procedure and remain in the door itself.
8 Plug in the electrical connectors and place the panel in position in the door. Press the door panel into place until the clips are seated and install the armrest/door pulls. Install the manual regulator window crank.

13 Door - removal, installation and adjustment

Refer to illustrations 13.4, 13.5 and 13.7

1 Remove the door trim panel (see Section 12). Disconnect any electrical connectors and push them through the door opening so they won't interfere with door removal.
2 Place a floor jack under the door or have an assistant on hand to support it when the hinge bolts are removed. **Note:** *If a jack is used, place a rag between it and the door to protect the door's painted surfaces.*
3 Scribe around the door hinges.
4 Disconnect the check strap by prying the clip out of the end of the retaining, then sliding the pin out **(see illustration)**.
5 Remove the hinge-to-door nuts and carefully lift off the door **(see illustration)**.
6 Installation is the reverse of removal.
7 Following installation of the door, check the alignment and adjust

it if necessary as follows:
a) Up-and-down and forward-and-backward adjustments are made by loosening the hinge-to-body nuts and moving the door as necessary.
b) The door lock striker can also be adjusted both up-and-down and sideways to provide positive engagement with the latch mechanism. This is done by loosening the mounting bolts and moving the striker as necessary **(see illustration)**.

14 Trunk lid - removal, installation and adjustment

Refer to illustrations 14.3, 14.7a and 14.7b

1 Open the trunk lid and cover the edges of the trunk compartment with pads or cloths to protect the painted surfaces when the lid is removed.
2 Disconnect any cables or electrical connectors attached to the trunk lid that would interfere with removal.
3 Make alignment marks around the hinge bolts **(see illustration)**.
4 While an assistant supports the lid, remove the lid-to-hinge bolts on both sides and lift it off.
5 Installation is the reverse of removal. **Note:** *When re-installing the trunk lid, align the lid-to-hinge bolts with the marks made during removal.*
6 After installation, close the lid and make sure it's in proper align-

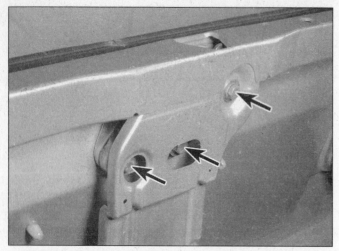

14.7a Loosen the latch bolts (arrows) and move the latch to adjust the trunk closing position

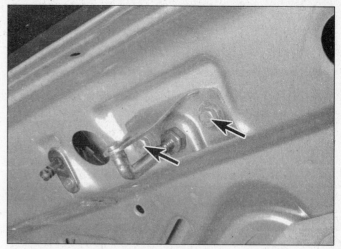

14.7b Adjust the trunk latch striker after loosening the bolts (arrows)

when closed can be changed by loosening the lock and/or striker bolts, repositioning the striker and tightening the bolts **(see illustrations)**.

15 Latch, lock cylinder and handles - removal, installation and adjustment

Refer to illustration 15.2

1   Remove the door trim panel and plastic shield (see Section 12).

Latch

2   Disconnect the operating rods from the latch **(see illustration)**.
3   Remove the three latch retaining screws located in the end of the door.
4   Detach the latch assembly and lift it from the door.
5   Installation is the reverse of removal.

Lock cylinder

6   Detach the linkage.
7   Use a screwdriver to slide the retaining clip off and withdraw the lock cylinder from the door.
8   Installation is the reverse of removal.

Inside handle

9   Disconnect the operating rod from the handle.
10 Remove the retaining screws and lift the handle from the door.
11 Installation is the reverse of removal.

Outside handle

12 Lift up the handle for access, remove the two retaining screws, then detach the handle from the door.
13 Installation is the reverse of removal.

16 Door window glass - removal and installation

1   Disconnect the cable from the negative terminal of the battery.
Caution: *If the radio in your vehicle is equipped with an anti-theft system, make sure you have the correct activation code before disconnecting the battery, Refer to the information on page 0-7 at the front of this manual before detaching the cable.* **Note:** *If, after connecting the battery, the wrong language appears on the instrument panel display, refer to page 0-7 for the language resetting procedure.*
2   Remove the door trim panel and the plastic water shield (see Section 12).
3   Pry the door inner and outer weatherstrips from the door.

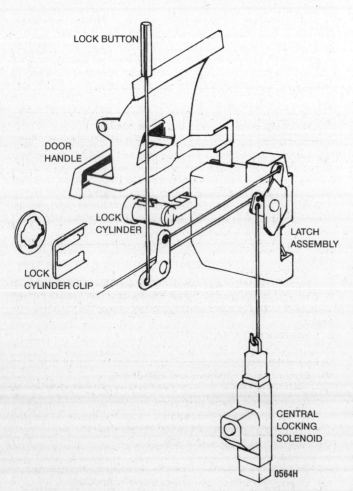

15.2 Typical door latch, lock cylinder and handle details

ment with the surrounding panels. Fore-and-aft and side-to-side are controlled by the position of the hinge bolts in the slots. To make an adjustment, loosen the hinge bolts, reposition the lid and retighten the bolts.
7   The height of the lid in relation to the surrounding body panels

Chapter 11 Body

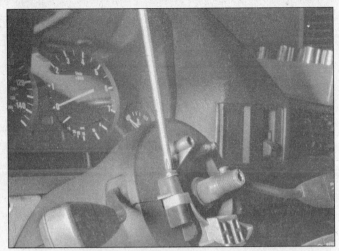

19.3 Use a Phillips screwdriver to remove the upper column cover screws

19.4 The lower screws are located under the tilt lever

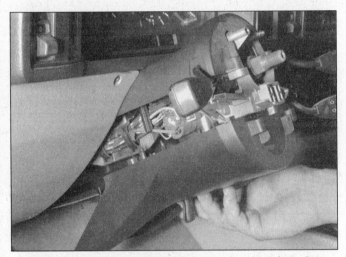

19.5a Pull the tilt lever down and lower the cover from the steering column

Front door
4   Raise the window so that the mounting bolts can be reached through the access hole. On power windows, temporarily reconnect the battery cable to accomplish this.
5   Support the glass and remove the retaining bolts securing the glass to the regulator.
6   Lift the window glass up and out of the door window slot, then tilt it and remove it from the door.
7   Installation is the reverse of removal.

Rear door
8   Perform steps 1 through 5.

3-Series models
9   Disengage the glass from the front guide roller and pry the rubber window guides out.
10   Remove the door handle screws and allow the handle assembly to hang out of the way.
11   Remove the rear window frame bolts, push the frame into the door and remove the window glass.
12   Installation is the reverse of removal.

5-Series
13   Loosen the rear window guide mounting bolt.
14   Detach the glass from the front guide roller, disengage the glass from the guide rails, then slide the glass to the rear and remove by lifting from above.
15   The fixed glass can be removed by removing the guide rail bolts, then pulling the rail down to the rear and detaching the glass.
16   Installation is the reverse of removal. The rear window glass can be adjusted by loosening the adjustment bolts, then raising the glass to within one inch of the top of the door opening. Adjust the glass-to-opening gap evenly, then tighten the bolts securely.

17 Window regulator - removal and installation

1   Disconnect the cable from the negative terminal of the battery. **Caution:** *If the radio in your vehicle is equipped with an anti-theft system, make sure you have the correct activation code before disconnecting the battery, Refer to the information on page 0-7 at the front of this manual before detaching the cable.* **Note**: *If, after connecting the battery, the wrong language appears on the instrument panel display, refer to page 0-7 for the language resetting procedure.*
2   Remove the door trim panel and the plastic water shield (see Section 12).
3   Remove the door glass.
4   Remove the attaching bolts or nuts and lift the window regulator assembly out of the door (withdraw the regulator mechanism through the access hole). On power window models, unplug the electrical connector.
5   Installation is the reverse of removal.

18 Exterior mirror - removal and installation

1   Pry off the cover panel or speaker.
3   Unplug the electrical connector.
4   Remove the retaining screws and lift the mirror off
5   Installation is the reverse of removal.

19 Steering column cover - removal and installation

Refer to illustrations 19.3, 19.4, 19.5a and 19.5b
1   Disconnect the negative battery cable. **Caution:** *If the radio in your vehicle is equipped with an anti-theft system, make sure you have the correct activation code before disconnecting the battery, Refer to the information on page 0-7 at the front of this manual before detaching the cable.* **Note**: *If, after connecting the battery, the wrong language appears on the instrument panel display, refer to page 0-7 for the language resetting procedure.*
2   Remove the steering wheel (Chapter 10).

19.5b Rotate the upper cover up and off the steering column

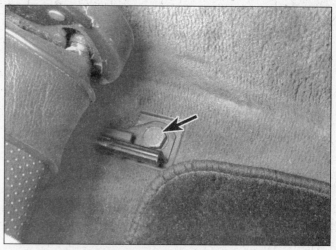

20.1 The front seats are held in place by bolts (arrow)

3 Remove the upper housing screws **(see illustration)**.
4 Remove the two screws from the underside of the column **(see illustration)**.
5 Detach the lower housing and lift the upper half off the column **(see illustrations)**.
6 Installation is the reverse of removal.

20 Seats - removal and installation

Front seat

Refer to illustration 20.1

1 Remove the four bolts securing the seat track to the floorpan and lift the seat from the vehicle **(see illustration)**.
2 Installation is the reverse of the removal steps. Tighten the retaining bolts securely.

Rear seat cushion

Refer to illustration 20.3

3 On models so equipped, removes the two retaining bolts, grasp the front of the cushion securely and pull up sharply **(see illustration)**.
4 Installation is the reverse of the removal.

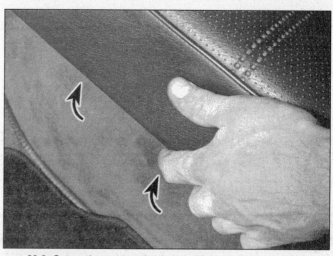

20.3 Grasp the seat at the front edge and pull up sharply

21 Seat belt check

1 Check the seat belts, buckles, latch plates and guide loops for obvious damage and signs of wear.
2 Check that the seat belt reminder light comes on when the ignition key is turned to the Run or Start position.
3 The seat belts are designed to lock up during a sudden stop or impact, yet allow free movement during normal driving. Check that the retractors return the belt against your chest while driving and rewind the belt fully when the buckle is unlatched.
4 If any of the above checks reveal problems with the seat belt system, replace parts as necessary.

22 Center console - removal and installation

1 Disconnect the negative battery cable.
2 Remove the lower trim panel below the steering column.
3 Remove the ashtray and lighter.
4 If equipped with a manual transmission, carefully pry off the shifter boot.
5 If equipped with an automatic transmission, remove the shift lever trim panel.
6 Remove the screws on each side of the center console.
7 On the 5-series, remove the retainers from both sides of the center console by turning the retainers 90 degrees (1/4 turn) clockwise.
8 Carefully pry up all the window switches and disconnect their harness connectors.
9 Remove the console mounting screws. There are two at the rear of the console, hidden under the plastic covers behind the handbrake lever, one under the rear handbrake lever brush, and two in the ashtray opening.
10 Remove the handbrake trim by sliding it to the rear and lifting it out. Disconnect any electrical connectors and remove the nut directly below the shift lever.
11 Slide the console to the rear revealing the two screws above the radio. Remove these screws and carefully push the radio through the opening.
12 Carefully slide the console out disconnecting any remaining electrical connectors.
13 Installation is the reverse of removal. Make sure all the console switches have been reconnected before final assembly.

Chapter 12 Chassis electrical system

Contents

Backup light check and adjustment ....See Chapter 7B	
Bulb replacement	15
Central locking system - description and check............................	20
Cruise control system - description and check	19
Electrical troubleshooting - general information	2
Fuses - general information	3
General information...	1
Headlights - adjustment	13
Headlights - replacement...	12
Headlight housing - removal and installation.................................	14
Ignition switch - removal and installation..............................	7
Instrument cluster - removal and installation.......................	10
Power window system - description and check............................	21
Rear window defogger - check and repair......................................	17
Relays - general information	4
Radio antenna - removal and installation..	9
Radio - removal and installation...	8
Service Indicator (SI) board - general information	11
Steering column switches - removal and installation....................	6
Supplemental Restraint System (SRS) - general information	18
Turn signal/hazard warning flasher - check and replacement	5
Windshield wiper motor - replacement..	16
Wiring diagrams - general information...	22

1 General information

The chassis electrical system of this vehicle is a 12-volt, negative ground type. Power for the lights and all electrical accessories is supplied by a lead/acid-type battery which is charged by the alternator.

This chapter covers repair and service procedures for various chassis (non-engine related) electrical components. For information regarding the engine electrical system components (battery, alternator, distributor and starter motor), see Chapter 5.

Warning: *To prevent electrical shorts, fires and injury, always disconnect the cable from the negative terminal of the battery before checking, repairing or replacing electrical components.*

Caution: *If the radio in your vehicle is equipped with an anti-theft system, make sure you have the correct activation code before disconnecting the battery. Refer to the information on page 0-7 at the front of this manual before detaching the cable.* **Note**: *If, after connecting the battery, the wrong language appears on the instrument panel display, refer to page 0-7 for the language resetting procedure.*

2 Electrical troubleshooting - general information

A typical electrical circuit consists of an electrical component, any switches, relays, motors, fuses, fusible links or circuit breakers, etc. related to that component and the wiring and connectors that link the components to both the battery and the chassis. To help you pinpoint an electrical circuit problem, wiring diagrams are included at the end of this book.

Before tackling any troublesome electrical circuit, first study the appropriate wiring diagrams to get a complete understanding of what makes up that individual circuit. Trouble spots, for instance, can often be isolated by noting if other components related to that circuit are often routed through the same fuse and ground connections.

Electrical problems usually stem from simple causes such as loose or corroded connectors, a blown fuse, a melted fusible link or a bad relay. Visually inspect the condition of all fuses, wires and connectors in a problem circuit before troubleshooting it.

The basic tools needed for electrical troubleshooting include a circuit tester, a high impedance (10 K-ohm) digital voltmeter, a continuity tester and a jumper wire with an inline circuit breaker for bypassing electrical components. Before attempting to locate or define a problem with electrical test instruments, use the wiring diagrams to decide where to make the necessary connections.

Voltage checks

Perform a voltage check first when a circuit is not functioning properly. Connect one lead of a circuit tester to either the negative battery terminal or a known good ground.

Connect the other lead to a connector in the circuit being tested, preferably nearest to the battery or fuse. If the bulb of the tester lights up, voltage is present, which means that the part of the circuit between the connector and the battery is problem free. Continue checking the rest of the circuit in the same fashion.

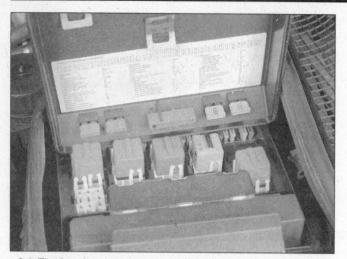

3.1 The fuse box is located in the engine compartment under a cover - the box also includes several relays

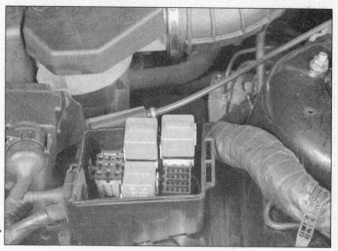

4.2 Engine compartment relays

When you reach a point at which no voltage is present, the problem lies between that point and the last test point with voltage. Most of the time the problem can be traced to a loose connection. **Note:** *Keep in mind that some circuits receive voltage only when the ignition key is in the Accessory or Run position.*

Finding a short circuit

One method of finding shorts in a circuit is to remove the fuse and connect a test light or voltmeter in its place. There should be no voltage present in the circuit. Move the electrical connectors from side-to-side while watching the test light. If the bulb goes on, there is a short to ground somewhere in that area, probably where the insulation has been rubbed through. The same test can be performed on each component in a circuit, even a switch.

Ground check

Perform a ground test to check whether a component is properly grounded. Disconnect the battery and connect one lead of a self-powered test light, known as a continuity tester, to a known good ground. **Caution:** *If the radio in your vehicle is equipped with an anti-theft system, make sure you have the correct activation code before disconnecting the battery, Refer to the information on page 0-7 at the front of this manual before detaching the cable.* **Note**: *If, after connecting the battery, the wrong language appears on the instrument panel display, refer to page 0-7 for the language resetting procedure.* Connect the other lead to the wire or ground connection being tested. If the bulb goes on, the ground is good. If the bulb does not go on, the ground is not good.

Continuity check

A continuity check determines if there are any breaks in a circuit - if it is conducting electricity properly. With the circuit off (no power in the circuit), a self-powered continuity tester can be used to check the circuit. Connect the test leads to both ends of the circuit, and if the test light comes on the circuit is passing current properly. If the light doesn't come on, there is a break somewhere in the circuit. The same procedure can be used to test a switch, by connecting the continuity tester to the power in and power out sides of the switch. With the switch turned on, the test light should come on.

Finding an open circuit

When diagnosing for possible open circuits it is often difficult to locate them by sight because oxidation or terminal misalignment are hidden by the connectors. Merely wiggling a connector on a sensor or in the electrical connector may correct the open circuit condition. Remember this if an open circuit is indicated when troubleshooting a circuit. Intermittent problems may also be caused by oxidized or loose connections. Electrical troubleshooting is simple if you keep in mind that all electrical circuits are basically electricity running from the battery, through the wires, switches, relays, fuses and fusible links to each electrical component (light bulb, motor, etc.) and then to ground, from which it is passed back to the battery. Any electrical problem is an interruption in the flow of electricity to and from the battery.

3 Fuses - general information

Refer to illustration 3.1

The electrical circuits of the vehicle are protected by a combination of fuses and circuit breakers. The fuse box is located in the left corner of the engine compartment and on some later models under the rear seat cushion **(see illustration)**.

Each of the fuses is designed to protect a specific circuit, and the various circuits are identified on the fuse panel itself.

Miniaturized fuses are employed in the fuse boxes. These compact fuses, with blade terminal design, allow fingertip removal and replacement. If an electrical component fails, always check the fuse first. A blown fuse is easily identified through the clear plastic body. Visually inspect the element for evidence of damage. If a continuity check is called for, the blade terminal tips are exposed in the fuse body.

Be sure to replace blown fuses with the correct type. Fuses of different ratings are physically interchangeable, but only fuses of the proper rating should be used. Replacing a fuse with one of a higher or lower value than specified is not recommended. Each electrical circuit needs a specific amount of protection. The amperage value of each fuse is molded into the fuse body.

If the replacement fuse immediately fails, don't replace it again until the cause of the problem is isolated and corrected. In most cases, the cause will be a short circuit in the wiring caused by a broken or deteriorated wire.

4 Relays - general information

Refer to illustration 4.2

Several electrical accessories in the vehicle use relays to transmit the electrical signal to the component. If the relay is defective, that component will not operate properly.

The various relays are grouped together in several locations under the dash and in the engine compartment for convenience in the event of needed replacement **(see accompanying illustration and illustration 3.1)**.

If a faulty relay is suspected, it can be removed and tested by a dealer or other qualified shop. Defective relays must be replaced as a unit.

Chapter 12 Chassis electrical system

5.1 The turn signal/hazard warning flasher is located on the steering column on most models - squeeze the tabs to detach it

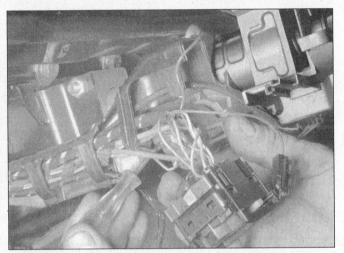

6.3 Follow the wire down the steering column to the connector

6.6 Squeeze the wiper/washer switch tabs and pull it directly out of the mount

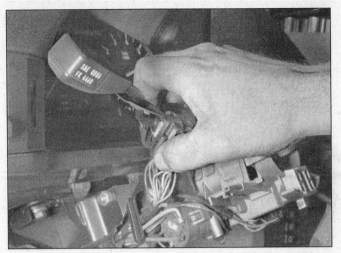

6.2 Squeeze the tabs to release the switch from the mount

lowing procedure should be left to a dealer service department, or other repair shop because of the special tools and techniques required.

1 The turn signal/hazard flasher is a small canister or box-shaped unit located in the wiring harness on or near the steering column (see illustration).
2 When the flasher unit is functioning properly, an audible click can be heard during its operation. If the turn signals fail on one side or the other and the flasher unit does not make its characteristic clicking sound, a faulty turn signal bulb is indicated.
3 If both turn signals fail to blink, the problem may be due to a blown fuse, a faulty flasher unit, a broken switch or a loose or open connection. If a quick check of the fuse box indicates that the turn signal and/or hazard fuse has blown, check the wiring for a short before installing a new fuse.
5 Make sure that the replacement unit is identical to the original. Compare the old one to the new one before installing it.
6 Installation is the reverse of removal.

6 Steering column switches - removal and installation

Warning: *Some later models are equipped with an airbag or Supplemental Restraint System. To avoid possible damage to this system, the manufacturer recommends that, on airbag-equipped models, the following procedure should be left to a dealer service department or other repair shop because of the special tools and techniques required.*

1 Disconnect the negative battery cable, remove the steering wheel (see Chapter 10) and steering column cover (see Chapter 11). **Caution:** *If the radio in your vehicle is equipped with an anti-theft system, make sure you have the correct activation code before disconnecting the battery. Refer to the information on page 0-7 at the front of this manual before detaching the cable.* **Note:** *If, after connecting the battery, the wrong language appears on the instrument panel display, refer to page 0-7 for the language resetting procedure.*

Turn signal/headlight control switch

Refer to illustrations 6.2 and 6.3

2 Depress the release clips and pull the switch out of the steering column mount (see illustration).
3 Trace the switch wires down the steering column to the electrical connector and unplug them (see illustration).
4 Installation is the reverse of removal.

Wiper/washer switch

Refer to illustration 6.6

5 Depress the release clip and detach the switch from the steering

5 Turn signal/hazard warning flasher - check and replacement

Refer to illustration 5.1
Warning: *Some later models are equipped with an airbag or Supplemental Restraint System. To avoid possible damage to this system, the manufacturer recommends that, on airbag-equipped models, the fol-*

Chapter 12 Chassis electrical system

6.9 Cruise control switch removal

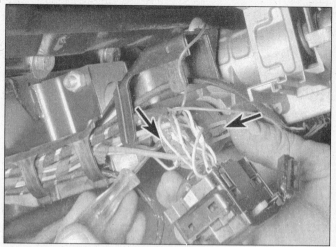

7.5 Insert a screwdriver into the openings (arrows) on each side of the switch to release the clip while pulling out

column mount.
6 Trace the switch wiring down the steering column to the electrical connector and unplug it.
7 Installation is the reverse of removal.

Cruise control switch

Refer to illustration 6.9
8 Remove the wiper/washer switch.
9 Squeeze the release tabs and withdraw the switch from the mount **(see illustration)**.
10 Disconnect the switch electrical connector from the harness at the base of the steering column.
11 Installation is the reverse of removal.

7 Ignition switch - removal and installation

Refer to illustration 7.5
Warning: *Some later models are equipped with an airbag or Supplemental Restraint System. To avoid possible damage to this system, the manufacturer recommends that, on airbag equipped models, the following procedure should be left to a dealer service department or other repair shop because of the special tools and techniques required.*
1 Disconnect the negative cable at the battery. **Caution:** *If the radio in your vehicle is equipped with an anti-theft system, make sure you have the correct activation code before disconnecting the battery, Refer to the information on page 0-7 at the front of this manual before detaching the cable.* **Note***: If, after connecting the battery, the wrong language appears on the instrument panel display, refer to page 0-7 for the language resetting procedure.*
2 Remove the steering wheel (see Chapter 10).
3 Remove the steering column cover (see Chapter 11).
4 Remove the turn signal/headlight control switch (see Section 6).
5 Detach the clips by inserting a small screwdriver into the opening on the sides while pulling out on the switch **(see illustration)**.
6 Unplug the electrical connector from the harness at the base of the steering column and remove the switch.
7 Installation is the reverse of removal.

8 Radio - removal and installation

1 Disconnect the negative cable at the battery. **Caution:** *If the radio in your vehicle is equipped with an anti-theft system, make sure you have the correct activation code before disconnecting the battery.* **Note***: If, after connecting the battery, the wrong language appears on the instrument panel display, refer to page 0-7 for the language reset-*

ting procedure.
2 The radios on these models are held in place by internal clips which are usually located at the sides or corners of the unit face plate. Removal requires a special tool which is inserted into the holes to release the clips so the radio can be pulled out. These tools can be fabricated from heavy wire or are available from your dealer or an automotive radio shop. On anti-theft radios, the clips are moved in and out by internal screws which require another type of tool.
3 Insert the tool into the holes until the clips release, then withdraw the radio from the dash panel.
4 Installation is the reverse of removal.

9 Radio antenna - removal and installation

1 Disconnect the negative cable at the battery. **Caution:** *If the radio in your vehicle is equipped with an anti-theft system, make sure you have the correct activation code before disconnecting the battery, Refer to the information on page 0-7 at the front of this manual before detaching the cable.* **Note***: If, after connecting the battery, the wrong language appears on the instrument panel display, refer to page 0-7 for the language resetting procedure.*
2 Use snap-ring pliers to unscrew the antenna-to-fender mounting nut.
3 Open the trunk and remove the left side trim panel.
4 Unplug the antenna power and radio lead connectors, remove the retaining bolts and remove the antenna and motor assembly.
5 Installation is the reverse of removal.

10 Instrument cluster - removal and installation

Refer to illustration 10.3, 10.4 and 10.5
Caution: *The instrument cluster and components are very susceptible to damage from static electricity. Make sure you are grounded and have discharged any static electricity before touching the cluster or components.*
1 Disconnect the negative cable at the battery. **Caution:** *If the radio in your vehicle is equipped with an anti-theft system, make sure you have the correct activation code before disconnecting the battery, Refer to the information on page 0-7 at the front of this manual before detaching the cable.* **Note***: If, after connecting the battery, the wrong language appears on the instrument panel display, refer to page 0-7 for the language resetting procedure.*
2 Remove the steering column covers (see Chapter 11).
3 Remove the screws holding the cluster to the instrument panel **(see illustration)**.

Chapter 12 Chassis electrical system

10.3 Use a Phillips head screwdriver to remove the instrument cluster retaining screws

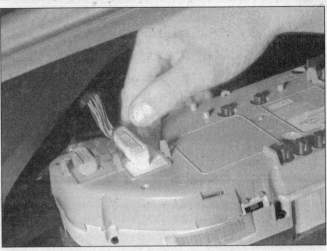

10.4 Push on the levers to detach the cluster electrical connectors

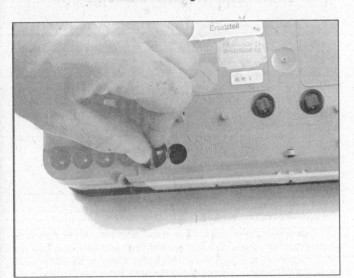

10.5 Turn the plastic knobs to release the back of the cluster (some model use screws)

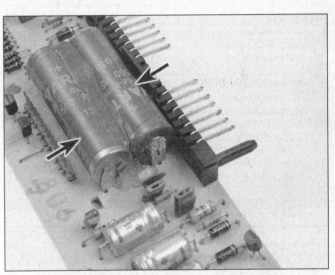

11.2 These batteries (arrows) power the Service Indicator (SI) board and will fail over time so they must be replaced

4   Tilt the top of the cluster back, reach behind it and detach the electrical connectors by pressing on the levers, then lift the cluster out of the dash opening **(see illustration)**.
5   For access to the cluster components, release the catches or remove the screws and separate the two halves **(see illustration)**.
6   Installation is the reverse of removal.

11 Service Indicator (SI) board - general information

Refer to illustration 11.2

All models that have service indicator lights are equipped with a Service Indicator (SI) board located in the instrument cluster. This board turns the lights on at the proper mileage intervals. The lights can only be turned off using a special tool which plugs into the engine check connector (see Chapter 1). The SI board is a self-contained computer which includes a chip and batteries.

The rechargeable SI board nickel cadmium (nicad) batteries maintain power to the computer memory in the event of a power drop (such as during starting) or complete power loss (such as a dead or disconnected battery) **(see illustration)**. This assures power so the computer can continue to keep track of mileage and turn the lights on at the proper interval.

The batteries have a life of approximately six years, at which time they must be replaced with new ones. Also, since they are recharged by the engine charging system so they can run down prematurely if power is cut off for some reason, such as a blown fuse, a fault in the wiring or extended storage of the vehicle. Excessive heat or cold can also shorten battery life, with heat the greatest enemy. Extreme heat can cause the batteries to actually split open, allowing acid to drip into the instrument cluster.

Several instruments controlled by the SI board can be affected by low or discharged batteries. Symptoms of low or dead SI board batteries can include inconsistent tachometer and temperature gauge readings, background radio noise and the inability to turn the service lights off with the special tool.

Although only complete SI boards are available from the manufacturer, batteries are available separately from aftermarket sources. While it is possible for the home mechanic to replace the batteries, they are soldered to the board, so unless you are skilled at this and have the proper tools, this job should be left to an experienced electronics technician. Considerable savings can be realized by removing the instrument cluster (see Section 10) and taking it to an electronics shop. **Caution:** *the instrument cluster and components are very susceptible to damage from static electricity. Make sure you are grounded and have discharged any static electricity (to a water pipe or metal furniture) before touching the cluster components.*

Chapter 12 Chassis electrical system

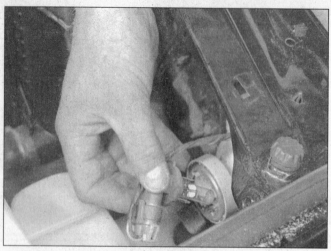

12.8 Rotate the bulb holder counterclockwise and pull it straight out of the housing

12.9 Squeeze the tabs and detach the bulb from the holder

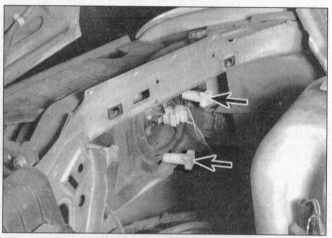

13.1 The headlight adjustment screws (arrows) are accessible from the back of the headlight on 3-Series models

12 Headlights - replacement

1 Disconnect the negative cable from the battery. **Caution:** *If the radio in your vehicle is equipped with an anti-theft system, make sure you have the correct activation code before disconnecting the battery. Refer to the information on page 0-7 at the front of this manual before detaching the cable.* **Note:** *If, after connecting the battery, the wrong language appears on the instrument panel display, refer to page 0-7 for the language resetting procedure.*

Sealed-beam type

2 Remove the grille (see Chapter 11).
3 Remove the headlight retainer screws, taking care not to disturb the adjustment screws.
4 Remove the retainer and pull the headlight out enough to allow the connector to be unplugged.
5 Remove the headlight.
6 To install the headlight, plug the connector in, place the headlight in position and install the retainer and screws. Tighten the screws securely.
7 Install the grille. Connect the negative battery cable.

Halogen bulb type

Refer to illustrations 12.8 and 12.9
Warning: *Halogen gas-filled bulbs are under pressure and may shatter if the surface is scratched or the bulb is dropped. Wear eye protection and handle the bulbs carefully, grasping only the base whenever possible. Do not touch the surface of the bulb with your fingers because the oil from your skin could cause it to overheat and fail prematurely. If you do touch the bulb surface, clean it with rubbing alcohol.*
8 Reach behind the headlight assembly, grasp the bulb holder and turn it counterclockwise to remove it **(see illustration)**. Lift the holder assembly out for access to the bulb.
9 Depress the tabs and detach the bulb assembly from the holder **(see illustration)**.
10 Insert the new bulb into the holder until it seats securely.
11 Install the bulb holder in the headlight assembly.
12 Connect the negative battery cable.

13 Headlights - adjustment

Refer to illustration 13.1
Note: *The headlights must be aimed correctly. If adjusted incorrectly they could momentarily blind the driver of an oncoming vehicle and cause a serious accident or seriously reduce your ability to see the road. The headlights should be checked for proper aim every 12 months and any time a new headlight is installed or front end body work is performed. It should be emphasized that the following procedure is only an interim step which will provide temporary adjustment until the headlights can be adjusted by a properly equipped shop.*

1 Headlights have two adjusting screws, one controlling up-and-down movement and one controlling left-and-right movement **(see illustration)**. It may be necessary to remove the grille (see Chapter 11) for access to these screws.
2 There are several methods of adjusting the headlights. The simplest method requires a blank wall 25 feet in front of the vehicle and a level floor.
3 Position masking tape vertically on the wall in reference to the vehicle centerline and the centerline of both headlights.
4 Position a horizontal line in reference to the centerline of all headlights. **Note:** *It may be easier to position the tape on the wall with the vehicle parked only a few inches away.*
5 Adjustment should be made with the vehicle sitting level, the gas tank half-full and no unusually heavy load in the vehicle.
6 Starting with the low beam adjustment, position the high intensity zone so it is two inches below the horizontal line and two inches to the right of the headlight vertical line. Adjustment is made by turning the adjusting screw to raise or lower the beam. The other adjusting screw should be used in the same manner to move the beam left or right.
7 With the high beams on, the high intensity zone should be vertically centered with the exact center just below the horizontal line.
Note: *It may not be possible to position the headlight aim exactly for*

Chapter 12 Chassis electrical system

14.4 Remove the screws (arrows) and detach the headlight housing

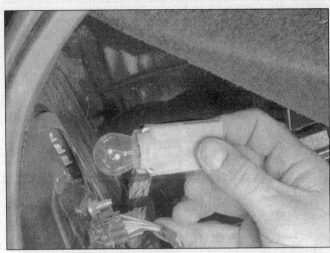

15.3a The tail light bulbs on later 5-Series models are in self-grounding holders which can be simply pulled out of the housing - the bulb is then removed from the holder

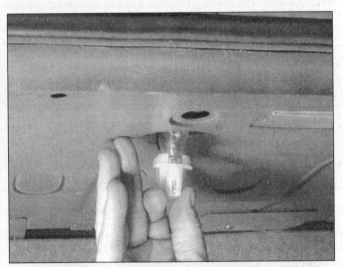

15.3b On models with high-mounted center brake lights, the self-grounding holder is accessible from the trunk - pull the holder out . . .

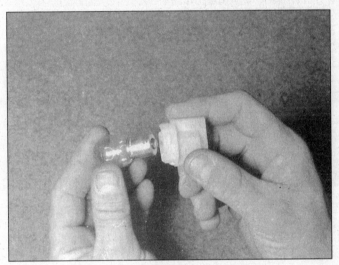

15.3c . . . then pull the bulb from the holder

both high and low beams. If a compromise must be made, keep in mind that the low beams are the most used and have the greatest effect on driver safety.

8 Have the headlights adjusted by a dealer service department or service station at the earliest opportunity.

14 Headlight housing - removal and installation

Refer to illustration 14.4

1 Disconnect the negative battery cable. **Caution:** *If the radio in your vehicle is equipped with an anti-theft system, make sure you have the correct activation code before disconnecting the battery. Refer to the information on page 0-7 at the front of this manual before detaching the cable.* **Note:** *If, after connecting the battery, the wrong language appears on the instrument panel display, refer to page 0-7 for the language resetting procedure.*

2 Remove the grille (see Chapter 11).
3 Unplug the headlight (sealed beam-type) or remove the bulb (halogen bulb-type).
4 Remove the screws and detach the housing **(see illustration)**.
5 Installation is the reverse of removal.

15 Bulb replacement

Refer to illustrations 15.3a, 15.3b, 15.3c, 15.4a, 15.4b and 15.5

1 The lenses of many lights are held in place by screws, which makes it a simple procedure to gain access to the bulbs.
2 On some lights the lenses are held in place by clips. The lenses can be removed either by unsnapping them or by using a small screwdriver to pry them off.
3 Several bulbs are mounted in self-grounding holders and are removed by pushing in and turning them counterclockwise **(see illustration)**. The bulbs can then be removed **(see illustrations)**.
4 The tail lights on 3-Series models are accessible after removing the housing, then removing the bulbs **(see illustrations)**.
5 To gain access to the instrument panel lights, the instrument cluster will have to be removed first **(see illustration)**.

16 Windshield wiper motor - replacement

Refer to illustrations 16.2a and 16.2b

1 Disconnect the negative cable at the battery. **Caution:** *If the radio in your vehicle is equipped with an anti-theft system, make sure you*

Chapter 12 Chassis electrical system

15.4a On 3-Series models the entire tail light housing assembly is self-grounding - loosen the plastic screw and pull the housing back . . .

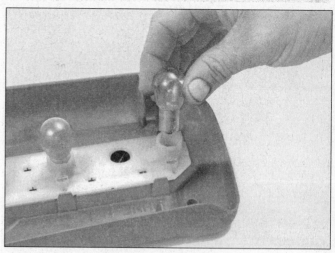

15.4b . . . then remove the bulb from the housing

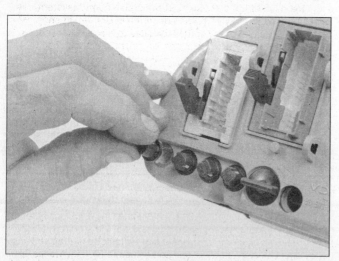

15.5 After removing the instrument cluster (see Section 10), turn the bulb holder counterclockwise to remove the bulb

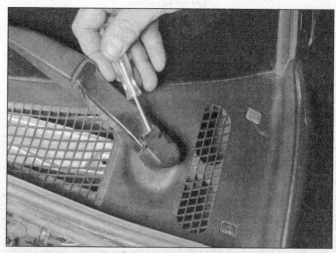

16.2a Use a small screwdriver to detach the wiper arm nut cover

16.2b After removing the nut, use a magnet to lift out the metal washer

have the correct activation code before disconnecting the battery, Refer to the information on page 0-7 at the front of this manual before detaching the cable. **Note**: *If, after connecting the battery, the wrong language appears on the instrument panel display, refer to page 0-7 for the language resetting procedure.*

2   Remove the covers and nuts, then detach the wiper arms **(see illustrations)**.
3   Pry out the retaining clips and detach the cowl grille for access to the wiper link assembly.
4   Remove the screws or nuts and detach the wiper cover located on the engine compartment firewall.
5   Unplug the electrical connector and detach the wiper linkage.
6   Mark the relationship of the wiper shaft to the linkage. Detach the wiper link from the motor shaft by prying carefully with a screwdriver.
7   Remove the three retaining bolts and remove the wiper motor from the vehicle.
8   Installation is the reverse of removal. When installing motor, plug in the connector and run the motor briefly until it is in the neutral (wipers parked) position.

17 Rear window defogger - check and repair

1   The rear window defogger consists of a number of horizontal elements baked onto the glass surface.

Chapter 12 Chassis electrical system

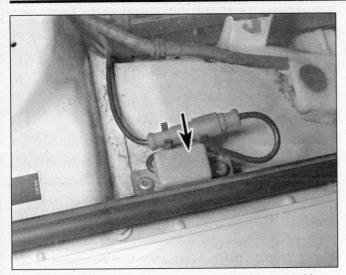

18.3 The SRS system crash sensors (arrow) are located in the engine compartment - check to make sure the wires are undamaged

2 Small breaks in the element can be repaired without removing the rear window.

Check

3 Turn the ignition switch and defogger system switches On.
4 Place the positive lead of a voltmeter to heater element nearest to an incoming power source.
5 Wrap a piece of aluminum foil around the negative lead of the voltmeter on the positive side of the suspected broken element and slide it slowly toward the negative side. Watch the voltmeter needle and when it moves from zero you have located the break.

Repair

6 Repair the break in the line using repair kit recommended specifically for this purpose, such as BMW repair kit No. 81 22 9 (or equivalent). Included in this kit is plastic conductive epoxy.
7 Prior to repairing a break, turn off the system and allow it to de-energize for a few minutes.
8 Lightly buff the element area with fine steel wool, then clean it thoroughly with rubbing alcohol.
9 Use masking tape to mask off the area of repair.
10 Mix the epoxy thoroughly, according to the instructions on the package.
11 Apply the epoxy material to the slit in the masking tape, overlapping the undamaged area about 3/4-inch on either end.
12 Allow the repair to cure for 24 hours before removing the tape and using system.

18 Supplemental Restraint System (SRS) - general information

Refer to illustration 18.3

Later models are equipped with a Supplemental Restraint System (SRS), more commonly called an airbag. This system is designed to protect the driver from serious injury in the event of a head-on or frontal collision. It consists of an airbag module in the center of the steering wheel, two crash sensors mounted at the front of the vehicle and a crash safing sensor located inside the passenger compartment.

The airbag module contains a housing incorporating the airbag and the inflator units. The inflator assembly is mounted on the back of the housing over a hole through which gas is expelled, inflating the bag almost instantaneously when an electrical signal is sent from the system. This signal is carried by a wire which is specially wound with several turns so the signal will be transmitted regardless of the steering wheel position.

The SRS system has three sensors: two at the front, mounted on the inner fender panels and a safing sensor located inside the passenger compartment. The crash sensors are basically pressure sensitive switches that complete an electrical circuit during an impact of sufficient force **(see illustration)**. The electrical signal from the crash sensors is sent to third sensor which then completes the circuit and inflates the airbag.

The module containing the safing sensor monitors the system operation. It checks the system every time the vehicle is started, causing th AIRBAG light to go on, then off if the system is operating correctly. If there is a fault in the system, the light will go on and stay on. If the AIRBAG light does go on and stay on, take the vehicle to your dealer immediately for service.

19 Cruise control system - description and check

The cruise control system maintains vehicle speed with a vacuum actuated servo motor located in the engine compartment, which is connected to the throttle linkage by a cable. The system consists of the servo motor, clutch switch, brake switch, control switches, a relay and associated vacuum hoses.

Because of the complexity of the cruise control system, repair should be left to a dealer service department or a repair shop. However, it is possible for the home mechanic to make simple checks of the wiring and vacuum connections for minor faults which can be easily repaired. These include:
a) Inspect the cruise control actuating switches for broken wires and loose connections.
b) Check the cruise control fuse.
c) The cruise control system is operated by vacuum so it's critical that all vacuum switches, hoses and connections are secure. Check the hoses in the engine compartment for tight connections, cracks and obvious vacuum leaks.

20 Central locking system - description and check

The central door locking system operates the door lock actuators mounted in each door. The system consists of the switches, actuators and associated wiring. Diagnosis can usually be limited to simple checks of the wiring connections and actuators for minor faults which can be easily repaired. These include:
a) Check the system fuse and/or circuit breaker.
b) Check the switch wires for damage and loose connections. Check the switches for continuity.
c) Remove the door panel(s) and check the actuator wiring connections to see if they're loose or damaged. Inspect the actuator rods to make sure they aren't bent or damaged. Inspect the actuator wiring for damaged or loose connections. The actuator can be checked by applying battery power momentarily. A discernible click indicates that the solenoid is operating properly.

21 Power window system - description and check

The power window system operates the electric motors mounted in the doors which lower and raise the windows. The system consists of the control switches, the motors, glass mechanisms (regulators) and associated wiring.

Diagnosis can usually be limited to simple checks of the wiring connections and motors for minor faults which can be easily repaired. These include:
a) Inspect the power window actuating switches for broken wires and loose connections.
b) Check the power window fuse/and or circuit breaker.
c) Remove the door panel(s) and check the power window motor wires to see if they're loose or damaged. Inspect the glass mechanisms for damage which could cause binding.

22 Wiring diagrams - general information

Refer to illustration 22.3

Since it isn't possible to include all wiring diagrams for every model year covered by this manual, the following diagrams are those that are typical and most commonly needed.

Prior to troubleshooting any circuit, check the fuses and circuit breakers to make sure they're in good condition. Make sure the battery is fully charged and check the cable connections (see Chapter 1). Make sure all connectors are clean, with no broken or loose terminals.

Refer to the accompanying table for the wire color codes applicable to your vehicle **(see illustration)**.

BK	Black
BL	Blue
BR	Brown
GR	Green
GY	Grey
OR	Orange
PK	Pink
R	Red
TN	Tan
V	Violet
W	White
Y	Yellow

22.3 Wiring diagram color codes

Chapter 12 Chassis electrical system

Key to Motronic engine control system wiring diagram

No	Description
1	Electronic Control Unit (ECU)
2	Speed control relay
3	Temperature switch
4	Air conditioner
5	Car wire harness connection
6	Throttle switch
7	Airflow sensor
8	Speed sensor
9	Reference mark sensor
10	Relay 1
11	Relay 2
12	Oil pressure switch
13	Temperature transmitter
14	Diagnosis connection
15	Engine plug
16	Battery
17	Spark plugs
18	Distributor
19	Ignition coil
20	Starter
21	Alternator
22	Position transmitter
23	Plug disconnected for automatic transmission
24	Coolant temperature sensor
25	Fuel injector
26	Solenoid
27	Electric power distributor
28	Oil pressure
29	Temperature gauge
30	Electric fuel pump
31	Service indicator
32	Drive motor
33	Temperature switch

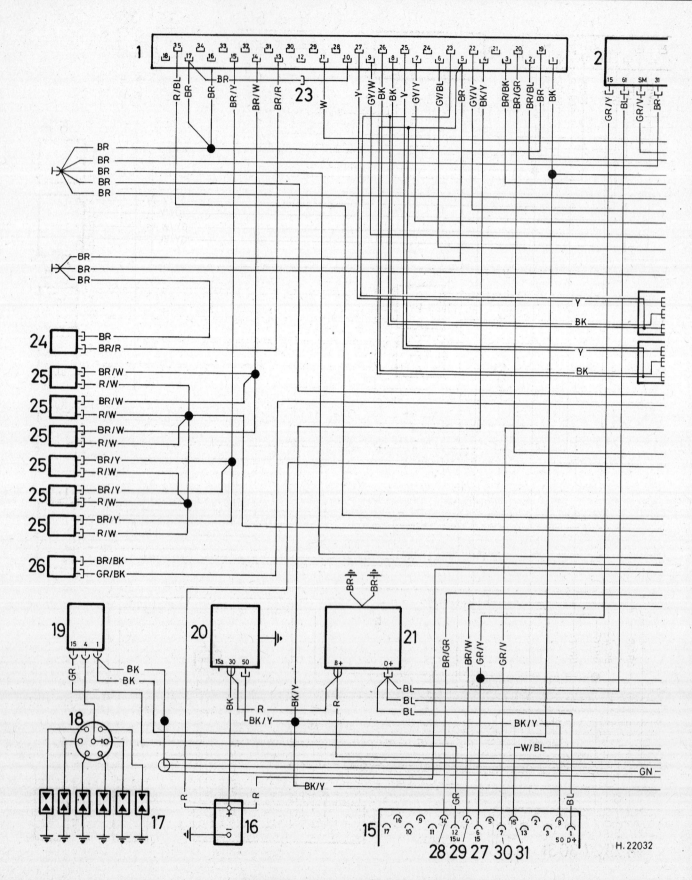

Typical Motronic system wiring diagram (1 of 2)

Chapter 12 Chassis electrical system

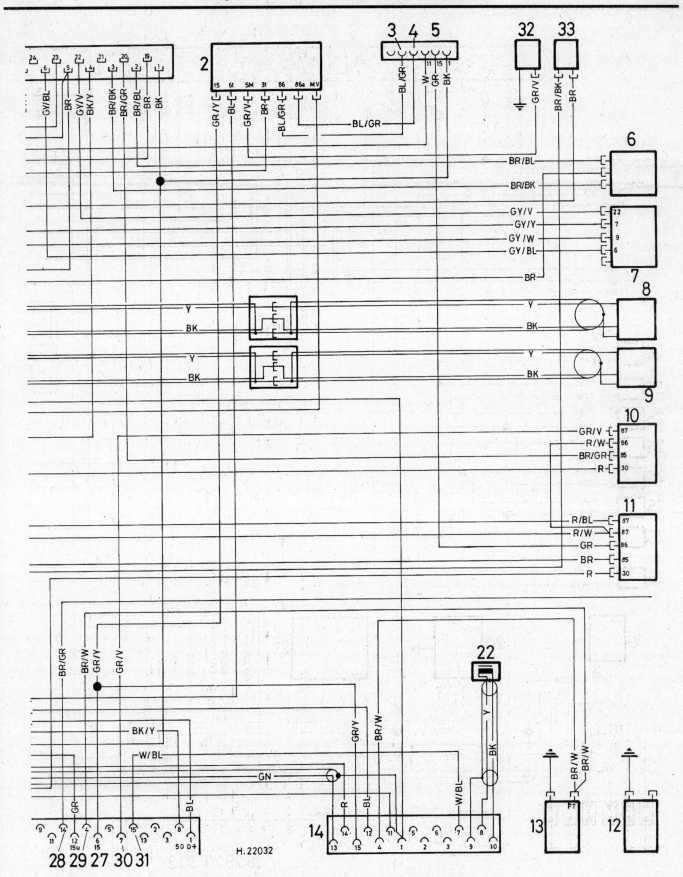

Typical Motronic system wiring diagram (2 of 2)

12-14 Chapter 12 Chassis electrical system

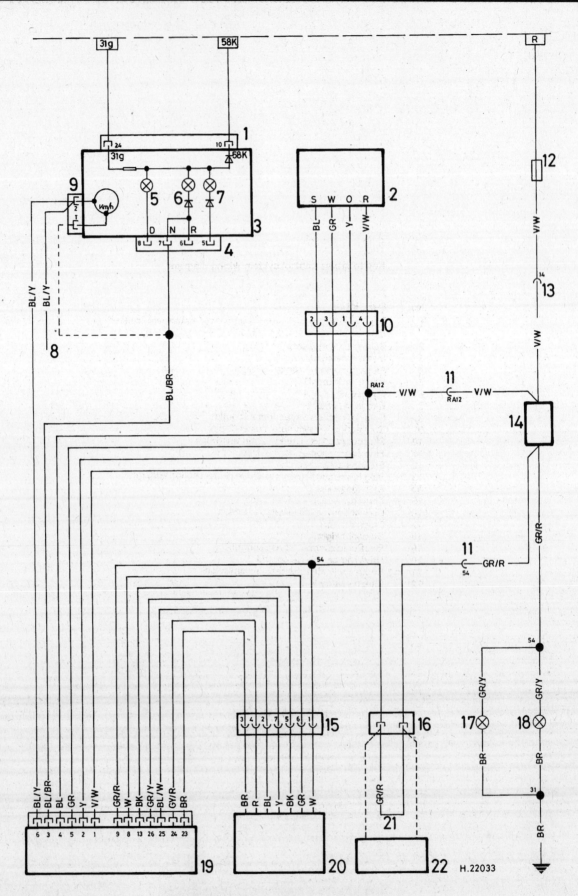

Typical cruise control system wiring diagram

Chapter 12 Chassis electrical system

Key to cruise control system wiring diagram

No	Description
1	Plug connection – center section to instrument cluster (26-pin)
2	Steering column switch
3	Instrument cluster
4	Plug connection – range indicator
5	Range indicator D
6	Range indicator N
7	Range indicator R
8	Plug connection – speedometer outlet
9	Connection – instrument cluster (2-pin)
10	Plug connection – steering column switch
11	Plug connection – special equipment
12	Steering column switch
13	Plug connection – rear section to center section (29-pin)
14	Stoplight switch
15	Plug connection – drive motor
16	Connection – clutch switch to bridge
17	Stoplight left
18	Stoplight right
19	Electronic control – cruise control
20	Drive motor – cruise control
21	Bridge (only for automatic transmission)
22	Clutch switch

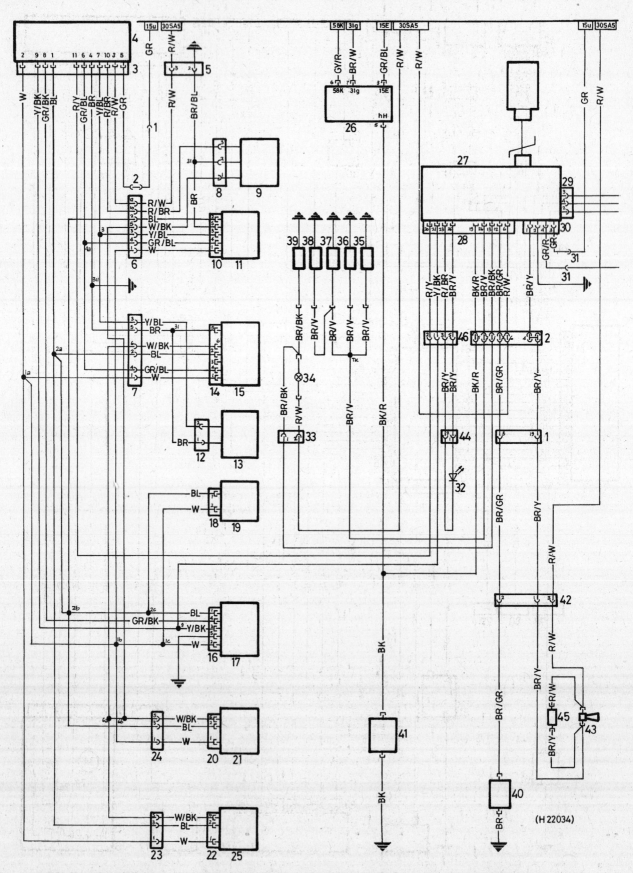

Typical wiring diagram for the power door locks, burglar alarm, on-board computer and digital clock (1 of 2)

Chapter 12 Chassis electrical system

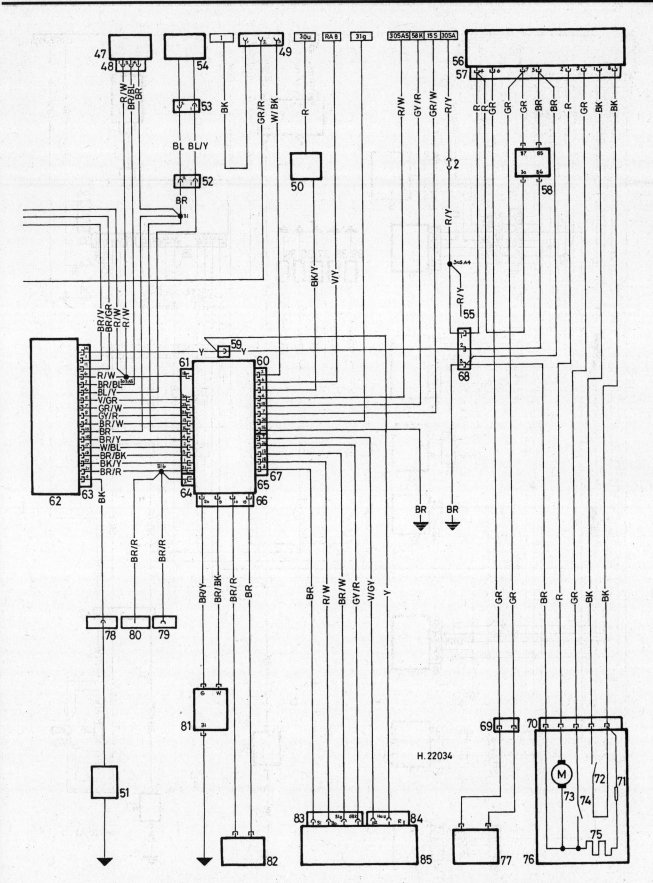

Typical wiring diagram for the power door locks, burglar alarm, on-board computer and digital clock (2 of 2)

Wiring diagram for central locking, burglar alarm, Electronic Control Unit (ECU) and digital clock

No	Description	No	Description
1	Plug – rear section to center section	45	Diode
2	Connection for special equipment plug	46	Plug – burglar alarm wire to central lock connecting wire
3	Connection for central lock control unit	47	Chime (left of steering column)
4	Central lock electronic control unit (A pillar end plate)	48	Connection for chime
5	Plug – driver's door wire to rear section	49	Plug – center section to LE-Jetronic wire harness
6	Plug – central lock connecting wire to driver's door wire (13-pin)	50	Ignition switch
7	Plug – central lock connecting wire to passenger's door wire	51	Remote control switch for on-board computer
8	Plug – driver's door central lock wire to switch	52	Plug – on-board computer to outside temperature sensor wire
9	Central lock switch/unblocking arrest (driver's door, on lock)	53	Plug – outside temperature sensor wire to outside temperature sensor
10	Connection for central lock motor to driver's door (6-pin)	54	Outside temperature sensor (lower front panel)
11	Central lock motor – driver's door	55	Plug – extra heater wire to power antenna
12	Plug – passenger's door wire to microswitch	56	Parked car heating electronic control unit (on parked car heater underneath right seat)
13	Microswitch (passenger's door, on lock)	57	Connection for electronic control unit
14	Central lock motor – passenger's door	58	Relay for parked car heater (on heater)
15	Central lock motor – passenger's door	59	Plug – on-board computer wire to extra heater wire
16	Connection for central lock motor to trunk lid (6-pin)	60	Plug – center section to instrument cluster
17	Central lock motor – boot lid	61	Connection for instrument cluster
18	Connection for central lock motor to fuel filler door (6-pin)	62	Electronic Control Unit (right of instrument cluster)
19	Central lock motor – fuel filler door	63	Connection for Electronic Control Unit
20	Connection for central lock motor to left rear door (6-pin)	64	Connection for instrument cluster II
21	Central lock motor – left rear door	65	Instrument cluster
22	Connection for central lock motor to right door (6-pin)	66	Plug – rear section to instrument cluster
23	Plug – central lock connecting wire to right rear door (7-pin)	67	Plug – digital clock wire to instrument cluster
24	Plug – central lock connecting wire to left rear door (7-pin)	68	Plug – extra wire to heater wire
25	Central lock motor – right rear door	69	Plug – heater wire to fuel pump wire
26	Rear window heater switch	70	Connection for heater
27	Burglar alarm electronic control unit (left of steering column)	71	Ballast resistor in heater
28	Connection for burglar alarm electronic control unit I (26-pin)	72	Thermoswitch (parked car heater)
29	Connection for relay box (4-pin)	73	Heater motor
30	Connection for burglar alarm electronic control unit II (4-pin)	74	Overheating switch (parked car heater)
31	Plug 150 (in main wire harness)	75	Heater plug for parked car heater
32	Light diode for burglar alarm	76	Heater
33	Plug for boot light	77	Fuel pump
34	Boot light	78	Plug – on-board computer to remote control
35	Door contact switch front left	79	Plug – speed dependent loudness control
36	Door contact switch front right	80	Plug – wire for cruise control
37	Door contact switch rear left	81	Fuel level transmitter
38	Door contact switch rear right	82	Speed transmitter
39	Trunk lid contact	83	Plug – digital clock wire to digital clock (4-pin)
40	Hood contact	84	Plug – digital clock wire to digital clock (2-pin)
41	Rear window heater	85	Digital clock
42	Plug – center section to wire for on-board computer/burglar alarm		
43	Horn		
44	Plug for light diode		

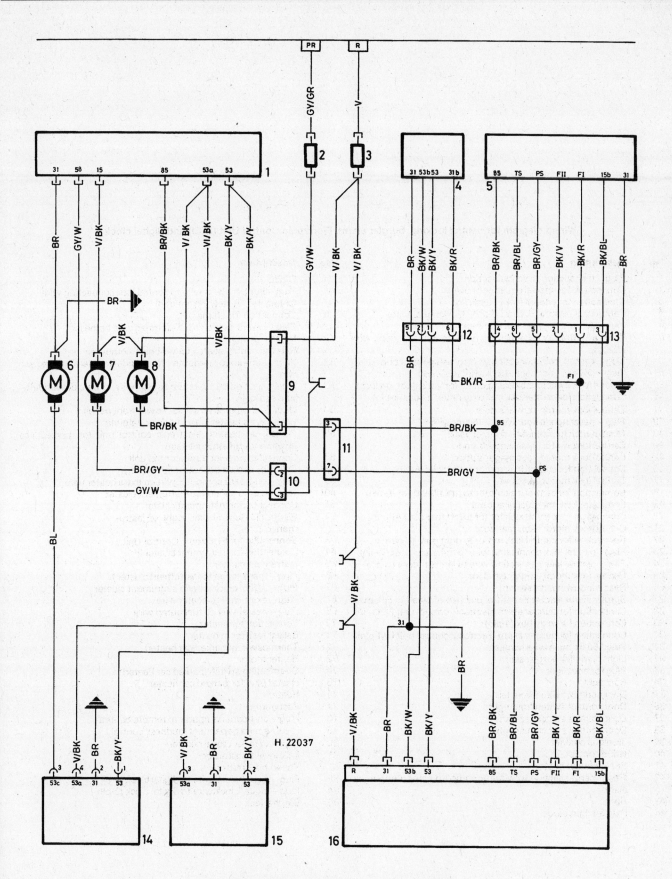

Typical headlight washer system wiring diagram

Key to headlight washer system wiring diagram

No	Description
1	Control unit for headlight cleaners (on fluid reservoir)
2	Fuse – overnight, tail and parking lights
3	Fuse – horns, wash/wipe control unit and headlight cleaners
4	Motor – windshield wipers
5	Wiper switch
6	Pump – headlight cleaning system
7	Pump – intensive cleaning fluid
8	Pump – windshield washing system
9	Plug – headlight cleaner wire to front section I (washer fluid pump)
10	Plug – Headlight cleaner wire to front section II (plug for headlight cleaners)
11	Plug – center section to front section (7-pin)
12	Plug for wiper motor
13	Plug – center section to wiper switch
14	Motor – windshield wipers
15	Motor – left headlight wiper
16	Wash/wipe interval control unit

Key to power window system wiring diagram

No	Description
1	Plug for rear section to driver's door (6-pin)
2	Plug for rear section to center section (27-pin)
3	Plug for window control and central lock wire to driver's door (13-pin)
4	Plug for window control and central lock wire to special equipment plug
5	Window switch rear left
6	Window switch rear left
7	Window switch rear right
8	Plug for left rear door wire to window motor rear left
9	Plug for right door wire to window motor rear right
10	Window motor rear left
11	Window motor rear right
12	Plug for window control and central lock wire to left rear door
13	Plug for window control and central lock wire to right rear door (7-pin)
14	Power safety switch
15	Child safety switch
16	Window motor front left
17	Window motor front right
18	Plug for driver's door wire to window motor front left
19	Plug for passenger's door wire to window motor of passenger's door
20	Relay
21	Plug for window control and central lock wire to passenger's door (13-pin)
22	Window switch front left
23	Window switch rear right
24	Window switch front right

Chapter 12 Chassis electrical system

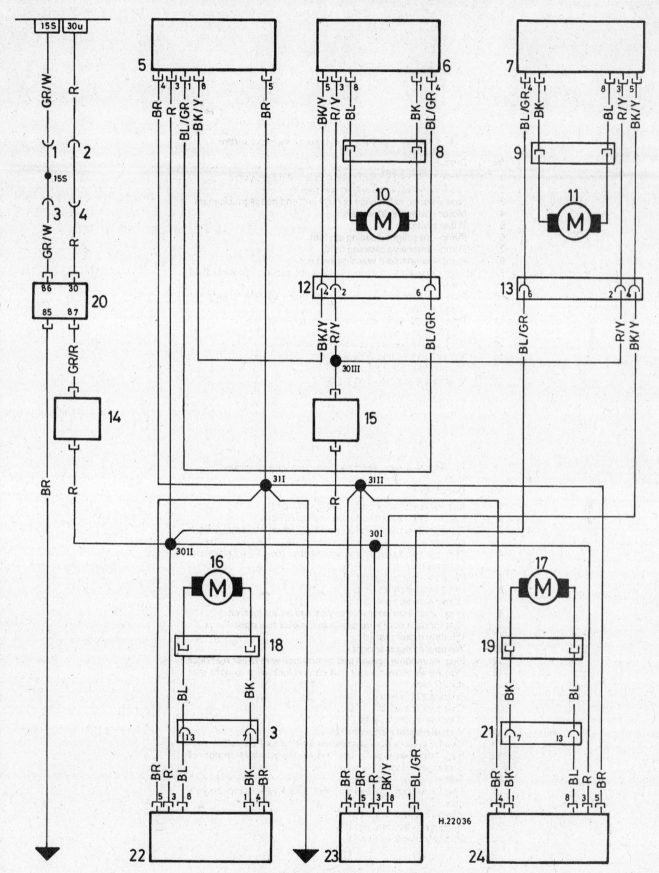

Typical power window system wiring diagram

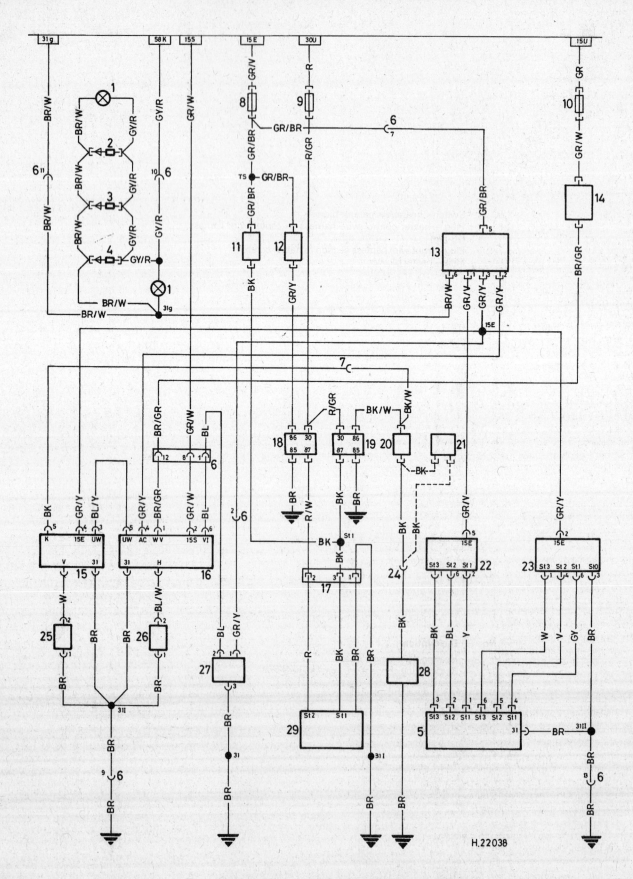

Typical heater and air conditioning system wiring diagram

Chapter 12 Chassis electrical system

Key to air conditioning system wiring diagram

No	Description
1	Light for heater controls
2	Light diode III
3	Light diode II
4	Light diode I
5	Switch – heater/evaporator blower
6	Plug – heater control wire harness to center wire harness (13-pin)
7	Plug – front wire harness section to heater controls
8	Fuse – heater blower
9	Fuse – extra fan stage II
10	Fuse – ind. lamp, reversing lights, tachometer and mirrors (power distributor)
11	Temperature switch 91°C – stage I
12	Temperature switch 99°C – stage II
13	Switch – air conditioner
14	Water valve
15	Evaporator temperature regulator
16	Air conditioner control unit (heater controls)
17	Plug – extra fan motor (on extra fan motor)
18	Relay – extra fan stage II (on power distributor)
19	Relay – extra fan stage I (on power distributor)
20	Switch – high pressure pressostat (drier)
21	Switch – temperature 110°C (only for 524 td)
22	Motor – heater blower
23	Motor – evaporator blower
24	Plug – high pressure pressostat to electromagnetic coupling
25	Evaporator temperature sensor (in evaporator)
26	Heater temperature sensor (in heater)
27	Inside temperature sensor (lower trim panel left)
28	Electromagnetic coupling for compressor
29	Motor – extra fan

Key to wiring diagram for heated seats

No	Description
1	Heating – passenger's seat
2	Seat heating connection – passenger's side
3	Seat heating switch – passenger's side
4	Plug for heated seat wire (driver's side) to special equipment plug (58K)
5	Plug for heated seat wire (driver's side) to passenger's side
6	Plug for heated seat wire (driver's side) to special equipment plug (15E and 30SA4)
7	Seat heating relay
8	Seat heating switch – driver's side
9	Heating – driver's seat
10	Seat heating connection – driver's side

12-24 Chapter 12 Chassis electrical system

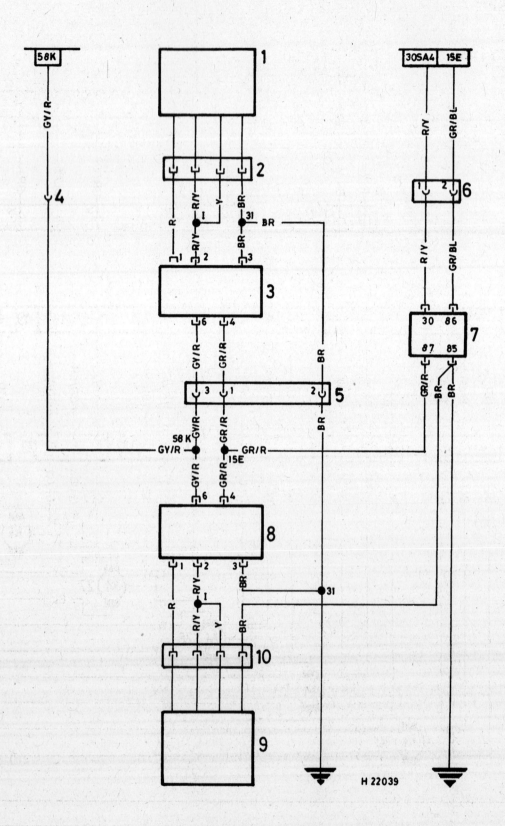

Typical heated seats wiring diagram

Chapter 12 Chassis electrical system

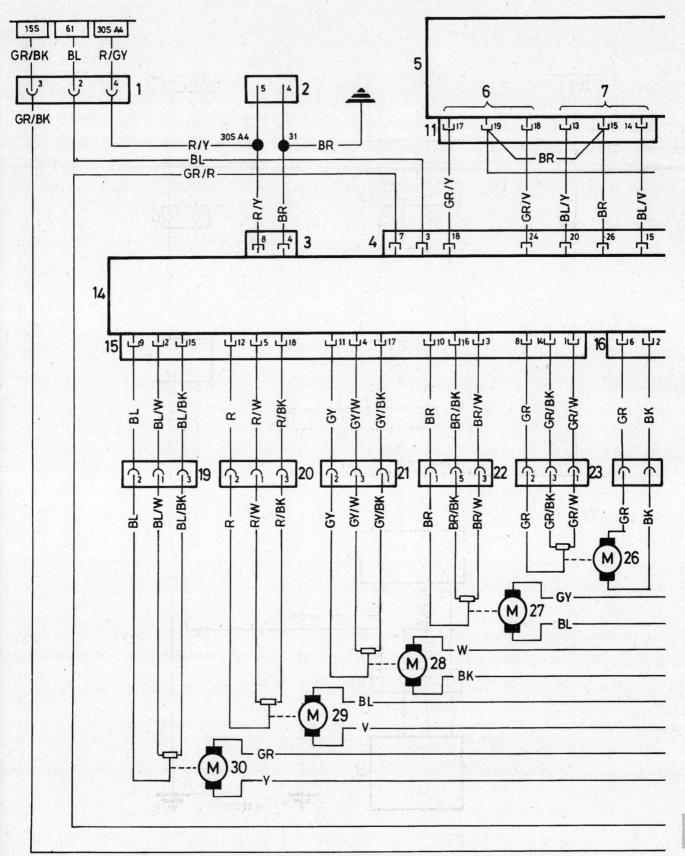

Typical wiring diagram for power seats with memory (1 of 2)

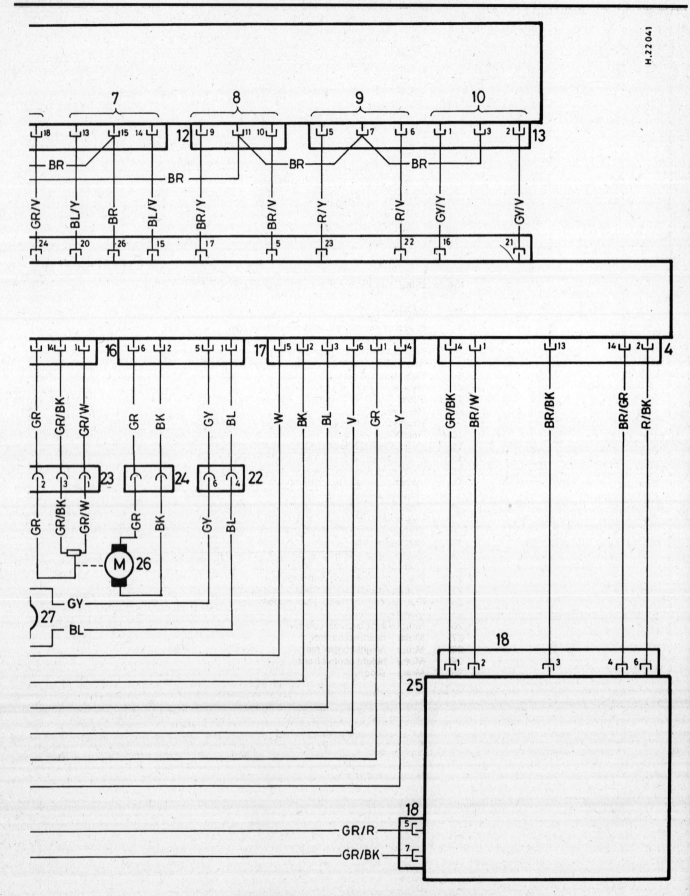

Typical wiring diagram for power seats with memory (2 of 2)

Chapter 12 Chassis electrical system

Key to wiring diagram for memory power seats

No	Description
1	Plug connection for special equipment plug
2	Plug connection of wire for passenger's seat
3	Plug connection of wire for seat control with memory
4	Connection of wire for seat control with memory
5	Seat control switch
6	Backrest
7	Slide
8	Headrest
9	Height front
10	Height rear
11	Plug connection for seat backrest/slide control
12	Plug connection for seat headrest control
13	Plug connection for seat height control
14	Electronic control unit (underneath seat)
15	Plug connection of wire for seat control with memory
16	Plug connection of wire for seat with memory
17	Plug connection for seat control drive
18	Plug connection for memory switch
19	Plug connection for slide potentiometer
20	Plug connection for front height potentiometer
21	Plug connection for rear height potentiometer
22	Plug connection for headrest motor
23	Plug connection for backrest potentiometer
24	Plug connection for backrest motor
25	Memory switch
26	Motor – seat backrest control
27	Motor – headrest control
28	Motor – height control rear
29	Motor – height control front
30	Motor – slide

12-28 Chapter 12 Chassis electrical system

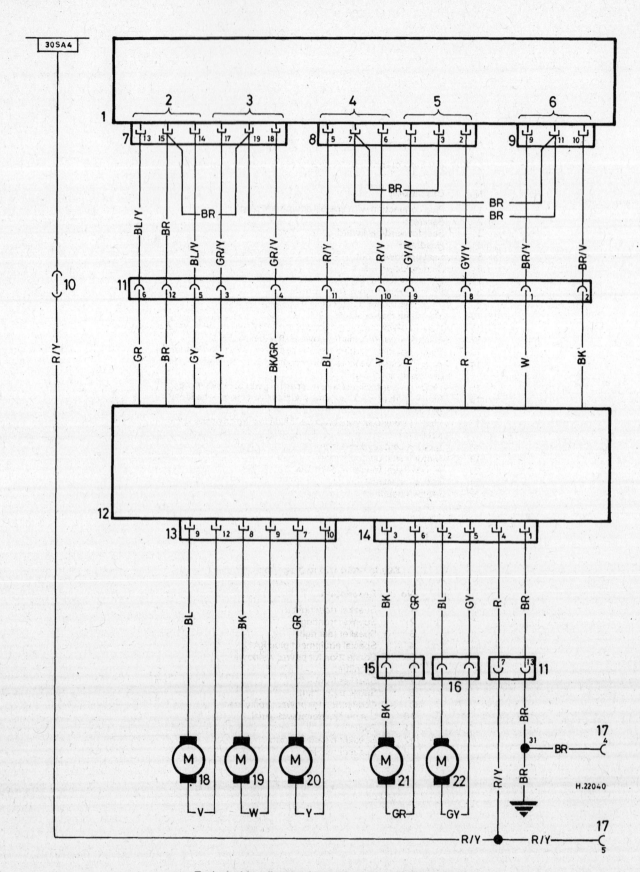

Typical wiring diagram for power seats without memory

Key to wiring diagram for power seats without memory

No	Description
1	Plug connection with special equipment plug
2	Backrest
3	Seat forward/backward
4	Headrest
5	Seat up/down front
6	Seat up/down rear
7	Plug – switch for backrest/seat control
8	Plug – switch for headrest control
9	Plug – switch for front/rear seat up/down control
10	Switch for power seats
11	Plug – power seat wire to power seat electronic control unit
12	Electronic control unit for power seats (below seats)
13	Plug – power seat drive to power seat electronic control unit
14	Plug – power backrest and headrest wire to power seat electronic control unit
15	Plug – power backrest and headrest wire to backrest motor
16	Plug – power backrest and headrest wire to the headrest motor
17	Plug – power seat wire on driver's side to wire on passenger's side
18	Motor – seat up/down front
19	Motor – seat up/down rear
20	Motor – seat forward/backward
21	Motor – backrest
22	Motor – headrest

Key to radio wiring diagram

No	Description
1	Speaker door right
2	Speaker front right
3	Speaker rear right
4	Special equipment plug RA12
5	Connection for power windows
6	Amplifier
7	Speaker front left
8	Speaker door left
9	Connection for power supply lead
10	Connection for power aerial
11	Radio
12	Speaker balance control
13	Speaker rear left

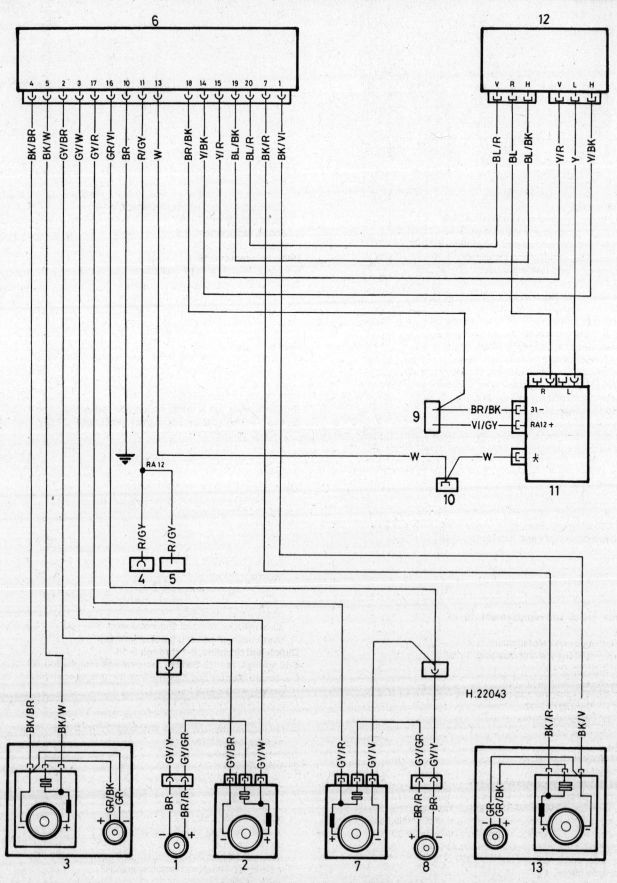

Typical radio wiring diagram - early model shown

Index

A

About this manual, 0-5
Accelerator cable, check, adjustment and replacement, 4-11
Air conditioning
    compressor, removal and installation, 3-9
    condenser, removal and installation, 3-10
    evaporator core removal and installation, 3-10
    receiver-drier, removal and installation, 3-10
Air conditioning and heating system, check, 3-8
Air filter housing assembly, removal and installation, 4-9
Air filter replacement, 1-21
Airbag (Supplemental Restraint System [SRS]), general information, 12-9
Alternator brushes, check and replacement, 5-13
Alternator, removal and installation, 5-12
Antenna, removal and installation, 12-4
Anti-lock Brake system (ABS), general information, 9-3
Anti-theft audio system, general information, 0-7
Antifreeze, general information, 3-2
Automatic transmission, 7B-1 through 7B-6
    diagnosis, general, 7B-2
    fluid and filter change, 1-26
    fluid level, check, 1-12
    general information, 7B-2
    neutral start switch, check and replacement, 7B-4
    oil seals, replacement, 7B-5
    removal and installation, 7B-5
    shift lever linkage, adjustment, 7B-3
    throttle Valve (TV) cable and kickdown, adjustment, 7B-3
Automotive chemicals and lubricants, 0-16

B

Balljoints, check and replacement, 10-10
Battery
    cables, check and replacement, 5-2
    check, maintenance and charging, 1-15
    jump starting, 0-15
    removal and installation, 5-2
Blower motor, removal and installation, 3-7
Body, 11-1 through 11-12
    general information, 11-1
    maintenance, 11-1
    repair
        major damage, 11-3
        minor damage, 11-2
Booster battery (jump) starting, 0-15
Brakes, 9-1 through 9-18
    Anti-lock Brake system (ABS), general information, 9-3
    booster (hydraulic), description, removal and installation, 9-14
    booster (vacuum), check, removal and installation, 9-13
    caliper, removal, overhaul and installation, 9-5
    disc, inspection, removal and installation, 9-7
    fluid, level check, 1-7
    general information, 9-2
    hoses and lines, inspection and replacement, 9-17
    hydraulic system, bleeding, 9-18
    light switch, check, adjustment and replacement, 9-16
    master cylinder, removal and installation, 9-12
    pads, replacement, 9-3
    parking brake assembly, check, removal and installation, 9-15
    parking brake cable(s), replacement, 9-14
    pedal adjustment, 9-16
    shoes, replacement, 9-9
    system, check, 1-24
Bulb replacement, 12-7
Bumpers, removal and installation, 11-8
Buying parts, 0-6

C

Cable
    accelerator, check, adjustment and replacement, 4-11
    parking brake, adjustment, 9-15
    parking brake, replacement, 9-14
Catalytic converter, 6-9
Center bearing (driveshaft), check and replacement, 8-8
Central locking system, description and check, 12-9
Charging system
    alternator brushes, check and replacement, 5-13
    alternator, removal and installation, 5-12
    check, 5-12
    general information and precautions, 5-11
    voltage regulator, replacement, 5-13
Chassis electrical system, 12-1 through 12-10
Chemicals, 0-16
Clutch
    components, removal, inspection and installation, 8-4
    description and check, 8-3
    general information, 8-2
    hydraulic system, bleeding, 8-4
    master cylinder, removal and installation, 8-3
    pedal, adjustment, 8-3
    slave cylinder, removal and installation, 8-4
    start switch, check and replacement, 8-3
Clutch and driveline, 8-1 through 8-16
Coil springs, rear (3-Series), removal and installation, 10-12
Cold start injector and thermo-time switch, check and replacement, 4-19
Compression check, 2B-4
Condenser, removal and installation, 3-10
Constant velocity (CV) joints and boots, check and replacement, 8-11
Conversion factors, 0-18
Cooling system
    antifreeze, general information, 3-2
    check, 1-22
    coolant level check, 1-7
    general information, 3-2
    radiator, removal and installation, 3-3
    servicing, 1-27
    temperature sending unit, check and replacement, 3-6
    thermostat, check and replacement, 3-2
    water pump
        check, 3-5
        removal and installation, 3-6

Index

Cooling, heating and air conditioning systems, 3-1 through 3-10
Crankshaft
   inspection, 2B-18
   installation and main bearing oil clearance check, 2B-20
   removal, 2B-13
Cruise control system, description and check, 12-9
Cylinder head
   cleaning and inspection, 2B-9
   disassembly, 2B-7
   reassembly, 2B-11
   removal and installation, 2A-11
Cylinder honing, 2B-15

D

Diagnosis, 0-19
Differential
   housing side seals, replacement, 8-13
   housing, removal and installation, 8-14
   lubricant level check, 1-20
   lubricant, change, 1-28
Disc brake
   caliper, removal, overhaul and installation, 9-5
   inspection, removal and installation, 9-7
   pad replacement, 9-3
Distributor
   cap, rotor check and replacement, 1-19
   removal and installation, 5-5
   rotor check and replacement, 1-19
Door
   Central locking system, description and check, 12-9
   latch, lock cylinder and handles, removal, installation and adjustment, 11-10
   removal, installation and adjustment, 11-9
   trim panel, removal and installation, 11-8
   window glass, removal and installation, 11-10
   window regulator, removal and installation, 11-11
Driveaxle boot check, 1-24
Driveaxles, removal and installation, 8-10
Drivebelt check, adjustment and replacement, 1-14
Driveline, general information, 8-2
Driveplate, removal and installation, 2A-14
Driveshaft
   center bearing, check and replacement, 8-8
   constant velocity (CV) joint, replacement, 8-10
   flexible coupling, replacement, 8-8
   front centering guide, check and replacement, 8-9
   removal and installation, 8-6
   universal joints, check and replacement, 8-9
Drum brake shoes (318i models), replacement, 9-9

E

EFI system self-diagnosis capability (Motronic 1.1 systems only), 6-2
Electrical system (chassis), general information, 12-1
Electrical troubleshooting, general information, 12-1
Electronic control unit (ECU), general information and replacement, 6-3
Emergency battery (jump) starting, 0-15
Emissions and engine control systems, 6-1 through 6-10
Engine
   block
      cleaning, 2B-14
      inspection, 2B-14
   crankshaft
      inspection, 2B-18
      installation and main bearing oil clearance check, 2B-20
      removal, 2B-13
   cylinder head
      cleaning and inspection, 2B-9
      disassembly, 2B-7
      reassembly, 2B-11
      removal and installation, 2A-11
   cylinder honing, 2B-15
   general information, 2A-2
   initial start-up and break-in after overhaul, 2B-23
   intermediate shaft
      installation, 2B-20
      removal and inspection, 2B-14
   main and connecting rod bearings, inspection, 2B-18
   mounts, check and replacement, 2A-15
   oil change, 1-10
   oil pan, removal and installation, 2A-12
   oil pump, removal, inspection and installation, 2A-13
   oil seals, front, replacement, 2A-10
   overhaul
      disassembly sequence, 2B-6
      general information, 2B-3
      reassembly sequence, 2B-19
   piston rings, installation, 2B-19
   pistons/connecting rods
      inspection, 2B-16
      installation and rod bearing oil clearance check, 2B-22
      removal, 2B-12
   rear main oil seal installation, 2B-21
   rebuilding alternatives, 2B-6
   removal and installation, 2B-5
   removal, methods and precautions, 2B-5
   repair operations possible with the engine in the vehicle, 2A-3
   timing belt and sprockets, removal, inspection and installation, 2A-8
   timing belt covers, removal and installation, 2A-7
   timing chain and sprockets, removal, inspection and installation, 2A-6
Engine cooling fan(s) and clutch, check and replacement, 3-4
Engine electrical systems, 5-1 through 5-16
Engine oil, level check, 1-7
Engines, 2A-1 through 2A-16
Evaporative emissions control (EVAP) system, 1-29, 6-7
Evaporator core, removal and installation, 3-10
Exhaust
   general information, 4-2
   manifold, removal and installation, 2A-5
   system
      check, 1-23
      servicing, general information, 4-22
Exterior mirror, removal and installation, 11-11

F

Fan(s) and clutch, check and replacement, 3-4
Fault finding, 0-19
Fixed glass, replacement, 11-3
Flexible coupling (driveshaft), replacement, 8-8
Fluid
   level checks, 1-7
Flywheel/driveplate, removal and installation, 2A-14
Front control arm (3-Series), inspection, bushing replacement and control arm removal and installation, 10-5
Front control arm and thrust arm (5-Series), inspection, removal and installation and bushing replacement, 10-6
Front hub and wheel bearing assembly, removal and installation, 10-10
Front stabilizer bar, removal and installation, 10-5
Front strut assembly, removal and installation, 10-8
Fuel and exhaust systems, 4-1 through 4-24
   filter replacement, 1-27
   injection system

Index

airflow meter, check, removal and installation, 4-16
check, 4-15
cold start injector and thermo-time switch, check and replacement, 4-19
general information, 4-12
Idle air stabilizer valve, check, adjustment and replacement, 4-21
injectors, check and replacement, 4-20
pressure regulator, check and replacement, 4-18
throttle body, check, removal and installation, 4-17
troubleshooting, 4-13
level sending unit, check, removal and installation, 4-6
lines and fittings, repair and replacement, 4-9
pressure relief procedure, 4-2
pump, check, removal and installation, 4-6
pump/fuel pressure, check, 4-3
system, check, 1-22
system, general information, 4-2
tank cleaning and repair, general information, 4-9
tank, removal and installation, 4-9
Fuses, general information, 12-2

G

General engine overhaul procedures, 2A-1 through 2A-24
Glass (door), removal and installation, 11-10
Glass (fixed), replacement, 11-3
Grille, removal and installation, 11-3

H

Hazard warning flasher, check and replacement, 12-3
Headlight
adjustment, 12-6
housing, removal and installation, 12-7
replacement, 12-6
Heater and air conditioning system
blower motor, removal and installation, 3-7
control assembly, removal and installation, 3-8
check, 3-8
Heater core, removal and installation, 3-8
Hinges and locks, maintenance, 11-3
Hood, removal, installation and adjustment, 11-6
Hub and wheel bearing assembly (front), removal and installation, 10-10
Hydraulic clutch system, bleeding, 8-4

I

Idle air stabilizer valve, check, adjustment and replacement, 4-21
Ignition
air gap (318i models), check and adjustment, 5-9
coil, check and replacement, 5-5
distributor, removal and installation, 5-5
impulse generator and ignition control unit, check and replacement, 5-8
speed sensors, check and replacement, 5-10
switch, removal and installation, 12-4
system, check, 5-3
system, general information and precautions, 5-3
timing, adjustment (318i models only), 5-4
Information sensors, 6-3
Instrument cluster, removal and installation, 12-4
Instrument panel language display, 0-7
Intake manifold, removal and installation, 2A-4

Intermediate shaft
installation, 2B-20
removal and inspection, 2B-14
Introduction to the BMW 3 and 5 Series, 0-5

J

Jacking, 0-14
Jump Starting, 0-15

L

Lubricant
differential
change, 1-28
level checking, 1-20
manual transmission
change, 1-28
level checking, 1-20
Lubricants, 0-16

M

Main and connecting rod bearings, inspection, 2B-18
Maintenance
introduction, 1-2
techniques, 0-7
Maintenance schedule, 1-6
Manifold, removal and installation
exhaust, 2A-5
intake, 2A-4
Manual transmission, 7A-1 through 7A-4
general information, 7A-2
lubricant change, 1-28
lubricant, check, 1-20
oil seals, replacement, 7A-2
overhaul, general information, 7A-4
removal and installation, 7A-4
shift lever, removal and installation, 7A-2
Master cylinder, removal and installation, 9-12
Mirror (outside), removal and installation, 11-11
Mounts, engine, check and replacement, 2A-15

N

Neutral start switch, check and replacement, 7B-4

O

Oil
engine, change, 1-10
level checks, 1-7
pan, removal and installation, 2A-12
pump and driveshaft, removal, inspection and installation, 2A-13
seal, manual transmission, replacement, 7A-2
seal, pinion, replacement, 8-15
seal, rear main, installation, 2B-21
seal, rear main, replacement, 2A-14
seals transmission, replacement, 7B-5
seals, front, replacement, 2A-10
Oxygen sensor replacement, 1-29

P

Pads (brake), replacement, 9-3
Pan, oil, removal and installation, 2A-12

Index

Parking brake
  assembly check, removal and installation, 9-15
  cable replacement, 9-14
  cable, adjustment, 9-15
Pinion oil seal, replacement, 8-15
Piston rings, installation, 2B-19
Pistons/connecting rods
  installation and rod bearing oil clearance check, 2B-22
  removal, 2B-12
Positive crankcase ventilation (PCV) system, 6-7
Power steering
  fluid level check, 1-12
  pump, removal and installation, 10-21
  system, bleeding, 10-21
Power window system, description and check, 12-9
Pressure (tire), checking, 1-9
Pump, power steering, removal and installation, 10-21

R

Radiator grille, removal and installation, 11-3
Radiator, removal and installation, 3-3
Radio
  antenna, removal and installation, 12-4
  removal and installation, 12-4
Rear coil springs (3-Series), removal and installation, 10-12
Rear main oil seal, installation, 2B-21
Rear main oil seal, replacement, 2A-14
Rear shock absorber/coil spring assemblies (5-Series), removal and installation, 10-12
Rear shock absorbers (3-Series), removal and installation, 10-11
Rear stabilizer bar, removal and installation, 10-12
Rear trailing arms, removal and installation
  3-Series, 10-13
  5-Series, 10-13
Rear wheel bearings, replacement, 10-14
Rear window defogger, check and repair, 12-8
Relays, general information, 12-2
Rocker arms, cleaning and inspection, 2B-9
Rotor (brake), inspection, removal and installation, 9-7
Routine maintenance schedule, 1-6

S

Safety, 0-17
Seat belt check, 11-12
Seats, removal and installation, 11-12
Service Indicator (SI) board, general information, 12-5
Service light resetting, 1-29
Shock absorber/coil spring assemblies, rear (5-Series) removal and installation, 10-12
Shock absorbers, rear (3-Series), removal and installation, 10-11
Spark plug
  check and replacement, 1-16
  wires, check and replacement, 1-19
Stabilizer bar, removal and installation
  front, 10-5
  rear, 10-12
Starter
  motor, in-vehicle check, 5-15
  motor, removal and installation, 5-15
  solenoid, removal and installation, 5-16
Starting system, general information and precautions, 5-15
Steering
  boots (3-Series), replacement, 10-17
  column cover, removal and installation, 11-11
  column switches, removal and installation, 12-3
  gearbox (5-Series), removal and installation, 10-19
  linkage (5-Series), removal and installation, 10-18
  rack-and-pinion (3-Series), removal and installation, 10-18

  system check, 1-23
  system, general information, 10-16
  tie-rod ends, removal and installation, 10-17
  wheel, removal and installation, 10-21
Stereo, removal and installation, 12-4
Strut assembly, removal and installation, 10-8
Strut or shock absorber/coil spring, replacement, 10-9
Supplemental Restraint System (SRS), general information, 12-9
Suspension and steering system, general information, 10-2
Suspension and steering systems, 10-1 through 10-22
Suspension check, 1-23
Switch
  ignition, removal and installation, 12-4
  steering column, removal and installation, 12-3

T

Thermo-time switch, check and replacement, 4-19
Thermostat, check and replacement, 3-2
Throttle body, check, removal and installation, 4-17
Throttle linkage, check and lubrication, 1-21
Throttle Valve (TV) cable and kickdown, adjustment, 7B-3
Tie-rod ends, removal and installation, 10-17
Timing belt and sprockets, removal, inspection and installation, 2A-8
Timing belt covers, removal and installation, 2A-7
Timing chain and sprockets, removal, inspection and installation, 2A-6
Timing chain covers, removal and installation, 2A-5
Timing, ignition, adjustment (318i models only), 5-4
Tire
  checking, 1-9
  rotation, 1-13
Tools, 0-10
Top Dead Center (TDC) for number one piston, locating, 2A-3
Towing, 0-14
Trailing arm, removal and installation
  3-Series, 10-13
  5-Series, 10-13
Transmission mounts, check and replacement, 7A-4
Transmission, automatic
  diagnosis, general, 7B-2
  fluid and filter change, 1-26
  fluid level check, 1-12
  general information, 7B-2
  lubricant change, 1-28
  neutral start switch, check and replacement, 7B-4
  oil seals, replacement, 7B-5
  removal and installation, 7B-5
  shift lever linkage adjustment, 7B-3
  Throttle Valve (TV) cable and kickdown adjustment, 7B-3
Transmission, manual
  general information, 7A-2
  oil seals, replacement, 7A-2
  overhaul, general information, 7A-4
  removal and installation, 7A-4
  shift lever, removal and installation, 7A-2
Trouble codes, 6-2
Troubleshooting, 0-19
Trunk lid, removal, installation and adjustment, 11-9
Tune-up general information, 1-7
Tune-up and routine maintenance, 1-1 through 1-30
Turn signal/hazard warning flasher, check and replacement, 12-3

U

Underhood hose check and replacement, 1-13
Universal joints, check and replacement, 8-9
Upholstery and carpets, maintenance, 11-1

Index

V
Valve clearance check and adjustment, 1-20
Valve cover, removal and installation, 2A-3
Valves, servicing, 2B-11
Vehicle identification numbers, 0-6
Vinyl trim, maintenance, 11-1
Voltage regulator, replacement, 5-13

W
Water pump
  check, 3-5
  removal and installation, 3-6

Wheel alignment, general information, 10-22
Wheel bearing and hub assembly (front), removal and installation, 10-10
Wheel bearings (rear), replacement, 10-14
Wheels and tires, general information, 10-21
Window regulator, removal and installation, 11-11
Windshield
  motor, replacement, 12-7
  washer fluid level check, 1-7
  wiper blade check and replacement, 1-25
Wiring diagrams, general information, 12-10
Working facilities, 0-13

HAYNES AUTOMOTIVE MANUALS

NOTE: New manuals are added to this list on a periodic basis. If you do not see a listing for your vehicle, consult your local Haynes dealer for the latest product information.

ACURA
- *1776 **Integra & Legend** all models '86 thru '90

AMC
- **Jeep CJ** - see JEEP (412)
- 694 **Mid-size models,** Concord, Hornet, Gremlin & Spirit '70 thru '83
- 934 **(Renault) Alliance & Encore** all models '83 thru '87

AUDI
- 615 **4000** all models '80 thru '87
- 428 **5000** all models '77 thru '83
- 1117 **5000** all models '84 thru '88

AUSTIN
- **Healey Sprite** - see MG Midget Roadster (265)

BMW
- *2020 **3/5 Series** not including diesel or all-wheel drive models '82 thru '92
- 276 **320i** all 4 cyl models '75 thru '83
- 632 **528i & 530i** all models '75 thru '80
- 240 **1500 thru 2002** all models except Turbo '59 thru '77
- 348 **2500, 2800, 3.0 & Bavaria** all models '69 thru '76

BUICK
- **Century (front wheel drive)** - see GENERAL MOTORS (829)
- *1627 **Buick, Oldsmobile & Pontiac Full-size (Front wheel drive)** all models '85 thru '93
  Buick Electra, LeSabre and Park Avenue; **Oldsmobile** Delta 88 Royale, Ninety Eight and Regency; **Pontiac** Bonneville
- 1551 **Buick Oldsmobile & Pontiac Full-size (Rear wheel drive)**
  Buick Estate '70 thru '90, Electra'70 thru '84, LeSabre '70 thru '85, Limited '74 thru '79
  Oldsmobile Custom Cruiser '70 thru '90, Delta 88 '70 thru '85,Ninety-eight '70 thru '84
  Pontiac Bonneville '70 thru '81, Catalina '70 thru '81, Grandville '70 thru '75, Parisienne '83 thru '86
- 627 **Mid-size Regal & Century** all rear-drive models with V6, V8 and Turbo '74 thru '87
- **Regal** - see GENERAL MOTORS (1671)
- **Skyhawk** - see GENERAL MOTORS (766)
- 552 **Skylark** all X-car models '80 thru '85
- **Skylark '86 on** - see GENERAL MOTORS (1420)
- **Somerset** - see GENERAL MOTORS (1420)

CADILLAC
- *751 **Cadillac Rear Wheel Drive** all gasoline models '70 thru '92
- **Cimarron** - see GENERAL MOTORS (766)

CAPRI
- 296 **2000 MK I Coupe** all models '71 thru '75
- **Mercury Capri** - see FORD Mustang (654)

CHEVROLET
- *1477 **Astro & GMC Safari Mini-vans** '85 thru '93
- 554 **Camaro V8** all models '70 thru '81
- 866 **Camaro** all models '82 thru '92
- **Cavalier** - see GENERAL MOTORS (766)
- **Celebrity** - see GENERAL MOTORS (829)
- 625 **Chevelle, Malibu & El Camino** all V6 & V8 models '69 thru '87
- 449 **Chevette & Pontiac T1000** '76 thru '87
- 550 **Citation** all models '80 thru '85

- *1628 **Corsica/Beretta** all models '87 thru '92
- 274 **Corvette** all V8 models '68 thru '82
- *1336 **Corvette** all models '84 thru '91
- 1762 **Chevrolet Engine Overhaul Manual**
- 704 **Full-size Sedans** Caprice, Impala, Biscayne, Bel Air & Wagons '69 thru '90
- **Lumina** - see GENERAL MOTORS (1671)
- **Lumina APV** - see GENERAL MOTORS (2035)
- 319 **Luv Pick-up** all 2WD & 4WD '72 thru '82
- 626 **Monte Carlo** all models '70 thru '88
- 241 **Nova** all V8 models '69 thru '79
- *1642 **Nova and Geo Prizm** all front wheel drive models, '85 thru '92
- 420 **Pick-ups '67 thru '87** - Chevrolet & GMC, all V8 and in-line 6 cyl, 2WD & 4WD '67 thru '87; Suburbans, Blazers & Jimmys '67 thru '91
- *1664 **Pick-ups '88 thru '93** - Chevrolet & GMC, all full-size (C and K) models, '88 thru '93
- *831 **S-10 & GMC S-15 Pick-ups** all models '82 thru '92
- *1727 **Sprint & Geo Metro** '85 thru '91
- *345 **Vans** - Chevrolet & GMC, V8 & in-line 6 cylinder models '68 thru '92

CHRYSLER
- *2058 **Full-size Front-Wheel Drive** '88 thru '93
- **K-Cars** - see DODGE Aries (723)
- **Laser** - see DODGE Daytona (1140)
- *1337 **Chrysler & Plymouth Mid-size** front wheel drive '82 thru '93

DATSUN
- 402 **200SX** all models '77 thru '79
- 647 **200SX** all models '80 thru '83
- 228 **B - 210** all models '73 thru '78
- 525 **210** all models '78 thru '82
- 206 **240Z, 260Z & 280Z** Coupe '70 thru '78
- 563 **280ZX** Coupe & 2+2 '79 thru '83
- **300ZX** - see NISSAN (1137)
- 679 **310** all models '78 thru '82
- 123 **510 & PL521 Pick-up** '68 thru '73
- 430 **510** all models '78 thru '81
- 372 **610** all models '72 thru '76
- 277 **620 Series Pick-up** all models '73 thru '79
- **720 Series Pick-up** - see NISSAN (771)
- 376 **810/Maxima** all gasoline models, '77 thru '84
- 368 **F10** all models '76 thru '79
- **Pulsar** - see NISSAN (876)
- **Sentra** - see NISSAN (982)
- **Stanza** - see NISSAN (981)

DODGE
- **400 & 600** - see CHRYSLER Mid-size (1337)
- *723 **Aries & Plymouth Reliant** '81 thru '89
- *1231 **Caravan & Plymouth Voyager Mini-Vans** all models '84 thru '93
- 699 **Challenger & Plymouth Saporro** all models '78 thru '83
- **Challenger '67–'76** - see DODGE Dart (234)
- 236 **Colt** all models '71 thru '77
- 610 **Colt & Plymouth Champ (front wheel drive)** all models '78 thru '87
- *1668 **Dakota Pick-ups** all models '87 thru '93
- 234 **Dart, Challenger/Plymouth Barracuda & Valiant** 6 cyl models '67 thru '76
- *1140 **Daytona & Chrysler Laser** '84 thru '89
- *545 **Omni & Plymouth Horizon** '78 thru '90
- *912 **Pick-ups** all full-size models '74 thru '91
- *556 **Ram 50/D50 Pick-ups & Raider and Plymouth Arrow Pick-ups** '79 thru '93
- *1726 **Shadow & Plymouth Sundance** '87 thru '93
- *1779 **Spirit & Plymouth Acclaim** '89 thru '92
- *349 **Vans - Dodge & Plymouth** V8 & 6 cyl models '71 thru '91

EAGLE
- **Talon** - see Mitsubishi Eclipse (2097)

FIAT
- 094 **124 Sport Coupe & Spider** '68 thru '78
- 273 **X1/9** all models '74 thru '80

FORD
- *1476 **Aerostar Mini-vans** all models '86 thru '92
- 788 **Bronco and Pick-ups** '73 thru '79
- *880 **Bronco and Pick-ups** '80 thru '91
- 268 **Courier Pick-up** all models '72 thru '82
- 1763 **Ford Engine Overhaul Manual**
- 789 **Escort/Mercury Lynx** all models '81 thru '90
- *2046 **Escort/Mercury Tracer** '91 thru '93
- *2021 **Explorer & Mazda Navajo** '91 thru '92
- 560 **Fairmont & Mercury Zephyr** '78 thru '83
- 334 **Fiesta** all models '77 thru '80
- 754 **Ford & Mercury Full-size,** Ford LTD & Mercury Marquis ('75 thru '82); Ford Custom 500,Country Squire, Crown Victoria & Mercury Colony Park ('75 thru '87); Ford LTD Crown Victoria & Mercury Gran Marquis ('83 thru '87)
- 359 **Granada & Mercury Monarch** all in-line, 6 cyl & V8 models '75 thru '80
- 773 **Ford & Mercury Mid-size,** Ford Thunderbird & Mercury Cougar ('75 thru '82); Ford LTD & Mercury Marquis ('83 thru '86); Ford Torino,Gran Torino, Elite, Ranchero pick-up, LTD II, Mercury Montego, Comet, XR-7 & Lincoln Versailles ('75 thru '86)
- *654 **Mustang & Mercury Capri** all models including Turbo. Mustang, '79 thru '92; Capri, '79 thru '86
- 357 **Mustang V8** all models '64-1/2 thru '73
- 231 **Mustang II** 4 cyl, V6 & V8 models '74 thru '78
- 649 **Pinto & Mercury Bobcat** '75 thru '80
- 1670 **Probe** all models '89 thru '92
- *1026 **Ranger/Bronco II** gasoline models '83 thru '93
- *1421 **Taurus & Mercury Sable** '86 thru '92
- *1418 **Tempo & Mercury Topaz** all gasoline models '84 thru '93
- 1338 **Thunderbird/Mercury Cougar** '83 thru '88
- *1725 **Thunderbird/Mercury Cougar** '89 and '90
- *344 **Vans** all V8 Econoline models '69 thru '91

GENERAL MOTORS
- *829 **Buick Century, Chevrolet Celebrity, Oldsmobile Cutlass Ciera & Pontiac 6000** all models '82 thru '93
- *766 **Buick Skyhawk, Cadillac Cimarron, Chevrolet Cavalier, Oldsmobile Firenza & Pontiac J-2000 & Sunbird** all models '82 thru '92
- 1420 **Buick Skylark & Somerset, Oldsmobile Calais & Pontiac Grand Am** all models '85 thru '91
- *1671 **Buick Regal, Chevrolet Lumina, Oldsmobile Cutlass Supreme & Pontiac Grand Prix** all front wheel drive models '88 thru '90
- *2035 **Chevrolet Lumina APV, Oldsmobile Silhouette & Pontiac Trans Sport** all models '90 thru '92

GEO
- **Metro** - see CHEVROLET Sprint (1727)
- **Prizm** - see CHEVROLET Nova (1642)
- *2039 **Storm** all models '90 thru '93
- **Tracker** - see SUZUKI Samurai (1626)

GMC
- **Safari** - see CHEVROLET ASTRO (1477)
- **Vans & Pick-ups** - see CHEVROLET (420, 831, 345, 1664)

(Continued on other side)

---

* Listings shown with an asterisk (*) indicate model coverage as of this printing. These titles will be periodically updated to include later model years - consult your Haynes dealer for more information.

Haynes North America, Inc., 861 Lawrence Drive, Newbury Park, CA 91320 • (805) 498-6703

HAYNES AUTOMOTIVE MANUALS

NOTE: New manuals are added to this list on a periodic basis. If you do not see a listing for your vehicle, consult your local Haynes dealer for the latest product information.

HONDA
- 351 Accord CVCC all models '76 thru '83
- 1221 Accord all models '84 thru '89
- 2067 Accord all models '90 thru '93
- 160 Civic 1200 all models '73 thru '79
- 633 Civic 1300 & 1500 CVCC '80 thru '83
- 297 Civic 1500 CVCC all models '75 thru '79
- 1227 Civic all models '84 thru '91
- *601 Prelude CVCC all models '79 thru '89

HYUNDAI
- *1552 Excel all models '86 thru '93

ISUZU
- *1641 Trooper & Pick-up, all gasoline models Pick-up, '81 thru '93; Trooper, '84 thru '91

JAGUAR
- *242 XJ6 all 6 cyl models '68 thru '86
- *478 XJ12 & XJS all 12 cyl models '72 thru '85

JEEP
- *1553 Cherokee, Comanche & Wagoneer Limited all models '84 thru '93
- 412 CJ all models '49 thru '86
- *1777 Wrangler all models '87 thru '92

LADA
- *413 1200, 1300, 1500 & 1600 all models including Riva '74 thru '91

MAZDA
- 648 626 Sedan & Coupe (rear wheel drive) all models '79 thru '82
- *1082 626 & MX-6 (front wheel drive) all models '83 thru '91
- 267 B Series Pick-ups '72 thru '93
- 370 GLC Hatchback (rear wheel drive) all models '77 thru '83
- 757 GLC (front wheel drive) '81 thru '85
- *2047 MPV all models '89 thru '93
- 460 RX-7 all models '79 thru '85
- *1419 RX-7 all models '86 thru '91

MERCEDES-BENZ
- *1643 190 Series all four-cylinder gasoline models, '84 thru '88
- 346 230, 250 & 280 Sedan, Coupe & Roadster all 6 cyl sohc models '68 thru '72
- 983 280 123 Series gasoline models '77 thru '81
- 698 350 & 450 Sedan, Coupe & Roadster all models '71 thru '80
- 697 Diesel 123 Series 200D, 220D, 240D, 240TD, 300D, 300CD, 300TD, 4- & 5-cyl incl. Turbo '76 thru '85

MERCURY
See FORD Listing

MG
- 111 MGB Roadster & GT Coupe all models '62 thru '80
- 265 MG Midget & Austin Healey Sprite Roadster '58 thru '80

MITSUBISHI
- *1669 Cordia, Tredia, Galant, Precis & Mirage '83 thru '93
- *2022 Pick-up & Montero '83 thru '93
- *2097 Eclipse, Eagle Talon & Plymouth Laser '90 thru '94

MORRIS
- 074 (Austin) Marina 1.8 all models '71 thru '78
- 024 Minor 1000 sedan & wagon '56 thru '71

NISSAN
- 1137 300ZX all models including Turbo '84 thru '89
- *1341 Maxima all models '85 thru '91
- *771 Pick-ups/Pathfinder gas models '80 thru '93
- 876 Pulsar all models '83 thru '86
- *982 Sentra all models '82 thru '90
- *981 Stanza all models '82 thru '90

OLDSMOBILE
- Bravada - see CHEVROLET S-10 (831)
- Calais - see GENERAL MOTORS (1420)
- Custom Cruiser - see BUICK Full-size RWD (1551)
- *658 Cutlass all standard gasoline V6 & V8 models '74 thru '88
- Cutlass Ciera - see GENERAL MOTORS (829)
- Cutlass Supreme - see GM (1671)
- Delta 88 - see BUICK Full-size RWD (1551)
- Delta 88 Brougham - see BUICK Full-size FWD (1627), RWD (1551)
- Delta 88 Royale - see BUICK Full-size RWD (1551)
- Firenza - see GENERAL MOTORS (766)
- Ninety-eight Regency - see BUICK Full-size RWD (1551), FWD (1627)
- Ninety-eight Regency Brougham - see BUICK Full-size RWD (1551)
- Omega - see PONTIAC Phoenix (551)
- Silhouette - see GENERAL MOTORS (2035)

PEUGEOT
- 663 504 all diesel models '74 thru '83

PLYMOUTH
- Laser - see MITSUBISHI Eclipse (2097)
- For other PLYMOUTH titles, see DODGE listing.

PONTIAC
- T1000 - see CHEVROLET Chevette (449)
- J-2000 - see GENERAL MOTORS (766)
- 6000 - see GENERAL MOTORS (829)
- Bonneville - see Buick Full-size FWD (1627), RWD (1551)
- Bonneville Brougham - see Buick Full-size (1551)
- Catalina - see Buick Full-size (1551)
- 1232 Fiero all models '84 thru '88
- 555 Firebird V8 models except Turbo '70 thru '81
- 867 Firebird all models '82 thru '92
- Full-size Rear Wheel Drive - see BUICK Oldsmobile, Pontiac Full-size RWD (1551)
- Full-size Front Wheel Drive - see BUICK Oldsmobile, Pontiac Full-size FWD (1627)
- Grand Am - see GENERAL MOTORS (1420)
- Grand Prix - see GENERAL MOTORS (1671)
- Grandville - see BUICK Full-size (1551)
- Parisienne - see BUICK Full-size (1551)
- 551 Phoenix & Oldsmobile Omega all X-car models '80 thru '84
- Sunbird - see GENERAL MOTORS (766)
- Trans Sport - see GENERAL MOTORS (2035)

PORSCHE
- *264 911 all Coupe & Targa models except Turbo & Carrera 4 '65 thru '89
- 239 914 all 4 cyl models '69 thru '76
- 397 924 all models including Turbo '76 thru '82
- *1027 944 all models including Turbo '83 thru '89

RENAULT
- 141 5 Le Car all models '76 thru '83
- 079 8 & 10 58.4 cu in engines '62 thru '72
- 097 12 Saloon & Estate 1289 cc engine '70 thru '80
- 768 15 & 17 all models '73 thru '79
- 081 16 89.7 cu in & 95.5 cu in engines '65 thru '72
- Alliance & Encore - see AMC (934)

SAAB
- 247 99 all models including Turbo '69 thru '80
- *980 900 all models including Turbo '79 thru '88

SUBARU
- 237 1100, 1300, 1400 & 1600 '71 thru '79
- *681 1600 & 1800 2WD & 4WD '80 thru '89

SUZUKI
- *1626 Samurai/Sidekick and Geo Tracker all models '86 thru '93

TOYOTA
- 1023 Camry all models '83 thru '91
- 150 Carina Sedan all models '71 thru '74
- 935 Celica Rear Wheel Drive '71 thru '85
- *2038 Celica Front Wheel Drive '86 thru '92
- 1139 Celica Supra all models '79 thru '92
- 361 Corolla all models '75 thru '79
- 961 Corolla all rear wheel drive models '80 thru '87
- *1025 Corolla all front wheel drive models '84 thru '92
- 636 Corolla Tercel all models '80 thru '82
- 360 Corona all models '74 thru '82
- 532 Cressida all models '78 thru '82
- 313 Land Cruiser all models '68 thru '82
- 200 MK II all 6 cyl models '72 thru '76
- *1339 MR2 all models '85 thru '87
- 304 Pick-up all models '69 thru '78
- *656 Pick-up all models '79 thru '92
- *2048 Previa all models '91 thru '93

TRIUMPH
- 112 GT6 & Vitesse all models '62 thru '74
- 113 Spitfire all models '62 thru '81
- 322 TR7 all models '75 thru '81

VW
- 159 Beetle & Karmann Ghia all models '54 thru '79
- 238 Dasher all gasoline models '74 thru '81
- *884 Rabbit, Jetta, Scirocco, & Pick-up gas models '74 thru '91 & Convertible '80 thru '92
- 451 Rabbit, Jetta & Pick-up all diesel models '77 thru '84
- 082 Transporter 1600 all models '68 thru '79
- 226 Transporter 1700, 1800 & 2000 all models '72 thru '79
- 084 Type 3 1500 & 1600 all models '63 thru '73
- 1029 Vanagon all air-cooled models '80 thru '83

VOLVO
- 203 120, 130 Series & 1800 Sports '61 thru '73
- 129 140 Series all models '66 thru '74
- *270 240 Series all models '74 thru '90
- 400 260 Series all models '75 thru '82
- *1550 740 & 760 Series all models '82 thru '88

SPECIAL MANUALS
- 1479 Automotive Body Repair & Painting Manual
- 1654 Automotive Electrical Manual
- 1667 Automotive Emissions Control Manual
- 1480 Automotive Heating & Air Conditioning Manual
- 1762 Chevrolet Engine Overhaul Manual
- 1736 GM and Ford Diesel Engine Repair Manual
- 1763 Ford Engine Overhaul Manual
- 482 Fuel Injection Manual
- 2069 Holley Carburetor Manual
- 1666 Small Engine Repair Manual
- 299 SU Carburetors thru '88
- 393 Weber Carburetors thru '79
- 300 Zenith/Stromberg CD Carburetors thru '76

Over 100 Haynes motorcycle manuals also available

5-94

* Listings shown with an asterisk (*) indicate model coverage as of this printing. These titles will be periodically updated to include later model years - consult your Haynes dealer for more information.

Haynes North America, Inc., 861 Lawrence Drive, Newbury Park, CA 91320 • (805) 498-6703